MULTIVARIATE
MORPHOMETRICS

MULTIVARIATE MORPHOMETRICS

By

R. E. BLACKITH

Department of Zoology,
Trinity College, Dublin, Ireland

and

R. A. REYMENT

Department of Historical Geology and Paleontology
Paleontological Institute, Uppsala, Sweden

1971

ACADEMIC PRESS • London and New York

ACADEMIC PRESS INC. (LONDON) LTD.
24-28 Oval Road,
London NW1 7DD

U.S. Edition published by
ACADEMIC PRESS INC.
111 Fifth Avenue,
New York, New York 10003

Library of Congress Catalog Card Number: 70-178219
ISBN: 0-12-103150-0

PRINTED IN GREAT BRITAIN BY:
The Whitefriars Press Ltd., London and Tonbridge

Preface

Less than a century ago in biology, and more recently still in geology, the acts of counting and measuring were regarded as something of an intrusion into pleasantly speculative ways. The rising tide of statistical works has so nearly drowned this attitude that there is, perhaps, not enough emphasis on the need to get the best out of statistical analyses, and there is a continuing need for statistical counsellors to prevent the use of inappropriate statistical techniques and to ensure that the results of the analysis are interpreted to best advantage in biological or geological terms.

Until relatively recently, few statistical texts reached further into multivariate analyses than a discussion of multiple regressions. Although an adequate number of mathematical expositions of multivariate analyses is now available, there is at present almost no way other than the hard school of experience by which a natural scientist can learn how to choose and interpret his multivariate analysis. He will be faced in such texts as exist with a perfectly understandable preoccupation with significance testing, natural enough as the pabulum for mathematical investigations, but quite secondary in importance for the natural scientist to that of estimation.

In multivariate work more perhaps than in any other branch of statistical counselling, the biometrician must be thoroughly familiar with the biology or geology of the problem before giving advice. Granted that this is more easily said than done, we trust that this book will, by providing numerous case histories, help to bridge the gap between counsellor and counselled. This help is all the more necessary in the rapidly growing field of morphometrics because many of those coming into the subject have a highly developed insight into biological or geological problems, but little familiarity with mathematical ideas, and a suspicion that it is hardly worth their while to acquire one. To them we can only plead that each of us has published works in zoology as well as in geology, which makes us sympathetic to the words of Veitch (1963): "The primary purpose of this report is to show, by example, how important it is for the statistician to acquire a sufficiently detailed knowledge of the non-statistical aspects of the problem. It is only then that one can ensure that the statistical techniques are doing what is expected of them."

v

This neutral attitude is howbeit inappropriate. There is a real pleasure in watching a tangled mass of data taking shape; and useful and meaningful shape at that, after passing through a well-chosen analysis. As Herne (1967) remarks: "The practical man wants an understandable relationship that is comprehensible in his technology. The statistician cannot do this with statistical analysis alone; he must be a technician as well. There is no delight in just cooking, one must also taste and relish the result."

As this is, so far as we know, the first textbook on morphometrics, apart from the specialist works of Rao (1952), Sokal and Sneath (1963) and Seal (1964), each of which touches upon only a small part of the field we have tried to cover, there is no more elementary work to which readers can be referred. It is customary in advanced works to invite the less experienced reader to begin with a general textbook on statistics to acquire the necessary background information. Yet for the most part he will not find the central topics of this book mentioned. Nevertheless, these are subjects discussed by artists, physicians and scientists for millennia; they have found these ideas a perennial challenge and so do we. At long last we can test the historic hypotheses involved, thanks to the development of multivariate analysis; we are, finally, operational in the study of the forms of animals and plants and in many related topics.

Although morphometrics leans heavily on the techniques of multivariate analysis, this is not by any means a statistical textbook. There are numerous excellent general multivariate texts available and we refer our readers to one or more of these for the technical details of calculation.

Although the whole idea of morphometrics centres around the quantitative study of animals and plants, it is clearly possible to make use of the results and analytical methods in a more abstract manner. We have therefore not hesitated to bring in examples from Geology and Sociology, for instance, to illustrate some particular point or other. Because the text cuts across disciplines, a glossary of technical terms is provided at the end of the book. Generally, we have given recent references since these usually contain bibliographies that afford an entry into the subject in question. We recognize that this practice is less than just to the pioneers of the subjects, but the saving of space and time is substantial.

We wish to express our thanks to Knut and Alice Wallenberg's Stiftelse and to Uppsala University for generous financial assistance which allowed R.E.B. to visit Uppsala for our close collaboration on the text of the book. R.A.R. was given a term's study leave by

Universitetskanslersämbetet (Swedish Universities' Chancellery) for the purpose of working up many of the examples used here. We also wish to mention Mrs. Ruth Blackith and Mr. Hans-Åke Ramdén for invaluable assistance with many of the problems treated. Mrs. Dagmar Engström drew the figures and Mrs. Eva Eklind typed the manuscript. Finally, an acknowledgment to the Irish and Swedish students upon whom we have tested our text. For the curious: there is no senior author—our names stand in alphabetical order.

UPPSALA   R. E. BLACKITH
August, 1971   R. A. REYMENT

CONTENTS

1
The Historical Background to Morphometric Studies

Although the history of morphometrics is, inevitably,
with the much wider issue of the history of attempts
animals and plants, morphometrics antedates classificator
For there is a strain of thought in the consideration of t
living things which is much more a question of art than o
the modern senses of these words. The desire to abstrac
great variety of living organisms forms which are
harmonious, aesthetically superior to other forms, or else
abstraction as the archetypical form of some wider
shapes and sizes, is very deep seated in human behavior.
is ignored at the risk of serious misunderstandings, beca
evidence that at least a part of the current dif
communication between leading schools of thought ir
stems from a lack of historical insight into the
taxonomists, and into the extent to which different p
likely to be aesthetically satisfying.

In the earliest studies of the forms of living organisms
have any record, there is evidence that classification, ae
what we should now call functional analysis are interl
extent which makes any attempt to separate them fu
Babylonians, back into the beginnings of the second mi
and for the Greeks before Plato and Aristotle, that is t
about 350 B.C., there was a strong conviction that form
music as in anatomy, was capable of numerical expre
that the very essence of musical harmony and of aesth
resided in that theory of numbers which came to be on
points of the thought of the Pythagorean school of ph
the Greek-speaking, and at least partly, ethnically Gre
of southern Italy from around 600 B.C. to around
Stapleton (1956, 1958) has put it, the mathematics

1

and his successors must be regarded as the offspring of both music and mysticism. Pythagoras (580-497 B.C.) himself, the son of a Phoenician jewel engraver by a Greek mother, and reputed to have been an Olympic boxer, knew of the numerical relations between the lengths of strings (and pipes) emitting concordant notes (Levy, 1926). His immediate successors probably knew most of what we should now call Euclid book II and part of book VII (van der Waerden, 1963).

Activities which a modern numerical taxonomist would instantly recognize were part of the teaching of the Pythagorean school; for instance, the essential quality of an animal or plant was "captured" by drawing an outline sketch, almost a caricature, in the sand, and then "recording" this essence by counting the number of junctions between the lines of the sketch, keeping this number of lines to the minimum required to evoke the image of the animal. Both Theophrastos and Aristotle describe how Eurytos, at the end of the fifth century B.C., made a low relief sketch of a man in thick limewash on a wall, sticking coloured pebbles into the junctions of the lines. A later account by the pseudo-Alexandrian commentator adds that Eurytos would finish his sketched-in representation of man with pebbles equal in number to the units which in his view defined a man (Guthrie, 1962): the "number" of man was 250 and of a plant was 360. No doubt a numerical taxonomist would part company with this activity at the point where the mystical interpretation of the numbers began, as combinations of the male and female principals of the universe (3 and 2) or as symbols of the hand of the Creator (1 and 4). Nevertheless, no one who has followed the development of Palaeolithic art can fail to recognize the deep-seated need to "capture" the essence of certain animals of the hunt by sketching them, perhaps on the walls of caves, perhaps by making bone or ivory or stone models. It may, indeed, be that the constant practice of this ritual sketching led to the extraordinary disparity between the skill with which animals of the hunt were depicted by Palaeolithic man and his crude and fumbling attempts to portray the human form, a disparity which becomes ever more striking as more and more artefacts of many periods are discovered.

Nor can anyone who has watched a skilled cartoonist portray the essential qualities of a person or of an animal, with great economy of line, fail to appreciate how quickly the human mind responds to visual clues which evoke an emotional response.

The way in which Greek morphometrics began is not necessarily, or even probably, true of other civilizations. The classification of

animals in India, for example, was based, so far as we know, from about 600 B.C. to about 200 A.D., on the mode of reproduction of the animal (real or supposed), or towards the end of this period, on the number of senses that it was supposed to possess (Sinha and Shankarnarayan, 1955).

The early Greek attempts to discover the essence of form, and hence, as a secondary aim, to classify organisms, were cast in a geometrical rather than an algebraic mould, and although with hindsight we can see how the highly sophisticated branch of geometry known as topology would have enabled them to continue their trend of thought, it is clear that, in the state of knowledge then prevailing, geometry was at a dead end: the kind of problem soluble by geometrical techniques and epitomized by Euclidean geometry was of very limited use in describing the forms of organisms. By the fourth century B.C. the time was ripe for a new approach, based on the logical syllogism of Aristotle (384-322 B.C.), systematizing and amplifying the thought of Plato and Socrates.

This school was opposed to the Pythagoreans because, to quote Aristotle "they construct the whole universe out of numbers, not however strictly monadic numbers, for they supposed the units to have magnitude". To some extent there seems to have been an inadequate distinction, natural enough in the fifth century B.C., between the unit of number *per se* and that geometric point whose motion generated lines, which in turn generated planes, and they again generated solid structures. But it is also possible to argue that much of the disagreement stemmed from Aristotle's concept of number as an essentially scalar quantity, with ratios and proportions to represent form, whereas the Pythagoreans had an almost vectorial interpretation of numbers, which were conceived as "flowing". Their system of representing numbers by rows of pebbles (hence "calculus") arranged in regular patterns (gnomons) instead of numerical symbols would have encouraged a way of thinking about the system of natural numbers not far removed from some of the simplest ways of thinking about a vector plane. Aristotle's inability to understand the Pythagoreans "flowing" numbers rankled, and he became quite petulant on the subject.

Few phenomena are more curious than the erection of Aristotelian ways of thinking into the epitome of classical teaching; a way of thinking about Aristotle which is still widespread. For he was in many ways a destroyer of the old Pythagorean habits of mind as well as an innovator of genius himself. As Bertrand Russell has commented, he was inept at quantitative work, and was thus forced to think in terms which were qualitative, and although he made tremendous advances in what we should now describe as taxonomic theory, these advances were made at the high cost of destroying the concern for what underlay the external form so characteristic of Pythagoras' school. This change was made necessary, in the conditions of Greece in the fourth century B.C., partly for historical reasons stemming from the growth of scepticism and partly because the Pythagorean philosophy was non-operational in the sense that one could applaud its teachings with the heart but fail to see how the reason could act on its precepts. At this period too the Pythagorean ideas of the heliocentric (or at least fire-centred) solar system, the spherical earth, the atomic nature of matter, propounded by Pythagoras' contemporary Democritos in the Macedonian city of Abdera, and indeed the widespread practice of partial democracy faded into the mists from which they were to re-emerge only in comparatively recent times. Pythagorean thought was eclipsed in the intellectual revolution.

Although Aristotle was willing (indeed he had little option in the matter) to abandon to the straight-jacket of the syllogism the descriptive function of classification, he was also the heir, and of even more concern to us, the principal instrument for bequeathing to later generations the concept of pure form. In the immediate past, Plato had discussed the idea that behind the everyday appearance of things there was an ideal form which possessed, unsullied by the imperfections of ordinary life, the quintessential qualities of the things studied; that immanent in the bulls of the fields around Athens, one could create in the mind a concept of a bull perfect in its power, its virility, its aesthetic proportions. This idea was to be carried still further by Aristotle, mainly by way of his influence on Plotinus whose transcendental philosophy at once diminished the operational nature and hence the emotional impact of Aristotelian biological theory and increased the chances that Aristotelian philosophy would survive the harrowing by the mediaeval Church. This survival was a near thing. Even by the time of the foundation of the University of Paris (1215), the fear of Aristotle was so great as to induce the insertion into the statutes of a ban on carrying Aristotle's work on pain of excommunication. Gradually, as classical learning revived, so did the syllogistic framework of classification, and the ideal of pure form which, to varying extents, influenced the taxonomic concept of a type in its formative stages.

We need only to look at two very different workers in the

biological field, Linnaeus and Goethe. Linnaeus was taught syllogistic logic at school, and it would have seemed to him natural enough to employ Aristotle's syllogisms to form the intellectual framework of the Linnean classification. Moreover, operating within a strictly logical and theocentric framework, it would also have seemed natural to Linnaeus to find one good character by which species could be differentiated, for if each species is thought of as having been sent by the Creator for man's benefit, it would be attributing some defects to the logic of the Creation were man not to be in a position to see the distinguishing marks. We owe much to Cain's historical research for our understanding of the mainsprings of Linnaeus' thought (Cain, 1958, 1962).

With the genesis of the dichotomous key in syllogistic logic, and that of the monothetic classification in the theology of Linnaeus' times, we turn to Goethe.

In some far from trivial ways Goethe's conception of form was nearer that of Aristotle than that of Plotinus (Wilkinson, 1951). He distrusted Plotinus' idea that form is quintessential, perfect only when it does not descend into matter; but if his mental image of the ideal form transcended the outward manifestations of an organism, it also embraced its outward form, since he was deeply concerned with function in relation thereto. To a considerable extent Goethe was a link between the Linnean approach and that of the taxonomists of the school now known as Adansonian, for his concern with "gestalt" (the whole organism) was much more in the line of Pythagorean ideals.

The general consequences of the introduction of evolutionary theory were surprisingly unhelpful to taxonomy. Linnean systematics accommodated itself so easily to evolutionary modes of thought that almost no changes of practice were necessary, only the description of that practice needed alteration. This fact in itself is witness to the superficial nature of biology at the classificatory level. What was worse, the ready acceptance of, and indeed demand for, phylogenetic speculation, lowered the standards of classificatory work. Agnes Arber has commented, "The whole attitude of many post-Darwinian botanists . . . has been distorted through trying to compel the study of form to observe phylogenetic ends" (Arber, 1950, p. 7). Now that standards in conventional as well as in quantitative taxonomy are sharply rising, we are entitled to insist that the one great, generalizing principal in biology, evolutionary theory, should not be prostituted to this end. Very little serious work has been done on the inheritance of form, and Grafius' work in this field stands out as a promising line of study (Grafius, 1961, 1965).

Pythagorean ideals, if not ideas, emerged strongly in the work of D'Arcy Wentworth Thompson (1942). It is a measure of the problems facing the earlier Greek philosophers in this field that Thompson was, in his own way, as remote from an operational solution to the comparison of forms as they were.

Thompson seems to have thought in terms of gradients of growth whose end-product was the final form of the organism. With his remarkable classical knowledge and sense of artistic appreciation, he was an instigator of the revival in studies on growth and form even though he was unable to progress beyond a restatement of the problem in terms of his transformation grids. He thought that mathematical methods had been slow to aid morphology because of the hostility, or rather want of understanding of the nature of the problem, implicit in Bergson's (1911) comment "Organic creation . . . the evolutionary phenomenon which properly constitutes life, we cannot in any way subject to a mathematical treatment". The problem was consistently tackled on an inappropriate basis, attempting to describe the form of a given species mathematically, instead of describing the contrasts of form between two affine species in vector terms.

After Thompson, or rather toward the close of his life, which included 64 years in university chairs, came the development of studies of allometric growth, with which the names of Huxley and Teissier are closely associated (Huxley and Teissier, 1935). Although almost entirely confined to the studies of the joint variation of two characters, it became apparent at an early stage that the generalization of the allometric relationship was going to mark a milestone in the history of growth studies. This step has been taken mainly as a result of the work of French-speaking biologists, amongst whom we may mention Teissier (1938) for one of the earliest applications of properly multivariate methods to invertebrate studies; Cousin (1961) for an extension of the use of factor analyses to studies of inheritance, already investigated using allometric methods by Bocquet (1953); and Jolicoeur (1963d) for placing the multivariate generalization on an acceptable mathematical footing.

It should not be thought that Pythagorean ideas disappeared entirely in the third century B.C.: ideas which meet a human need are tenacious of life, and these were partly modified, partly continued, by Persian and Arabic-speaking philosophers and then by mathematicians such as Leonardo Fabonacci of Pisa (twelfth century) and artists such as Albrecht Durer (1471-1528) (see Durer, 1613) and Leonardo da Vinci (1452-1519) whose passionate concern for the harmony of form led them to make sketches which stem from Pythagorean preoccupations with form yet anticipate the ideas of D'Arcy Thompson. Medicine afforded a further line of intellectual descent, since Galen of Pergamon (131-201 B.C.) taught that the incidence and course of disease was dependent on the body-build of the patient, a perennial subject of medical interest now known as somatotypy.

Whereas, up to about 1750, it had seemed proper to select characters for taxonomic study on Aristotelian *a priori* principles, connected with their supposed importance in determining the way of life of the animal or plant, Adanson (1759) produced a radically new departure, allotting equal *a priori* weighting to all characters. In this he was, as Cain (1959) remarks, much ahead of his time, and we refer the reader to Cain's work for an appreciation of the state of taxonomic theory in Adanson's times. Adanson's earlier progress towards equal weighting came from his studies on molluscs; it came to fruition in his great botanical treatise (Adanson, 1763).

Independently of Adanson, Scopoli (1777) who also studied plants as well as animals was to publish very similar views on equal weighting, although his contribution has received much less notice during the revival of what is now often called Adansonian taxonomy. Scopoli's work was brought to our attention by Mr. R. G. Davies of Imperial College, London.

This reinforcement of Adanson's ideas was not, however, sufficient. The claims of Aristotelian logic, the belief that by taking thought one could construct the natural system, without recourse to mundane observation, remained too strong. A whole series of *a priori* weighting systems came to the fore, only to fall as the next fashion came on the scene. This is not the place to discuss the circularity of much *a priori* thinking, which has been cogently argued by Cain (1959), Sokal and Sneath (1963) and Colless (1967a) amongst others. Darwin's *a priori* system was, however, based on observations linked to evolutionary criteria. It has seemed to many workers during the last century so satisfactory that two decades ago few would have predicted that it would be the subject of serious dispute; but so it is and the issue is still open. The swing away from the "New Systematics" lies at the heart of the "new taxonomy" initiated by the coming of numerical taxonomy; although the two issues of numerical methods and empirical theory are in fact logically distinct, they are closely linked to deeply ingrained attitudes of mind: there is also a natural alliance between a defence of conventional techniques and a defence of phylogenetic weighting. As Sylvester-Bradley (1968) remarks, perhaps the new taxonomy has not been born, but at least it has been conceived.

2
The Mathematical Framework of Morphometric Analyses

An essential problem in morphometrics is to measure the degree of similarity of two forms. Although much can be, and has been, written about each stage of the analysis, it is possible to strip the problem down to the following exposition.

Consider the two forms shown in Fig. 1. To help the biologist to visualize the idea of a degree of similarity, the smaller of the two forms has been drawn inside the larger one. Imagine the smaller form inflated by an internal pressure (growth, in effect) so that the difference in volume between the two forms is reduced. The growth pattern may be of any kind, so that as each of the marked points on the left-hand side of the smaller form approaches its homologue on the larger, we permit each marked point to travel at a rate independent of that of any other. It is intuitively obvious that the two forms become identical when the two sets of marked points coincide, unless of course there are features of either organism which are unlike but which have not been allocated a marked point. This would be a failure of homology. We will assume for the moment that the marked points are sufficient to follow the growth of all salient features of the organisms. The volume occluded between the two forms may be regarded provisionally as one measure of the dissimilarity of the forms, and if we consider the expansion of the smaller form as being accompanied by a displacement of every marked point, Gauss' divergence theorem can be used to establish the relationship

$$\Delta = \operatorname{div} \{ \mathbf{x_i x_j} \} \tag{2.1}$$

where Δ is the fractional expansion at any point on the surface, and $\mathbf{x_i x_j}$ is the vector of displacements of the marked points on the smaller form towards the homologous points on the larger form

(Bickley and Gibson, 1962). Essentially, we have set our analysis in motion in terms of the vector $x_i x_j$, and this is the key to all truly multivariate studies of form.

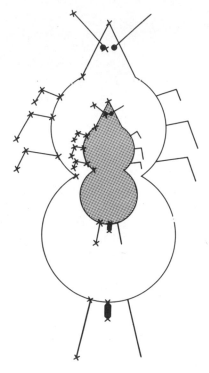

Fig. 1. Affine forms of a hypothetical animal: the marked points on the smaller, shaded, form move to their homologues on the larger, unshaded, form as the smaller form "grows".

Formally, instead of considering the measurement of the volume displaced by the expansion of the smaller form, we can set up a p-dimensional set of Cartesian coordinates in which the axes are the directions of displacement of the marked points. Each of the two forms can then be located within this framework of reference by noting the amount of the displacement of each marked point along its axis of variation. To find the geometrical distance between the two forms is a simple matter of applying the p-dimensional equivalent of Pythagoras' equation

$$d^2 = \sum_0^p x_i x_j \qquad (2.2)$$

The distance as measured in this Euclidean space has indeed been suggested as suitable for use in taxonomic work by Sokal (1962), despite a potentially serious flaw in his analysis. For the multivariate extension of Pythagoras' theorem to hold, the axes of the chart must be at right angles to one another, i.e. the space in which the calculation is performed must be Euclidean. However, the distance so calculated will only be correct if the movement of any one marked point is independent of the movement of any other marked point, there being no correlation between the measured characters. In general, we can expect substantial correlations to exist, seriously biasing our measure of distance.

To eliminate the effects of correlated characters we could, in principle, so set the angles between the axes of our chart that the cosine of the angle between any two axes equalled the coefficient of correlation between the characters whose displacement they represent. To act in this way would be impossibly cumbersome, so that we take a step which, although calling for much imagination for its comprehension, enables us to compute the distance whilst taking into account the correlations between characters. This step is to insert into the calculation a metric tensor, the "fundamental tensor", descriptive of a space in which the correlations between the characters are removed by distorting the space to a calculated extent. We now have

$$D^2 = \sum g^{ij} x_i x_j, \tag{2.3}$$

where D^2 is the generalized distance between the two forms, adjusted for any correlation there may be between the measured characters and g^{ij} is the metric tensor, which represents the inverse of the matrix (of dispersions) of the characters. This matrix is sometimes referred to as the "covariance matrix" and the "variance-covariance matrix". As before $x_i x_j$ is the vector of differences between the characters on the smaller form and those on the larger one.

The rôle of g^{ij} in this equation is brought out by replacing it with the generalized Kronecker delta, a unit matrix of the same rank with all leading diagonal terms unity and all pre- and post-diagonal terms zero. This matrix evidently describes a space in which the characters are uncorrelated, and its use in the equation for D^2 reduces the latter to the Pythagorean equation. In fact, g^{ij} describes the extent to which the Riemannian hyperspace has to be distorted in order to accommodate the interrelationships existing between the characters when measured in Euclidean space (Adke, 1958). The generalized distance has the character of a geodesic, that is to say, the line of shortest distance between the two forms in a curved space; it reduces, as we have seen, to a straight line in Euclidean space.

With the advent of computers, few would now wish to undertake large-scale matrix computations by hand, though they may serve a useful purpose in helping the analyst to understand the mathematical basis of the work. However, just as it is no longer essential to know how an internal combustion engine works in order to drive a car, so one may cause multivariate analyses to be performed on data without being able to perform the calculations oneself, or even without understanding their mathematical basis. The effort to understand, even if the understanding is incomplete, is, we feel, rewarding, not only in the utilitarian sense of providing some control over the calculations, but also of affording a glimpse of the workings of a technique of staggering generality. We feel it is in place to sound a note of warning about the use of the electronic computer by workers who lack insight into the workings of the techniques involved and in the way in which computers work (and do not work). Multivariate statistical programmes are becoming increasingly available at most computation centres. With very little effort it is possible for somebody with a collection of data to have these processed. Unfortunately, even many well-tried programmes contain faults (so-called "bugs"), and unless the user has a fair knowledge of what should be happening inside the machine, he runs a strong chance of being saddled with incorrect results, at the best partly wrong, at the worst woefully off key.

Calculations which involve the use of metric tensors to represent spatial distortions play a large part in elasticity theory, in the general theory of relativity and in the thermodynamics of irreversible processes (Branson, 1953).

The need for these calculations in statistical work produced a striking example of how a crucial problem, long pondered over, is suddenly solved, in different ways and independently, in different countries. In England, Fisher developed the idea of the discriminant function, in America Hotelling generalized the t-test for use in the multivariate case, and in India Mahalanobis was responsible for the evolution of the generalized distance between two multivariate populations, recognizing at an early stage of this evolution the close relevance to relativity theory arising from the fact that, in relativity and in the calculation of D^2, we can choose to transform from a set of correlated variables to an uncorrelated set in a constant field. Although Hotelling's generalizations of the t-test appeared in 1931, key papers from all three authors appeared within a few months of one another in 1936. The D^2 had been in use for a decade in Mahalanobis' anthropological papers, but with a theoretical background which prevented its full exploitation.

In seeking for methods of greater and greater generality, we emerge into a field where, as Einstein himself has remarked, "the coordinates, by themselves, no longer express metric relations, but only the 'neighbourliness' of the things described" (Einstein, 1950). Rashevsky (quoted by Vinck, 1962) has also expressed the view that much of interest to the biologist can only be discussed profitably in a non-metric space with the aid of set theory.

We should emphasize that, whereas much of the development of multivariate theory has been concerned with significance testing, such tests play a minor, though assuredly not unimportant, role in morphometrics. Pearce (1965) has noted that whereas significance testing follows experimentation, for many purposes much multivariate exploration is concerned with the construction rather than the testing of hypotheses and therefore precedes experimentation. The D^2 concept thus lies at the heart of morphometrics. In the following sections we shall show how this has lead to many useful developments, and has become an indispensable tool (Rao, 1960).

3
The Biological Background to Multivariate Analysis

For the benefit of readers coming to the problems of analysing multiple measurements for the first time, or with rather limited experience of these topics, we offer the following simple account of what each of the techniques we discuss actually does, expressed in the familiar terms of studies of the shapes and sizes of human beings. Teaching experience suggests that the problems of growth and form require for their understanding the capacity to conceptualize the mathematical manipulations in concrete terms, and that the only model whose growth patterns are sufficiently familiar to all readers is that of the human frame. Believing firmly that until a student knows what the mathematical techniques are capable of doing for him he is most unlikely to master them, we have not hesitated to sacrifice rigour for clarity.

Discriminant Functions

We will suppose that patients referred to an obesity clinic need to be assessed for "fatness" at the time of admission and again after some remedial treatments. How can a scale of "fatness" be created? It is not enough to weigh the patients because of the large differences of stature in human beings.

Some measurements are clearly going to be more sensitive to "fatness" than others. Girths of the upper arm, the trunk, and thighs will be such characters, whereas height, lengths of limbs, and cranial measurements will be less so. We cannot create a proper representation of "fatness" (or, more accurately, of the polarity between "fatness" and "thinness") unless we can set up a combination of these characters which represents the stature of the patient as well as his adiposity; to do this we can avoid the use of

14

ratios, shown in Chapter 4 to be dangerously unsatisfactory, by making up a "linear compound" of the characters

$$x_1 + x_2 + x_3 + x_4 + x_5 + x_6. \tag{3.1}$$

This linear compound, however, equally well reflects the changes of stature as the changes of "fatness", so that we need to calculate a set of weights to be attached to the characters which will properly reflect our concern with the polarity between fat and thin people (Brožek and Keys, 1951). By calculating the dispersion matrix between the six (in this instance) characters we have chosen, inverting this matrix and multiplying by the vector of distances between the means (difference mean vector) we provide the right-hand sides for a set of six simultaneous linear equations which have for their left-hand sides the coefficients required. Calculated in this way the coefficients maximize the separation of fat and thin people for the six characters, relative to the variation within each group.

We thus have as our new linear compound the six characters each of which is multiplied by the elements of the vector of coefficients we have just calculated.

$$\beta_1 x_1 + \beta_2 x_2 + \beta_3 x_3 + \beta_4 x_4 + \beta_5 x_5 + \beta_6 x_6 \tag{3.2}$$

If we had simply compounded the characters, without taking any steps to allow for the correlations between them, the linear compound would have increased the apparent separation between fat and thin people indefinitely as more and more characters were added. However, a little thought shows us that such additional separation is spurious, and arises from the fact that the information about the two groups common to each character is being added afresh with each new character. The object of the calculation of the inverse of the dispersion matrix for the six characters is to disentangle the correlations between them so that each new character brings only that amount of information which is unique to it and is not already carried by the earlier ones.

If the first three characters are, as stated above, the three fat-sensitive girth measurements, and the last three measurements are the stature-sensitive skeletal frame measurements, we should expect the β-coefficients to be large and positive for the first three, and smaller, perhaps negative, for the last three characters.

A negative sign for a coefficient simply means that its character develops much less than those which have the large positive coefficients as the form of the organism traces a path along the vector describing the contrasts of form. Evidently, if our vector is essentially describing a fatness-thinness polarity, we should expect positive high coefficients of fat-sensitive characters, when the vector is describing changes towards the "fatness" pole, whereas, when the vector is describing changes towards the "thinness" pole, it will have all its signs reversed, with substantial negative coefficients for the fat-sensitive characters and smaller ones of either sign for the stature-sensitive ones.

As one might expect, once the vector of coefficients has been reoriented to describe the contrast between people of large and small stature (by using such groups as the poles of the discriminant function), it will have large coefficients for the stature-sensitive characters and smaller ones of either sign for the fat-sensitive ones.

By the term "reorient" we mean that the vector has been rotated to a new direction so that it links two new groups of forms. It still consists of the coefficients of the same set of characters, but the proportions of the coefficients will have changed. In fact, the sum of cross-products of the "standardized" coefficients provides a measure of the angle between them, so that vectors whose coefficients, when multiplied together, give small values for the sum of cross-products will make large angles; in fact when the sum of cross-products is zero the vectors are at right angles to one another. We can in this way give mathematical expression to the intuitively attractive idea of considering the fatness-thinness polarity in human beings as a sequence of shape changes proceeding in a different direction to that of the tallness-shortness polarity that is the essence of changes of stature. Naturally, these contrasts do not exhaust the possibilities of human variation of form. The vector could be oriented yet again to cope with the polarity between stocky and slender people, and between men and women, and this list by no means exhausts the differences of form which have been studied. We have, therefore, a powerful means of describing, simply by changing the coefficients of our discriminant function, any contrasts of form that are of interest. Once they have been described by the objective calculation of the coefficients, we can proceed to compare the different polarities by calculating the angles between the vectors. It is sometimes argued that discriminant functions require that the polar groups should be specified *a priori* and generally this is so, but the process of selecting end-members of a series often works quite well despite possible risks through using truncated distributions. The polarity can then be refined iteratively (as an approximate though useful measure).

Generalized Distances

Once the description of the fatness-thinness polarity is possible in terms of the vector of coefficients in the appropriate discriminant function, we need to know how far these two groups are, in fact, separated by the function. What is the distance between any given group of fat people and another group of thin people? Very large numbers of measures of similarity have been proposed, and about 30 are known to us in the literature. However, there is one which stems from the same intellectual approach as the discriminant function and is readily calculable from it: this is the Generalized Distance of Mahalanobis (1936), which is found by multiplying the vector of coefficients that constitutes the discriminant function by the vector of differences between the means for the two groups. Because it is based on the discriminant function, the generalized distance allows each character to carry only its proper amount of information about the separation of the groups, and eliminates the effects of correlation between the characters.

The Construction of a Scale of "Fatness-Thinness"

Once we have the β-coefficients for any particular discriminant function, all we need to do to obtain the score of an individual along this vector is to multiply each β-coefficient by the value of its proper character for that individual, with due regard for signs, and sum the products. If desired, we can then proceed to express the score as a percentage of the total distance along the vector, say from "thinness" to "fatness". An obese patient might be scored as 86% on admission but as only 42% after treatment.

The coefficients in the discriminant function are, indeed, derived from a calculation designed to maximize the separation of the two poles of the function (the "fat" and "thin" groups), but, once established, the particular nature of the groups chosen for the initial orientation does not restrict the use to which the discriminant can be put; as Blackith (1957) has remarked, "by concentrating on the direction rather than the magnitude of the separation afforded, one evades the problem of whether the material studied exhibits the full range of morphological plasticity of which it is capable". Thus, if individuals fatter than the "fat" group, or thinner than the "thin" group, were subsequently to require a score on the scale of fatness-thinness, their measurements could be fitted into the discriminator without any difficulty.

Canonical Variates and Discriminatory Topology

Once we have established the morphometric relationship between "fatness" and "thinness", and that of "tallness-shortness" our intellectual curiosity, even if not the needs of the work, will seek a way of representing the two relationships. As we have seen, something can be done by comparing the angles between the discriminant functions. This approach can become cumbersome and does not help us to incorporate our knowledge of the distances between the defining groups of people. (We should not, in this context, speak of the length of the vector, which is not at all the same quantity as the extent to which that vector separates two groups of people.)

It is possible to make a geometrical representation of the arrangement of the various groups "fat"; "thin"; "tall"; "short" by calculating the generalized distances between each of the groups, choosing an arbitrary scale so that any one of these distances can conveniently be drawn on a sheet of paper, and then drawing arcs of a radius determined by the other generalized distances until the positions of the remaining groups have been localized, as Fig. 2 suggests. Sometimes the attempt to construct such a chart will be unsuccessful; one possible reason is that the chart should in fact be three (or more) dimensional: if it is three-dimensional, a solid model may be made, but if it is four-dimensional, or of higher dimensionality, only very imperfect representations can be made. A quite different reason for the failure of the construction may be that variation, as expressed by the dispersion matrices differs substantially from one group to another.

We can see from the hypothetical generalized distance chart of Fig. 2 that there are two distinct dimensions of variation, corresponding to the two polarities which we are investigating. Rao (1952) recommends that instead of constructing the generalized distance chart first, and then putting in, as it were, the underlying dimensions of variation by inspection, we should rather calculate the underlying dimensions of variation and then use these as the axes of charts on which the mean positions of the groups may be plotted (Fig. 3). Such underlying axes of variation are called the *canonical variates* and the arrangement of the groups in the space defined by these variates is called *discriminatory topology*. This technique has the advantage that, should the arrangement of the groups require three or more dimensions for its proper expression, the canonical variates can be taken two at a time so that different aspects of the

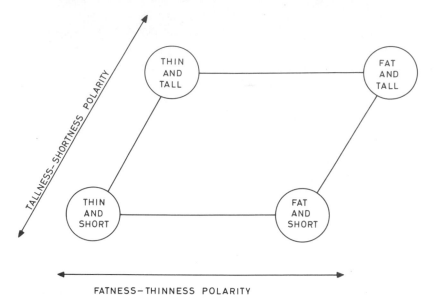

Fig. 2. Hypothetical generalized distance chart showing four groups and two polarities, or axes of variation.

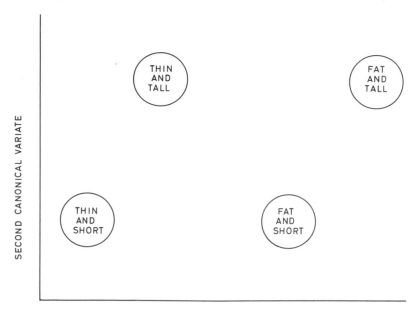

Fig. 3. Hypothetical canonical variate chart showing four groups and two polarities, or axes of variation.

truth can be examined in detail even though the overall picture is inaccessible to us, bound as we are in a three-dimensional world.

The necessary calculation, for which the aid of a computer is highly desirable although, when there are not too many variables measured, say less than six, the calculation can be done on a desk machine at the expense of several days of hard work, serves the same purposes as the calculation of discriminant functions and generalized distances. The canonical variates are necessarily at right angles to one another, the underlying axes of variation representing the polarities may not be orthogonal in this way, so that these canonical variates act as a formal framework for the groups, whose interrelationship can be studied as before once their positions have been plotted.

Canonical Correlation

Pursuing further our example of the fat and thin people we might wish to investigate the relationship between obesity and the external factors causing this unenviable condition. An ideal model would be to balance the morphometric characters we have selected as being descriptive of adiposity against the things underlying it. One would therefore have, on the one pan of the scale, a *set* of morphometric variables, and on the other scale-pan variables such as measurable genetic factors, intake of nourishment, metabolic factors and the like.

A reasonable method of analysis would then require, firstly, a measure of the correlation between the two sets and, secondly, the contribution of each variable to the correlation, albeit in approximate terms.

The multivariate statistical method of canonical correlation was developed by Hotelling (1936) to carry out this kind of analysis. It is in some respects a generalization of multiple regression.

Homogeneity of Dispersion Matrices

The topic of the homogeneity of the dispersion matrices of two or more samples (for tests) or universes (for discussions in general terms) is obviously an important one. Many of the multivariate statistical procedures used in this book have the theoretical requirement of homogeneity in the dispersion matrices in order to be legitimately applicable. This is the first point, and one to which we shall return in other connections. It is clearly a technical matter.

The second point relates to the geological and biological

significance of heterogeneity in dispersion matrices. Particularly in connection with palaeontological morphometric analyses it is frequently of importance to be able to pick up significant heterogeneity in covariance matrices as it may be a reflection of sexual dimorphism in the material, changes in breadth of variation deriving from different ontogenetic phases (particularly in arthropods), and sorting of hard parts by geological agencies. Fortunately, most of the multivariate methods used in this book are quite robust even to quite considerable deviations from homogeneity.

As will be elaborated upon in a subsequent section, there is no fully satisfactory method of testing homogeneity. The only test available is sensitive to departures from multivariate normality, as well as differences in the variances and covariances due to heterogeneity, and it therefore requires not a little care and forethought if a correct analysis of heterogeneity in dispersions is to be made. There is as yet no definite method for testing multivariate normality of dispersions. As in all statistical investigations, it is highly desirable that a graphical appraisal of the data be made (Healy, 1968). Even in a multivariate study, pronounced non-normality in one or more variables will be readily picked up from a graph (and if deemed necessary, univariate tests of normality) and in the stead of a multivariate testing procedure, these are adequate for most purposes.

Principal Components: Variation Within a Group

So far we have been concerned with the contrasts of form between two distinct groups (fat and thin). But the mass of the people will fall between the two extremes, being of what we should commonly call normal build. There is thus no very clear distinction between studying contrasts of form between two selected types and studying the variation of form to be found within any one group of people; inside the group the variation will still be canalized along vectors of various kinds. The situation in which we try to distinguish these canalized patterns of variation in a supposedly homogeneous group, and that in which we are deliberately selecting individuals which show one or other extreme form (to act as the terminal groups for discriminant functions) is in practice a matter of degree and, more still, a matter of the intellectual preferences of the investigator. Nevertheless, from the point of view of a statistician wishing to set up valid mathematical models, there is all the difference in the world between the two cases, since he cannot get started until a decision as to one or other approach has been made.

Pearce (1959) has stressed the fact that in growth studies the residual variation will often be canalized along certain dimensions of biological interest. The canalized patterns of growth and form which exist within a supposedly homogeneous population can be discovered from the dispersion matrix by setting it in a characteristic equation whose roots are the elements of the vector of λ-coefficients (the latent roots)

$$|S - \lambda I| = 0, \qquad (3.3)$$

where I is a unit matrix, of the same rank as S, that is to say a matrix having ones in the leading diagonal positions and zero elsewhere. The solution of this equation, which is also best done on a computer, but can for 4-5 variables be performed on a desk machine, consists of a set of latent roots, as many as there were characters in the dispersion matrix; each of these latent roots generates a corresponding latent vector, yet another set of coefficients for our original choice of characters. Generally, the first latent vector, corresponding to the largest latent root of the characteristic equation, reflects the variation in size of the organisms, the polarity of stature as between tall and short individuals that we have discussed above. It is quite likely that the second latent vector would reflect the fatness-thinness polarity, although much depends on the other possible sources of variation in this respect among the patients as there would be in a already been selected for fatness, as would inevitably occur amongst the patients at an obesity clinic, there might not be so much variation in this respect among the atients as there would be in a relatively unselected population. In such a case, the second latent vector might represent the polarity between people of stocky build and those of slender build, leaving to the third latent vector the job of representing fatness-thinness. Again, much would depend on whether the clinic accepted patients of both sexes, since some measurements would clearly be sensitive to the differences between men and women.

There are other ways of extracting vectors descriptive of canalized variation from the dispersion covariance matrix. When the technique briefly sketched above is used, the latent vectors are known as principal components, proper vectors or eigenvectors: such components are always at right angles (orthogonal) to one another. Various techniques of "factor analysis" have been devised so as to reorient the vectors more closely with whatever sources of variation the experimenter is interested in: the arbitrary nature of these "rotations" has led to much confusion.

We are not suggesting that when the fatness-thinness polarity is represented by a discriminant function, by a principal component, and by a canonical variate, all three vectors will lie parallel to one another. There will, normally, be a general tendency for the three types of representation to lie roughly parallel, but the extent to which this happens in any particular case will be variable: at least it should be possible to recognize what each vector is representing by observing the nature of the organisms (or non-biological material as the case may be) at each end of the vector. This comparison is sometimes known as reification.

Factor Analysis: Variation Along Axes Determined by an *a priori* Theoretical Analysis

In all the techniques for analysis which have been discussed so far, the choice of a suitable method, determined by the experimenter's needs, is the only choice that has to be made apart from technical matters like the numbers of characters, and the numbers of organisms on which to assess them. Once the technique has been chosen, all analysts will arrive at the same result, though they may differ in their ability to interpret it. With factor analytical methods, however, a new element of choice enters into the calculation, for in factor methods the experimenter deliberately sets out to orient his axes of variation according to his theoretical knowledge, real or assumed, about the nature of variation in his material. As Kendall has cogently put it, in principal component analysis we work from the data towards a hypothetical model, whereas in factor analysis we work the other way round. But Kendall (1957) also wisely remarks that at different stages of the development of a subject, the emphasis may well shift from one aspect of this dialectic to another; by virtue of this feedback arrangement, so typical of scientific activity, much progress may be made at the expense of some logical tidiness.

Naturally, the nature of the *a priori* hypotheses will change from one application of factor analysis to another; it would be foolish to expect a theory of the intellectual capacities of the human mind to bear any useful relationship to theories about the growth of insects, to choose only two examples. Yet methods of analysis which have built-in psychometric theories are quite commonly applied to morphometric problems without much consideration of their applicability. We might feel able to excuse this practice by saying that morphometric methods are so robust that it matters little what prior assumptions are made, and later on (p. 297) evidence for this proposition will be considered. It does seem, however, eccentric, and possibly unwise, to use a more complicated model than is necessary, solely on the grounds that its complexities are likely to be irrelevant to the problem. For the essence of factor analysis is the freedom that it gives to the analyst to rotate his calculated axes of variation into positions predicted by his theoretical model. In order to do this the analyst must have some theory on which to work, and the theories built into factor methods are those of psychometricians (such as Thurstone's Simple Structure) whose applicability to morphometric problems in biology is, to say the least, tenuous (Sokal *et al.*, 1961). For simple structure arrangements of the axes of variation imply that the underlying causes of that variation affect only a few of the multiplicity of characters assessed; indeed it forces the rotation of the axes to positions in which the "factor loadings" or correlations between the character and the factor are large for some characters and negligibly small for others. If such a situation does, in the experimenter's belief, actually underly the data he is analysing, then factor analysis is evidently the preferred course for him to adopt; but in the overwhelming proportion of published factor analyses, the experimenter has quite patently no such idea in mind, and in many instances there is no evidence that these problems have ever been considered prior to the application of the analytical procedure. Similar arguments also apply to forms of factor analysis in which rotations to other than "simple structure" are adopted. It seems likely that factor methods have often been chosen on such understandable if trivial grounds as the availability of a book on the subject, or of a computer programme or quite simply on the "follow my leader" principle.

Not infrequently, people claim to be doing a factor analysis when they are in effect making a principal components analysis. When the latent vectors of such an analysis are then rotated to some kind of structure, it is difficult to grasp the rationale of the underlying logic, for rotation can only reasonably be advocated for a true factor-analytic structure. That in actual practice (for example in many fields in geology) a factor analysis yields intelligible results does not justify its incorrect application, and it may be reasonably assumed that the use of the "correct" procedure would produce even more intelligible results. Moreover, there will be cases in which the incorrect application of factor analysis will lead to a wrong outcome.

One of the most remarkable aspects of multivariate morphometric analysis is that what are, conceptually, some of the most difficult techniques should have long antedated the conceptually simpler ones. The reason for this improbable inversion of the normal sequence lay in the genius of Spearman (1904) who thought of the advantages of matrix operations to disentangle the underlying vectors describing a multivariate situation. Spearman was however, dealing with what, perhaps, is the most difficult of all experimental situations, the unravelling of the structure of the human mind. He found it scarcely possible to proceed with the unravelling process until he had imposed some degree of order on the observations, imposed, in effect an *a priori* structure on the raw data, in terms of what he called the factors of the mind. This is not the place to discuss the correctness or otherwise of Spearman's psychological ideas, which formed the basis of a passionate discussion which has yet to reach any substantial asymptotic approach to finality. What is of concern to us now is the fact that Spearman felt that he could not proceed without this imposed structure, and that, throughout the history of factor analysis, there is this essential distinction between the factor model which is imposed by the experimenter and the ability of the factor model to permit of the extraction from the data of its intrinsic components of variation.

It is generally agreed that in both factor and principal component analyses (and other methods discussed earlier in this book) the objective is to reduce a large number of correlated variables to some smaller number of underlying variates whose nature one hopes to be able to identify by the process of "reification". In principal components, however, the axes are orthogonal because this is the logical arrangement in order to determine the independent components of variation. Orthogonality is the condition of independence. Factor analysts, many of whom do not seem to understand the bases of their subject, are wont to say that principal components analyses are inappropriate because an experimenter who resorts to this technique "assumes" that the factors of the mind are orthogonal. This statement is simply untrue.

Orthogonality in principal components and other truly multivariate analyses plays exactly the same role as it does in univariate analyses; namely, the preservation of the statistical independence of the components of variation. No one would suggest that a balanced two-way analysis of variance assumes that the factors studied are necessarily independent in nature, and the suggestion is no less out of place in multivariate studies.

At this stage we need to enquire what services factor methods can perform that other multivariate methods (in particular that of principal components) cannot. Is there, in fact, any situation in which we should actively recommend factor analysis? Clearly some form of factor analysis is called for when the experimenter has a theory, which can be expressed in quantitative terms, about the behaviour of his material, and feels that this theory should determine the way in which the results are interpreted. Great restraint is needed in such cases to avoid using the results so acquired as a justification for the underlying theory. If the theory is built into the mathematical model, it will necessarily rearrange the data to conform with its requirements, leading to a circular justification of the data by the model and the model by the data.

What of the cases where no underlying theory is to be invoked in the analysis? One such arises when the characters assessed are not the same in two different experiments but the experimenter wishes to compare some notional axis of variation. In psychometric work this situation seems to be common: the test subjects are given a "battery" of tests, differing from one laboratory to another, and notional factors such as those related to intelligence or skills are to be extracted from the matrices of correlations between the performances. An analogous situation in morphometrics would arise when two experimenters wished to compare vectors of, say, size and shape, but had for some reason been unable or unwilling to measure the same characters of the organisms. There can be little doubt that certain factor analytical techniques can arrive at an assessment of these axes of variation in terms which minimize the nuisance caused by the different suites of characters used. Against this advantage has to be set the lack of any way of testing whether factors so uncovered are significantly discrepant, but in the unusual circumstances that would be a lot to ask of any technique. Despite the frequency with which such situations arise in psychometric work, they are not common in morphometrics so far as our experience is a guide.

4

Some Simple Forerunners
of Multivariate Morphometrics

The earliest attempts at a quantitative assessment of shapes were almost all conducted with the aid of ratios of characters; often there was, underlying this use, the idea that one of the characters could be regarded as indicative of some feature of primary interest whereas the other effectively standardized its variation by providing a measure of absolute size. The use of ratios continues, partly because of a misconceived dogma that one cannot handle more than two characters at once and that ratios are the method of choice for two characters (neither part of this belief being well founded), but mainly because their use constitutes what Blackwelder (1964) has called "an acceptance", that is, a proposition or other system of sentences which cannot be shown to be true but which serves to meet a need, and hence generally passes unquestioned because it is thought of as a "standard" technique.

In fact the use of ratios implies certain prior knowledge about the material under examination which, if set out explicitly, would almost certainly be denied by any experienced worker in the field. The weaknesses of ratios include:

(1) The fact that a ratio will not be constant for organisms of the same species unless these are also of the same size, by virtue of the almost universal occurrence of allometric growth.

(2) As generally used, ratios contain only two characters and thus afford a poor appreciation of what may turn out to be an involved contrast between forms.

(3) To compound two characters into a ratio implies that there is only one contrast of form to be studied, and that that unique contrast is well assessed in terms of two characters of equal weights, but opposite in sign. The assumption that only one contrast of form accounts for all the observable variation in the material is particularly misconceived because almost all properly conducted experiments

designed to test this assumption have shown it to be untrue, as the illustrations given later in this book show. The use of ratios of characters has been discussed by Barraclough and Blackith (1962), Christensen (1954) and by Jeffers (1967a). Nevertheless, one should not lose sight of the usefulness of some quite simple techniques for shape measurement. For example, the seeds of *Menyanthes*, found in rocks of the order of 10^6 years old in Japan, become flattened by the internal pressure in the sediments. They thus become a built-in indicator of the age of the sediments, and Kokawa (1958) has constructed a nomogram to relate the age of the sediment to the length and thickness of the seeds it contains. In general terms, the thickness of the *Menyanthes* seeds decreases exponentially with the square of the age of the sedimentary rock.

Traditionally, morphometric analysis has centred on the study of allometric growth. If much of the recent emphasis in growth studies has moved away from allometry, as expressed in the classical equations, towards multivariate methods, there is still a substantial amount of research devoted to the exploration of the allometric relation. This work has been well summarized by Gould (1966). Recently, Laird *et al.* (1965) have shown how the equation for allometric growth can be related to, and deduced from, the relativity relations of biological time. The work of Jolicoeur (1963d; 1968) and of Jolicoeur and Mosimann (1968) has taken the allometric relationship far along the multivariate road, so that from the allometric relationship we proceed to its generalization in terms of the major axis of the dispersion matrix, and thence to the "reduced major axis" of the matrix, which is in effect a ratio of two measures of dispersion. Although recommended by Kermack and Haldane (1950), the use of the reduced major axis seems to Jolicoeur to afford an apparent gain of precision which is illusory, since one can estimate its slope with precision, even when there is no significant association between the variables, as a non-zero quantity. Now that it is possible to compute the confidence limits for the slope of the unreduced major axis of the bivariate dispersion matrix Jolicoeur considers it preferable to do so rather than to rely on calculations based on the reduced axis.

There are some remarkably difficult, even paradoxical, situations in the theory of allometric growth which have been most thoughtfully reviewed by Martin (1960). Sprent (1968) has noted that Jolicoeur's model, using essentially the first principal component as a growth vector, is acceptable to workers who believe that the remaining components define the error space. If one believes that this defines a space relevant to the shape of the organisms the problems sharpen acutely.

Some of the problems with which biologists, and in particular palaeontologists, are familiar continue to require attention in the multivariate field as they did in the univariate one. Burnaby (1966a) has remarked that biological taxonomists are often reluctant to employ multivariate methods in cases where the organism continues to grow throughout life. He also pointed out that growth is not the only generator of "nuisance factors" in taxonomy, but that other components of variation, in which the investigator may be quite uninterested, may occur. Burnaby has prepared a general procedure capable of eliminating such unwanted components of variation from the several multivariate techniques that we shall be concerned with.

Sometimes, these nuisance components interfere with the analysis of only some of the parts of the organisms under investigation. Hopkins (1966) reported that in rats, the liver, heart, spleen, lungs, skin and adrenal weights develop in a pattern consistent with a single allometric pattern, whereas the growth of the kidneys, genitalia and brain developed in a manner which was size-sensitive.

Although the allometry model put forward by Hopkins in the cited paper seems to us to be logical and useful, in actual practice the differences between coefficients found by principal components seem to differ very little from Hopkins' factor-analytic values. The differences obtained by Hopkins for the rat data could well be an artifact of the method of factor analysis used, namely, the centroid method of extraction of the "principal factor". Our attempts to duplicate his results indicate, moreover, that the dispersion matrix he used is not positive definite.

Ideas for the separation of shape and size components of the various contrasts of form with which one may have to deal date back to Penrose (1954). The extent to which it is preferable to eliminate such components of variation rather than to keep them in the calculations, and hence in the representation of the results, and so to study them on the same footing as other components of, perhaps, more obvious concern, is debatable. The present authors tend towards the practice of retaining all discoverable components of variation in the analysis at all stages.

From Ratios to Discriminants

Teaching experience suggests that biology students are often "left standing" at the beginning of a course on multivariate techniques, because the very idea of adding and subtracting characters in a discriminant function seems out of touch with their needs. The idea of standardizing the dimensions of an animal by dividing by some measured representative of the size of the animal is easily accepted, but adding and subtracting dimensions seems bizarre.

Middleton (1962) sets out a simple relationship, which helps biologists and geologists who find it easy to think in terms of ratios of characters to make the conceptual jump required to manipulate discriminant functions and other linear compounds. We offer this simple example whose logic applies just as much to the measurements of animals and plants as it does to the constituents of rocks. Middleton points out that knowledge of the ratio $y = Na_2O/K_2O$ in the analysis of a sandstone often enables the geologist to say what its tectonic origin (eugeosynclinal, taphrogeosynclinal or exogeosynclinal) may be. Omitting irrelevant constants this equation may be written

$$\log y = (\log Na_2O) - (\log K_2O) \qquad (4.1)$$

in which the potash content is subtracted from the soda content. If we write this equation in the more general form

$$\log y = b_1 (\log Na_2O) + b_2 (\log K_2O) \qquad (4.2)$$

we can see that the relationship remains unchanged only if $b_1 = +1$ and $b_2 = -1$. To make this equation as powerful a discriminator as possible we can change it in one of four ways.

Firstly, we can calculate the values of b_1 and b_2 which do the best job of discriminating between the broad categories of sandstone. Secondly, we can, if we have the necessary data, enlarge the equation by including the magnesia content, the lime content and so on in the hope that each component will help us to distinguish the nature of the sedimentation processes (eugeosynclinal or taphrogeosynclinal) that given rise to the sandstone. To add more elements in such a linear compound is easy for we then have

$$\log y = b_1 (\log Na_2O) + b_2 (\log K_2O)$$
$$+ b_3 (\log CaO) + b_4 (\log MgO) \ldots \qquad (4.3)$$

whereas to incorporate this extra information in the original ratio would be cumbersome to say the least. Thirdly, we can use our new-found freedom from the restraints of ratios by enquiring whether the logarithms of the characters are the most appropriate transformations to use. We might do just as well, or better, with the original arithmetic values or with some other transformation.

Middleton included the content of eight oxides in his analysis, finding that, when the simple arithmetic values were used, the soda content was most useful for distinguishing eugeosynclinal sandstones from exogeosynclinal ones, whereas the potash content best separated exogeosynclinal sandstones from taphrogeosynclinal ones; this last distinction was substantially improved in the case where logarithmic transformations of the data were used, by including the magnesia content of the rock.

Fourthly, there is no necessity to restrict ourselves to a linear discriminant function, since some or all of the characters may be raised to higher powers. A function of the type

$$y = b_1 x_1 + b_2 x_1^2 + b_3 x_2 + b_4 x_2^2 + \ldots \tag{4.4}$$

is a quadratic discriminant, and has been exploited for morphometric purposes by Burnaby (1966b). There is also an excellent theoretical paper by Cooper (1965).

5
Some Practical Considerations

The Choice and Number of Characters for Analysis

As soon as the material for a morphometric analysis is arrayed, the decision has to be taken as to the nature and number of characters to be studied.

Where the organisms are fairly closely related, and all the measurements are quantitative, few problems concerning the nature of the characters are likely to arise. Homologies are generally evident in such cases, and the main problem will be to try to spread the measurements as evenly as possible over the animal or plant, ensuring that at least something is measured on most of the major physically available components of the organism. Quite small numbers of characters are needed. Satisfactory analyses have been made with as few as three to six characters, although ten might be regarded as near optimal in a preliminary experiment, where not very much is known about the discriminating power of the various characters. As with many other aspects of morphometrics, it is hard to give advice in the abstract, but certain rules are helpful in particular instances. For example, with animals it is undesirable to measure nothing but the lengths of parts, some breadths should be included, or there is a risk that the discriminant will measure little but pure size. It may be undesirable to measure parts that are susceptible to shrinkage or expansion in specimens preserved in different ways, for example, dried or pickled insects. If in a preliminary investigation only a small fraction of some potentially larger body of material is used, it is inconvenient to employ characters that may be difficult or impossible to measure on material that one may want to include at a future date.

Where the organisms are less closely related, problems of homology arise. Sometimes, as in the Basidiomycetes, homologies may be so difficult to determine that satisfactory characters outside the stable sexual ones may be almost impossible to find (Kendrick and Weresub, 1966): the situation varies greatly from group to group. The character "length of wing" in Diptera (flies) is not homologous with the same nominal measurement in Strepsiptera (stylopids). Key (1967) has produced an operational definition of homology which seems entirely satisfactory, so long as one is prepared to accept a purely phenetic approach to the subject. This definition is "Feature a_1 of organism A is said to be homologous with feature b_1 of organism B if comparison of a_1 and b_1 with each other, rather than with any third feature, is a necessary condition for minimizing the overall difference between A and B." A more detailed discussion of homologies is given by Jardine (1969).

Since any numerical technique such as generalized distance analysis, or canonical variate analysis, involves the inversion of a matrix of rank equal to the number of measured characters, a certain amount of care is required not to exceed the capacity of the computer, whether human or machine. Considerable experience of inversion of matrices on desk machines suggests that a 13×13 matrix is about as much as can reasonably be tackled by hand. Larger matrices involve such voluminous calculations that slips in the arithmetic are likely to creep in as fast as they are cured, although a determined and persistent computer might manage a few more attributes. With electronic computers the problem is mainly that of care in the programming. Since matrix operations have a propensity to accumulate rounding-off errors, there is a risk that, if the capacity of the computer is stretched too far, some of the "results" may consist mainly of rounding-off errors. This situation can arise with certain types of desk calculation as well as with electronic machines. Where possible, the use of double-precision arithmetic on large computers is a worthwhile insurance.

Most University computing centres today dispose of a library of reasonably well-tried multivariate statistical programmes. Unless you have had some experience of programming, we advise you to discuss your particular computing problems with one of the systems analysts at the centre. He will be able to advise you on eventual limitations and restrictions of the computer, and on the true capabilities of the programme in which you may be interested. Unless you are aware of the possible drawbacks you are likely to do yourself far more harm than good.

If many or all of the characters are qualitative (dichotomous; presence or absence) so that the analysis takes on some of the characteristics of an exercise in numerical taxonomy, the requisite number may be increased to around 100. This figure has been discussed by Sokal and Sneath (1963) as the upper limit of the range of numbers of characters over which there is an appreciable improvement in precision as the number of characters is increased.

The topic of "how many characters" is inseparable from the question of redundancy in the character-suite, which is discussed later (p. 300). Anticipating this discussion, we may suggest that for qualitative characters 50 must be close to the limit beyond which redundancy makes further accumulation of data less profitable; everything, however, depends on whether the new characters contribute new information; whether, in fact, the addition of some new suite of, say, internal anatomic characters would add information that could not have been recovered from the indefinite accumulation of external morphological characters. Drastically new characters, such as would arise when biochemical characters were added to morphological ones, are likely to differentiate the material along new axes of variation rather than to add to the differentiation achieved along the morphological axes of variation.

Generally, redundancy is invited when a great number of variables are instituted for parts that are closely bound to each other. For example, there is no particular difficulty in finding redundant variables for bones of the skull of mammals. If you have succeeded in gathering together a great number of purely morphological variables you may expect many of them to be redundant. If, however, you find it possible to quantify qualitative data as well and to include dichotomous variables these will almost always enrich your total suite of variables with characters that convey additional information.

Palaeontology poses quite a difficult problem at times. Whereas zoologists often have to worry about which of a multitude of characters should be selected for measurement in order to obtain greatest efficiency in an analysis, the palaeontologist is frequently hard put to find a sufficient number of variables so as to make his analysis worthwhile. This is admittedly a rather extreme situation and should be accepted as such, but may be brought home by considering such a genus as *Cytherella*, the species of which ostracod are mostly as smooth as an egg.

Some palaeontologists, in their attempts to find a reasonably large number of variables for a study, become more enthusiastic than critical and manufacture new "variables" by making ratios of some of their measurements.

The Effects of Correlated Characters

Although there are very many references in the literature to the fact that characters are often correlated, positively or negatively, and to the possibility of using such correlations to determine the degree of "primitiveness" of an organism (Sporne, 1960) there is remarkably little information about the effect of varying degrees of correlation on any form of quantitative taxonomy. Indeed, there appear to be few examples in the literature where a taxonomic distance has been computed with and without an appropriate allowance for the degree of correlation between the characters used.

Rohlf (1967) claims that, in a numerical taxonomic analysis of the pupae of 45 species of mosquitoes, the only effect of using characters which were chosen for their high correlation was that the generic clusters appeared to be unduly elongated. It is hard to see the basis for this claim, because the pupal mosquitoes were not also classified on the basis of less correlated attributes, and if they had been, the validity of the comparison would have depended on the non-specificity hypothesis (see p. 241). The effective basis of the argument must lie in the fact that the pupae were classified by these characters into a hierarchy not very different from that obtained by classifying the adults of the same species, on the basis of characters that appeared to be less strongly correlated. Most classifications, however, are remarkably robust, and emerge, perhaps slightly mutilated, from a wide variety of technical operations designed to give them birth. In any event the elongation remarked upon by Rohlf seems to be confined to one of the five generic clusters, that for *Anopheles*.

In the early stages of numerical taxonomy, there was a marked emphasis on the production of phenograms (dendrograms representing relationships based on phenetic similarity rather than on common descent, as in a conventional phylogenetic diagram). More recently, there has been a movement towards the use of distance measurements or of various types of factor, principal components, or canonical variate analyses because of the great distortion that may be suffered by the relationships when these are forced into a dendrogram in two dimensions. There is thus more need to consider how the correlations between characters influence distance estimates.

Bartlett (1965) gives a lucid theoretical discussion of the problem based on Cochran's analysis of the situation (Cochran, 1962). Taking the simple case of two variables in Euclidean space, we begin with the familiar Pythagorean distance formula:

$$D^2 = d_1^2 + d_2^2 \qquad (5.1)$$

and rearrange this to give

$$D^2 = d_1^2(1 + f^2) \qquad \text{where} \quad f = d_2/d_1. \qquad (5.2)$$

If there is a correlation ρ between the two variables, then

$$D^2 = d_1^2 + \frac{d_1^2(f-\rho)^2}{1-\rho^2} \qquad (5.3)$$

The effects of correlation will augment the D^2, provided that

$$(f-\rho)^2 > f^2(1-\rho^2), \qquad (5.4)$$

so that negative correlation always increases the D^2. The effect of positive correlation is more complicated, and, perhaps, rather different from what an intuitive appreciation of the situation would lead one to expect. Certainly, positive correlation diminishes the value of D^2, but only up to a certain limit; once the correlation coefficient $\rho > 2f/(1+f^2)$ the positive correlation will also augment the D^2. It is worth looking at some of the numerical consequences of this relationship. If we have a marked disparity between d_1 and d_2 so that $f = 0.1$, any correlation coefficient greater than 0.22 will augment the D^2. For small asymmetries, such as would make $f = 0.4$, only high correlations above 0.928 will have this effect. Where f exceeds 0.4212, no value of the correlation can have the effect of increasing the D^2.

Table 1 provides an experimental test of the effects of computing generalized distances, with or without the elimination of the effects of correlation between the characters.

The example is a morphometric analysis of the shapes of two species of European wasps, *Vespula germanica* and *V. rufa*. The original data, with the D (not D^2) values, has already been published (Blackith, 1958), together with a generalized distance chart which was a three dimensional model. The same set of D-values has now been calculated without eliminating the effects of inter-character correlation.

The main influence of the correlation has been greatly to exaggerate the distances between those groups which were already well separated; that is to say, the sexual dimorphism in each species has been effectively doubled. The distances from queens to workers have also been much exaggerated, whereas those between males and workers are uniformly reduced. The net effect of these changes

Table 1

Experimental test of effects of correlations on generalized distances

	Queens	*Vespula germanica* Workers (large)	Workers (small)
Vespula germanica			
Males	13.32 (24.87)	6.60 (2.74)	6.57 (5.05)
Queens	0	10.57 (29.37)	12.01 (29.45)
Workers (large)		0	3.05 (5.63)
	Males	*Vespula rufa* Queens	Workers
Vespula germanica			
Males	4.43 (4.15)	11.01 (21.24)	7.95 (5.17)
Queens	14.37 (27.49)	5.95 (7.49)	12.39 (29.48)
Workers (large)	8.09 (3.68)	8.12 (19.69)	4.68 (5.62)
Workers (small)	6.80 (2.85)	12.59 (25.30)	2.30 (1.57)
Vespula rufa			
Males	0	11.29 (22.97)	6.89 (2.52)
Queens		0	9.96 (24.88)
Workers			0

Generalized distances between groups of wasps and, in parentheses, the distances uncorrected for correlations between characters.

would be to give the impression that the workers are very much more like the males than is in fact the case. The reason for this distortion is that the sexual dimorphism is essentially a matter of size (from the point of view of the gross biometry of the insects), whereas the contrast between the workers and the males is mainly a matter of differential growth patterns leading to insects of the same size, broadly speaking, but distinct shapes. Where one is dealing with large size differences, the correlations between the attributes are almost sure to be all positive, but where there is allometric growth leading to organisms of the same size, but different shapes, the correlations will contain quite a number of negative elements. It is noteworthy that the difference between the two species is not very seriously changed by failure to eliminate the effects of correlation between the attributes.

In another example, the comparisons were made between the generalized distances and "Pythagorean" distances for the skull measurements published by Mukherjee *et al.* (1955). Taking contrasts between the Jebel Moya people and several other African groups, we note that in almost all cases, the Pythagorean distance is at least twice as great as the generalized distance. If this increase were uniform, the effect would not be serious, since it would act as if there had been a scale change to a chart on which all dissimilarities were multiplied by some such factor as three.

However, the augmentation is not uniform, there is a tendency for those groups which are least removed from the Jebel Moya people to seem more remote than they really are. This tendency takes the form of a systematic distortion of the chart. As Seal (1964) comments, Sokal's (1961) suggestion of the "Pythagorean" distance for use in taxonomy may lead to unjustifiable conclusions, although its convenience and utility in situations where the data are entirely qualitative is great.

The effect of eliminating correlations between characters has been considered more in anthropology than in any other branch of biometry, because of the popularity of Pearson's Coefficient of Racial Likeness (C.R.L.): this is identical with the Generalized Distance, except for constant factors, and the fact that the C.R.L. as it is known to anthropologists, does not allow for the correlations between characters. Mulhall, in Talbot and Mulhall (1962), reaches the general conclusion that "allowance for correlation alters the magnitude of the coefficients significantly but not enormously", and again, that "the C.R.L. is an unsatisfactory measure of group divergence, and that neglect of the mutual intercorrelations between the characters can result in false conclusions being drawn regarding racial affinity". He appends a table of distances between various groups of Nigerians computed with and without the allowance for correlations between characters.

Some Comments on the Place of Significance Tests in Morphometrics

A high proportion of the papers and books published on the theory of multivariate analysis concern tests of significance. In this book, tests of significance are rarely mentioned, and then only in passing. Some explanation of this difference of emphasis seems called for.

Discriminatory analysis was first introduced as a test of multivariate significance, and this aspect of the subject is the one that naturally attracts the attention of mathematical statisticians because it provides a ready source of problems to be solved.

However, in many of the situations with which an experimental scientist has to deal, tests of significance are superfluous or secondary to the problems of ascertaining the structure of the experiment in multidimensional spaces. If a group of organisms is not known *a priori* to consist of definable subgroups, then a principal components, or principle coordinates, analysis seems the appropriate tool for probing its morphometric structure. Once the group is known *a priori* to consist of definable subgroups, an analysis along canonical variates seems to be the tool of choice for such an investigation. Once the decision to use canonical variates has been taken, tests of multivariate significance lose much of their point, for we know in advance that the subgroups differ. Whether they differ significantly in the particular experimental material that comes to hand in the case examined by the experimenter seems, on the whole, rather trivial: the outcome of the test will be influenced by the sample sizes, by the wise or unwise choice of variables to be measured, by the way in which variation along particular axes of growth or development happens to be represented in the material to hand (see p. 98), to name only some obvious factors. It is as much a mark of immaturity in the experimenter to leave multivariate problems at the level of the significance test, without proper estimation and interpretation, as it is commonly agreed to be in the univariate analysis of variance.

The point at issue is not primarily a statistical one: an entomologist investigating the form of insects in a bisexual species would rarely be well advised to test the significance of the sexual dimorphism, for a glance at the genitalia will settle the question of sex in most instances. A palaeontologist concerned with the question of whether or not some sexual dimorphism exists in fossil brachiopods might well take a quite opposite view, if by dissecting living forms he could gain an insight into the nature of the fossil record.

Another instance of significance testing which can be taken too seriously occurs in the tests for the roots of determinantal equations. The point is perhaps best illustrated by an example. Consider an organism which exhibits three polarities, which we shall suppose for convenience to be represented along orthogonal canonical variates. If, in the material to hand, the three polarities are about equally well represented, each will take up approximately one-third of the total variance. Such a structure can be determined by the measurement of only three characters of the organism: we may not consider this practice wise, but it is certainly possible. In such a case, the third root of the determinantal equation can never be significant in one statistical sense, but it may well order the material in a meaningful fashion, and in extreme cases, where there are many *a priori* distinguishable groups, it may even be possible to show the ordering to be significant by the use of non-parametric tests: if the third root generates a vector distinguishing males from females, and a non-parametric test shows that the individuals have been ordered by the third vector into these two groups, then the third root is in a very real and relevant sense significant. Such a possibility may well lie at the heart of the dispute between those who continue to practice discriminatory topology when there are more groups than characters, and those who see in such practices nothing but a lack of statistical sophistication, or even wrongheadedness. Some sense of balance is required here, as an extreme indifference to significance testing is highly dangerous: an extreme insistence on the letter of the statistical law (model) may frustrate an experimenter who can sense that he is being unduly restricted, but is not articulate enough to explain his misgivings.

A third issue concerns the applicability of those tests which are currently available. So far as we know, there is no way of testing the significance of the differences between two groups along a specified vector, other than the discriminant function. For instance, in a canonical variate analysis, the application of Wilks' criterion will test whether there is significant variation along any one of the variates, but it will not enable one to compare the means of two clusters of points (out of several). A similar problem can arise in principal components analyses. One solution which has been suggested (Blackith, 1965) is to compute the scores of the individuals along the vector in question and to perform a univariate test on these scores. Of course there may be qualms about non-normality of the scores (a graphical appraisal should be made) but since the test is univariate, or at least quasi-univariate, the traditional robustness of the t-test and the analysis of variance is likely to be relevant.

It seems to us worth the trouble to ventilate these problems now because it is unrealistic to expect progress in statistical theory to cover fields that, intrinsically, hold almost no interest for those likely to be able to resolve such problems.

It is hoped that these remarks will not be taken as meaning that we wish to belittle the importance of significance testing as a whole. This is by no means our aim and we advocate that where the situation is clearly understood, and significance tests have a biological meaning, it is foolish not to carry out the necessary appraisals.

6
The Use of Angles
Between Vectors

For certain purposes, it is helpful to be able to pick out some kind of biological contrast, common to a number of species, and compare the contrast in one species with that in another. For instance, sexual dimorphism is often similar in several related species, the males often being consistently larger or smaller than the females (Reyment, 1969d). Or one may be interested in the change of habit of several species of plants growing either in their normal habitat or on top of a windswept, overgrazed, mountain, where the habit of growth may be quite different, leading to stunted, prostrate forms, at first sight difficult to reconcile with the normal appearance of the plant.

One can, of course, proceed to analyse all available measurements made on all the material, so as to be in a position to construct a generalized distance chart, canonical variate chart or other model in order to display the relationships between the entities studied. However, such an elaborate procedure may not meet the experimenter's needs in cases where he is concerned only with one kind of variation. Such a case arose when it became desirable to examine the effects of crowding on the shapes of adult locusts (Blackith, 1962). A peculiar feature of the locusts is that although several species are known which respond to crowding by changes of shape, physiology and behaviour leading ultimately to swarm formation, other species do not respond in this way. Moreover, some genera include both swarming and non-swarming species, whereas the swarming species are to be found in different subfamilies, so that they are not of necessity closely related.

By calculating the discriminant functions between the populations of any given species which had been reared in crowded conditions of crowded parents, or had been relatively isolated for two generations, it was possible to obtain a vector describing the resulting change of shape ("phase" change) for each species. If each vector is represented as before by an equation of the form:

$$y_1 = a_1 x_1 + a_2 x_2 + a_3 x_3 + a_4 x_4 + \ldots a_i x_i \ldots$$

with a second discriminant function being represented by

$$y_2 = b_1 x_1 + b_2 x_2 + b_3 x_3 + b_4 x_4 + \ldots b_i x_i \ldots$$

then by a well-known relationship the angle θ between the vectors is given by

$$\cos \theta = \frac{\sum (a_i b_i)}{\sqrt{\sum (a_i)^2 \cdot \sum (b_i)^2}}. \tag{6.1}$$

In general, for two discriminants which represent unrelated patterns of growth, the angle will approach 90°, whereas when the two discriminants are substantially parallel, the angle will approach zero. The angle thus measures, objectively, the extent to which two taxonomic comparisons are alike.

When the discriminants had been computed for the two phases of each of ten species of swarming locust, they were arranged as shown in Fig. 4 with the angles between them corresponding to those calculated. It can be seen that the entire group of vectors forms a fan whose arc subtends an angle of less than 20°, which confirms the remarkable homogeneity of this phenomenon. Furthermore, when the same vector was computed for a non-swarming species *(Anacridium aegyptium),* the new vector was practically at right

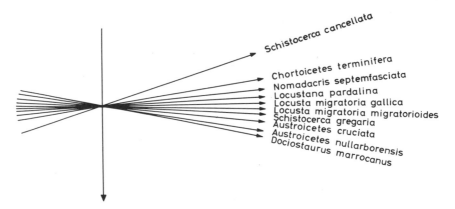

Fig. 4. Directions of vectors describing the response to crowding in ten species of swarming locust and one species of non-swarming grasshopper, based on three morphometric characters. Redrawn from Blackith (1962).

angles to the fan-shaped bundle of the "phase" vectors. This large "tree-locust" often forms aggregations in bushes in arid scrubland, but is not known to form true swarms, although a related species (*A. melanorhodon*) does so. However, *A. aegyptium* does differ in shape when reared in crowds or in isolation. Even more striking was the vector representing the effects of crowding in an insect that was not a locust at all, but a cotton stainer (a kind of bug), whose vector was found to project from the plane of the paper (not shown in the diagram).

This technique of "abstracting" vectors which describe some particular biological or geological quality might save time and enable the experimenter to concentrate his attention on what, to him, are the essentials of his problem.

A word of warning should perhaps be sounded in this connection. The angles between latent vectors or other discriminators will naturally be subject to quite considerable variability, which will be most pronounced where the sample sizes are small, say less than 30 observations per sample. In order to be reasonably reliable, angles calculated between vectors should therefore be a technique reserved for larger samples, except for work of a more exploratory kind.

Seal (1964) remarks that the comparison of a number of groups in a space defined by the number of characters, which will usually be greater than the number of groups, should be made in a hyperspace of dimensionality one fewer than the number of groups, and not in the original character-space. This comment is fair, but loses much of its point when one recalls that the object of the analysis is to reveal "structure" in a hyperspace which will often be defined by a few oligovariates; the dimensionality of the hyperspace so defined will, in many instances where a multivariate analysis has been a success, be so much less than that of either the character-space or the group-space than neither limiting dimensionality is relevant. Thus, although one should be aware of Seal's reason for concern about the use of angles between vectors as suggested above, it will only become a relevant part of the problem when this is strongly hyper-multivariate, as in some ecological experiments.

Reyment (1963a) has made graphical use of the relative orientations of the first latent vector of samples of various species of ostracods from different stratigraphic levels for the purpose of morphoevolutionary analysis and (Reyment, 1966a) the angles between first eigenvectors of morphometric dispersion matrices of related species of Palaeocene ostracods for morphometric comparisons in the same vein as outlined above for the locusts.

Another use of isolated vectors in morphometric work arises from the possibility of comparing observed vectors with hypothetical *a priori* ones. After, say, a principal components analysis has disclosed a general growth vector for a particular organism, this vector can be compared with one representing pure isometric growth which may be created synthetically as

$$1,1,1,1,1, \ldots 1,1,1$$

(Blackith, 1960; Jolicoeur, 1963b; Reyment and Sandberg, 1963) which defines the growth as being equal in respect of all measured characters.

Similarly, if the first four characters are length measurements and the second four (of eight) are breadth measurements, we can set up an hypothesis of attenuated growth as

$$1,1,1,1,0,0,0,0,$$

or in extreme cases as

$$1,1,1,1,-1,-1,-1,-1$$

(Jolicoeur, 1963b), and test any of these and other possible hypotheses against what has been observed by calculating the angle between observed and expected vectors.

By combining these vectorial representations with suitable transformations applied to the measured characters, many of the simpler kinds of allometric growth could find suitable expresssion. Jolicoeur (1963a), for instance, uses these synthetic, or *a priori* vectors, to good effect in testing hypotheses concerning the limb-bones and their robustness, a problem in multidimensional differential growth. Pearce and Holland (1961) also make use of the angles between vectors, and discuss how some simple vectors such as those mentioned above can be employed in growth studies.

While we are concerned with the general topic of calculating angles between vectors, there is a potential misunderstanding to which it may be desirable to call attention, arising out of the fact that vectors may have been calculated so as to be orthogonal in a space which has very different properties from the Euclidean space in which we live. An important property of this Euclidean space is that the sum of cross products of the coefficients which comprise two vectors is zero when they are uncorrelated. But if, for instance, the vectors are canonical variates, as discussed on p. 88, they will be orthogonal in the space from which the effects of correlations between the characters have been eliminated. Only in the improbable event of all the correlations between the characters being zero will the two canonical vectors be orthogonal also in the Euclidean space. Several cases have come to the writers' attention where biologists have tried to check the accuracy of canonical variate analyses this way, with unnecessarily disappointing results.

Case-histories Involving the Use of Multivariate Methods

For each of the multivariate methods described in the text we have included several "case-histories" to illustrate their use. To avoid a certain tendency for such illustrations to be copied from one book to another, we have chosen unfamiliar examples, or used original data, wherever possible; moreover, the fields from which the examples have been drawn illustrate, as far as is practicable, the range of topics which have actually been tackled by multivariate methods. It may be of more help to a potential user to have the problem stated in terms of situations with which he is familiar than to see what, in theory, may be a more comprehensive application but to a field in which the practical details are as strange to him as the biometrical ones.

There is also, we believe, a certain merit in illustrating the range of applications of morphometrics: these are spread over so many journals that the bibliography includes more than 150 of them, without any pretensions to completeness. Although the case-histories, combined with the bibliography, have not been designed primarily as a compilation of the uses of morphometrics, they should give a quick entry into the subject for many workers.

7
Formal Background of Discriminant Function Analysis

The statistical idea underlying the method of discriminant functions may best be discussed in terms of two universes. In the case of two universes, U_1 and U_2, reasonably well-known from samples drawn from them, a linear discriminant function is constructed, on the basis of k variables and two samples of size N_1 and N_2. On the grounds of measurements on the same k variables on a newly found individual the researcher wishes to assign this individual to one of the universes with the least chance of making a mistake. This presupposes that the individual actually does come from one of the universes. Let us just pause at this point in the narrative.

The supposition that the individual actually must come from either of the universes, is necessary from the purely statistical point of view, but it is one that makes rather poor biological sense and is even worse in the context of palaeontology and geology. As biologists, we would never be so bold as to make such a claim as the statistician tries to force us into. One could, naturally, with a tongue-in-the-cheek attitude, "say" that this is so, but usually the most logical decision would be that the specimen in question might come from one of the two universes, but that it is more likely to be *close* morphometrically to either of these theoretical universes. This brings us to a concept of *closeness* and it is in this context that discriminant functions are of particular service in morphometry.

Whilst we are fully aware of the fact that statisticians are not very happy about the biological use of discriminant functions, the closeness application is the one of most use in morphometric research, and the one we shall adhere to, in the main, in this book.

As a consequence, we reframe the two-sample concept in more biologically meaningful terms, as follows:

There are two universes for which a sample linear discriminant function has been constructed. A newly acquired specimen is of interest in that it may derive from either of the universes, or be morphometrically close to either of them. The biological usefulness of such a decision procedure is obviously far greater than the rigid concept of assigning the material to one of two populations.

As we have pointed out elsewhere in this text, the morphometric user of multivariate statistical analysis tends to be more preoccupied with the quantitative description and analysis of data, rather than with significance testing and the niceties of distribution theory. The application of discriminant functions made in this section offers a useful, albeit approximate, procedure for a graphical display of populations with respect to the degree of morphometric likeness between samples, drawn from various geographical and, or, stratigraphical locations.

The linear discriminant function between two samples may be defined as

$$Y = (\bar{x}_1 - \bar{x}_2)'S^{-1}X, \qquad (7.1)$$

where $\bar{x}_1$ and $\bar{x}_2$ are the mean vectors for the respective samples, S^{-1} is the reciprocal of the pooled sample dispersion matrix, and X is a vector of variables (for three dimensions it would be (X_1, X_2, X_3)).

The coefficients of the discriminant function are defined as

$$a = S^{-1}(\bar{x}_1 - \bar{x}_2), \qquad (7.2)$$

where a is the vector of coefficients. If the variances of the variables are almost equal, the discriminator coefficients give an approximate idea of the relative importance of each variable to the "discriminatory power" of the function.

The linear discriminant function is connected with the Mahalanobis' generalized distance by the relationship:

$$\begin{aligned} D^2 &= (\bar{x}_1 - \bar{x}_2)'S^{-1}(\bar{x}_1 - \bar{x}_2) \\ &= d'a, \end{aligned} \qquad (7.3)$$

where the vector d is the difference between the two sample mean vectors. D^2 is thus the inner vector product of the difference mean vector and the vector of discriminator coefficients. For those interested in a detailed exposé of discriminant functions, good books to study are those of Anderson (1958) and Rao (1952). Discriminant analysis with presence or absence data has been illustrated by Cochran and Hopkins (1961).

A generalized concept of what is done by a discriminant function analysis under ideal conditions is given in Fig. 5. This geometric interpretation of the procedure applies to two groups and two variables. The two variables, X and Y, are slightly correlated. A line

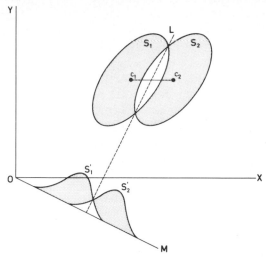

Fig. 5. Schematic diagram indicating part of the concept underlying discriminant functions. The line C_1-C_2 is a measure of the statistical distance between the two samples (developed from a diagram in Cooley and Lohnes, 1962).

through the points at which the ellipsoids of scatter S_1 and S_2 cut each other, denoted by L, and a perpendicular, M, from it to the origin of the coordinate system, O, are marked. The major axes of the ellipses are parallel and of equal length (i.e. the ellipses are "equally inflated"). The projection of the ellipsoids onto the perpendicular gives the two univariate (normally distributed) distributions S_1' and S_2' shown in the diagram. This serves to illustrate one of the uses of the discriminant function, in which the transformed values of the observational vectors may be used in a univariate comparison. The two distributions of our schematic example in Fig. 5 display a slight degree of overlap.

MULTIPLE DISCRIMINANT ANALYSIS

There are two ways of presenting this topic. The logical extension of the case of two groups, just discussed, is to a concept of linear discriminant functions between more than two groups. Thus, one would wish to construct g discriminant functions between each of g groups. This approach requires the assumption that the probability that a specimen being classified falls into one of the groups is $1/g$ for each of the groups. This is, however, an unreal assumption to make and in some studies, there may be good *a priori* reasons for weighting the probabilities so that the likelihood of a certain specimen falling into one or more of the groups is greater than for its landing in others. This is the method of multiple discriminant analysis reviewed in Anderson (1958).

A second line of attack is by means of the method of canonical variates. This canonical analysis consists of a transformation of the original observation space, which results in a reduction of the within-sample ellipsoids of scatter to spheres. As a consequence, the latent vectors of canonical variate calculations are not at right angles to each other (as they are in principal components). The canonical discrimination method thus produces a set of discriminant functions which are the latent vectors corresponding to the latent roots of the generalized determinantal equation of this canonical analysis. The coefficients of the variables (characters) in the discriminators are the corresponding elements of these latent vectors. The non-significance of a latent root does not necessarily prove that the associated canonical variate is non-existent, negligible, or useless as a discriminant function. What it does tell us is that when the individuals and the group means are projected on to the canonical variate, the differences among the group means are small in relation to the differences among individuals within a group. Experience shows that not infrequently canonical discriminators connected with quite small latent roots may be more efficient identifiers than those associated with larger roots.

As Kendall and Stuart (1966, p. 314) aptly underscore, the discriminant function identifies and does not classify. The classification problem is logically quite distinct from the identification concept and it is to be lamented that many statisticians speak of "classifying" when they work with discriminatory problems.

The Problem of Misidentification

Misidentification, that is, the incorrect identification of a specimen with a universe, is a subject that has greatly interested statisticians and there is now a very extensive corpus of literature on the subject. Its morphometric importance is not great and it is only in special situations that one would be genuinely interested in carrying out the tests for misidentification. The subject is taken up at some length in Kendall and Stuart (1966).

Homogeneity of Dispersion Matrices

Here, as in so many other connections in multivariate work involving two or more groups, the theory presupposes that the universes are multivariate normally distributed and that the sample dispersion matrices are statistically homogeneous. In many cases this requirement is met, particularly for biological data, but deviations from the ideal state are sufficiently common to warrant attention. Providing the heterogeneity is not grotesquely manifested, little harm is done in performing the calculations in the usual way. In most situations of a doubtful or marginal nature, a few exploratory empirical studies of the data will often be sufficient to make the experimenter realize that he should make use of the special procedures available for analyses involving heterogeneous dispersion matrices. Some aspects of problems arising in this connection are taken up in Chapter 29 on the study of sexual dimorphism in fossils.

CASE-HISTORIES INVOLVING DISCRIMINANT FUNCTIONS

Distinctions between Morphometrically Similar Beetles

Some species of animals are almost indistiguishable in external appearance; identifications, to be reliable, need to delve more deeply into the organism than is done in the conventional key. Lyubischev (1959) has studied such situations in the Chrysomelid beetles of the genus *Halticus*. Here the beetles may differ only in respect of the male genitalia, but the dissection and study of the genitalia is a lengthy task, and may be quite impracticable if, for example, museum regulations forbid the dismembering of specimens.

For two species of *Halticus* from the European part of the USSR, Lyubischev examined 21 characters and found that none of them gave frequency distributions for the two species (*H. oleracea* and *H. carduorum*) that did not overlap seriously. However, the four best of these characters, when combined into the discriminant function

$$y = 1.00x_5 + 0.40x_{14} + 0.25x_{17} + 0.40x_{18} \qquad (7.4)$$

avoided all overlap between the two species in the sample on hand, and reduced the potential overlap in very large samples to one specimen in every 33 examined.

With computer facilities available, one would be inclined to place all the characters into the preliminary discriminant function, to save the many eliminatory trials that Lyubischev carried out by hand. If a more concise function were needed, for routine identification, the less useful characters (those that carried the lowest coefficients in the discriminant) could be dropped, until the frequency of misidentification had reached a predetermined acceptable level.

A Discriminant Function to Distinguish Asiatic Wild Asses

Nine measurements of the skulls of two samples of wild asses, Khurs from the Little Rann of Cutch, and Khiangs from Ladakh and Western Tibet, were converted into a discriminant function by Groves (1963). Nine Khurs and twelve Khiangs were available, together with much smaller numbers of several other groups of Asiatic wild ass. This discriminant, despite the small samples and the distortion of some of the skulls because the animals had lived on unsuitable diets in captivity, clearly distinguished the two groups. Three of the attributes had positive signs in the discriminant, six had negative ones, so that the differences are not primarily of size but involve changes of shape. A certain amount of sexual dimorphism was noted in the Kiangs but not in the Khurs. On the other hand, the Khurs showed most distortion of the skull due to life in captivity, with only a small overlap of scores between the wild and the captive specimens. Once the framework of reference had been set by calculating the discriminant, the other isolated skulls could be fitted into the picture. There is always a risk that new material, brought into an analysis after the main calculations have been done, may differ from the earlier material in some dimension for which the earlier analysis cannot allow. This risk has to be balanced, in each case, against the certain loss of information if odd samples of what is, in the nature of things, likely to be rare material are rejected on the grounds that there is insufficient replication for them to be included in the main analysis. The fact is that a single specimen is incomparably more useful than no specimen at all, and can often be made to serve a useful purpose in a multivariate analysis if, as an isolated point, it is fitted into a conceptual framework constructed with other more plentiful material of the same general form.

Changes of Shape in Whitefly Growing on Different Host-plants

Whiteflies of the genus *Bemisia* are pests of cassava, cotton and tobacco plants in West Africa. It has been strongly suspected that the Cassava, Cotton, and Tobacco Whitefly are all members of the same species, but that they grow up to look different according to the host-plant. To clarify this issue, Mound (1963) measured seven characters of the fourth-stage larva drawn from populations on cassava and tobacco plants all larvae being originally descendants of a single parthenogenetically reproducing female. These seven characters were: length and breadth of pupal case (the fourth instar larva is commonly referred to as a pupa); length and breadth of vasiform orifice; length and breadth of the lingula tip; and length of the caudal furrow. The discriminant function which optimally separates the two populations on this basis is: 0.23; −0.38; 0.35; 2.69; 2.08; −3.99 and −0.73.

The corresponding variance ratio (Wilks' Criterion) for 562 degrees of freedom was 14.23, with a probability of the null hypothesis being true of less than 1 in 10,000, so that there can be no doubt of the reality of the host-correlated variation. The nature of the discriminant function shows that the form from tobacco is longer and narrower than that from cassava, and its vasiform orifice is much broader and slightly longer. The lingula tip is narrower and longer in the form from tobacco, and the caudal furrow shorter. The fact that three of the seven coefficients in the discriminant function are negative whereas four are positive makes it clear that a general change of shape rather than of size is involved. If all, or nearly all, the coefficients were of the same sign one would suspect that the difference between the two forms was mainly one of size, and hence probably due to differences in the nutritional status of the two populations. In fact, it is the growth pattern, rather than the amount of growth, that is changed by the host-plant.

In a somewhat similar way, Fraisse and Arnoux (1954) found that the shape of the silkworm cocoon varied systematically according to the diet of the developing larvae. These authors also used a discriminant function to assess the changes of form.

The Assessment of Ear-formation in Barley Breeding Experiments

In breeding plants for better yield, Mather and Philip (in Mather, 1949) note that just as a single character can be resolved into subcharacters, so can characters be compounded into supercharacters such as yield, ear conformation etc., which, taken together, form an estimator of the total merit of a variety. They in fact compounded the length of the ears (neglecting the awns); the maximum breadth; and the combined length of the central six internodes. Their objective was not so much the discrimination of one plant variety from another, but to maximize variation between plants, relative to that within plants, so as to produce a measure of ear conformation whose genetical variation is at a maximum compared with at least one important kind of non-heritable variation. This vector turned out to be 1, −5.3, 5.8 when the two varieties Spratt and Goldthorpe were contrasted. These varieties differ in the genetical architecture of the ears, but no simple Mendelian differences could be detected between them, variation being continuous in the F_2 generation.

Partition of Multivariate Variation and Shape and Size Changes in an Ostracod Species

Reyment (1966a, p. 118) used Dempster's (1963) method of principal variable analysis, a form of stepwise multivariate analysis of variance, for studying the variation in the carapace of the Palaeocene ostracod *Iorubaella ologuni,* in order to attempt the identification of possible environmental influences on the morphology of the shell. The method is based on an analysis of variance in steps, related to each of the latent roots. The analysis disclosed that significant differences in mean vectors have resulted from pure size differences as well as shape differences.

Fossils in Ice-transported Blocks—an Example of Multiple Discrimination (Reyment, 1968)

Blocks of limestone of various ages are widely spread around the Baltic area as a result of the transportational activities of the ice sheets of the last Ice Age. Blocks of limestone containing the Palaeozoic ostracod genus *Steusloffina* found in mid-eastern Sweden and Gotland were thought to be relatable to three infraspecific groups based on *in situ* occurrences. A multiple discriminant analysis was carried out to produce discriminators between the three groups. The variables measured on the ostracod carapaces were: length of carapace, distance of the lateral spine from the anterior margin and its distance from the dorsal margin. Five samples from five different blocks from the Baltic area were substituted into these discriminators. It was found that one of these could be identified with the first region of classification, two with the third. The results make good palaeontological sense.

Comparisons of Linear and Quadratic Discriminant Functions

The input spectrum from ten vocalized monosyllabic words (Bit, Bet, Bat, Bot, But, Bert, Beet, Boot, Book and Bought) was assessed in terms of the output from 35 band-pass filters by Smith and Klem (1961). A machine was programmed to recognize each sound by means of a discriminant function, with variables made up of these outputs. The machine calculated the likelihood that each input sound belonged to a particular one of the ten words and allocated the sound to the most probable word. Using a linear discriminant function the machine was able to allocate 87% of its input correctly. With a quadratic function it improved its performance to 94%, but it scored the most "difficult" word (But) correctly only 71% of all the times that this word was vocalized, irrespective of whether a linear or quadratic function was used. The substantial extra work involved in computing the quadratic function hardly seems justified for a 7% improvement in performance, especially as this improvement applies to recognition of words that are only rarely misclassified by the machine. One can well imagine a situation, such as space travel, however, where the additional expense would be quite trivial in relation to the cost of the project as a whole. Every case must be judged on its merits, and other authors such as Welch and Wimpress (1961) have chosen linear discriminants for vowel recognition.

Where the discriminant function is to be used in systematic work, however, the quadratic form is often markedly superior to the linear function (Burnaby, 1966b; Cooper, 1963; 1965; Lachenbruch, 1968; Lerman, 1965).

Where there are several parent populations to be compared even the arithmetical work need not be greater than for the linear discriminant (Burnaby, 1966a), but it is desirable to transform the data so that the dispersion matrices of the populations are equal.

Anthropometry of the Skulls of the Jebel Moya People of the Sudan

The Jebel Moya people of the Sudan were a negroid people who moved into the southern Sudan during the first millenium B.C. and whose affinities have long been a source of speculation. One approach was made by Mukherjee et al. (1955), who compared as many of the skulls as could be recovered more or less intact (from a large collection with an unfortunate history), with skulls drawn for candidate populations from which, at some historical or prehistorical era, the Jebel Moya people might have arisen. The method of comparison chosen was to compute the generalized distances between all the various groups of skulls, using seven morphometric characters. We have adapted the diagram presented in the publication of Mukherjee *et al.* to show more clearly the fact that the

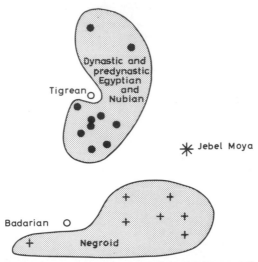

Fig. 6. Generalized distance chart showing the affinities of the Jebel Moya people of the Sudan, based on seven cranial characters. Adapted from Mukherjee *et al.* (1955).

populations chosen for comparison with the Jebel Moya skulls form distinct clusters according to their racial affinities, the two main groups being the dynastic and predynastic Egyptian and Nubian peoples, whose skulls were derived from tombs of at least approximately known data, and the peoples of negroid stock who include a population of Nilotic negroes from Egypt (Fig. 6).

The Jebel Moya people stand out as clearly distinct from all the candidate populations chosen for comparison, although presenting some negroid features. It is interesting that many of the skeletal features, and also some of the cultural traits, found in these ancient graves could be found in the local population of the area when the excavations were performed between 1911 and 1914. Two other peoples, the Tigréans and the Badarians, seem difficult to place on the chart. The Tigré-speaking people of Eritrea have some local traditions of connections with the classical peoples of Egypt, which seems to be supported by the morphometric findings of Mukherjee *et al.* who do not mention these traditions. Although much mixed ethnically and culturally nowadays, as an East African coastal population is bound to be, the Tigré seem to form a distinct entity. The Badari population comprises the earliest known collection of skulls of any substantial numbers from Egypt. Obviously they represent predynastic peoples, but little seems to be known of their origins. On the multivariate evidence they seem to be negroid, but on classical archeological grounds several of the populations which cluster with the Egyptian/Nubian populations have been considered negroid and the criteria perhaps leave something to be desired.

This investigation stands as a warning to biometricians not to take their material for granted: out of the remains of 3137 people excavated in the Sudan, only 15 complete sets of seven measurements could be obtained for the work reported here. Indeed, much of the multivariate effort in this work went into the correction of the earlier anthropologists' attempts to sex the bones, a subject discussed in more detail in Chapter 29.

Another anthropological investigation employing discriminant functions is that of Kaufmann *et al.* (1958), who made eight measurements on over 700 Swiss villagers. A wide range of anthropological and geological applications of the discriminant function is given by Griffiths (1966).

Phenotypic Flexibility of Shape in the Desert Locust

In a very full investigation of the form of the Desert Locust, as assessed by the three characters, head width, length of hind femur, and elytron length, Stower *et al.* (1960) acquired samples of locusts from a variety of East African localities. These localities included some that were particularly hot whilst the locusts were breeding and some that were cooler. In some places the locusts were breeding at high densities, in which case the adults have a characteristic body-shape, colour, and other characters which place them in the phase *"gregaria"*. At other places, the locusts were relatively isolated during their development, leading to adults of the phase *"solitaria"*. In some instances, it was possible to obtain samples from a parental population and from its progeny, which enabled the authors to examine the remarkable phenotypic plasticity of locusts, leading to changes of body-shape when the progeny are reared under conditions different from those in which the parental population had been reared.

By computing the generalized distances between each sample, and constructing charts showing the relationships, many interesting facts were established. Basically, there are two fundamentally distinct dimensions of variation, the one dominated by sexual dimorphism, the other dominated by "phase" variation reflecting the density at which the locusts were reared. Both these dimensions of variation, however, can reflect changes of shape associated with factors other than sexual dimorphism and phase variation. The "phase" axis, which is almost orthogonal to the "sex" axis, also carries the effects of temperature differences during the period of development of the young locusts, in such a way that those reared under hot conditions resemble *"solitaria"* locusts, whatever their rearing density. The effects of rearing density and rearing temperature are thus "symbatic" in the sense of Blackith and Roberts (1958).

It seems likely that the use of only three characters rather limits the discriminatory power of the technique, not so much in providing significant discrimination between populations, but in providing significant discrimination between axes of variation. The authors suggested the inclusion of a further character, length of pronotum, a suggestion which unfortunately, has not as yet been adopted.

This situation is instructive in showing how widely disparate are the points of view of people with different backgrounds; a numerical taxonomist would probably be disconcerted to see as few as four characters employed in a serious study. A biologist conditioned to accept the conventions of research into locust morphometrics, where, as in anthropometry, ratios of two characters are the rule as working tools rather than the exception, might regard the simultaneous handling of four variables as a radical departure from the framework of ideas within which he has learned to think. The extra power of the multivariate approach seems well worth the added trouble. Even with three characters, it enabled Stower *et al.* to determine the nature of populations of the phase *"congregans"* (i.e. parental population isolated, progeny crowded) and those of the phase *"dissocians"* (i.e. parents crowded, progeny isolated) from the fact that *congregans* populations have larger adults and *dissocians* populations have smaller adults, thus shifting the populations along the axis of variation representing sexual dimorphism because this axis also, to a large extent, reflects the size differences. A somewhat similar analysis using canonical variates (Symmons, 1969) shows that the first variate (of three examined) can be used to order locusts according to their "phase" status.

Later, Davies and Jones (unpublished) showed that a third

dimension of variation reflected differences between the Desert Locust *(Schistocerca gregaria)* and other species of the same genus such as *S. americana* and another subspecies of *S. gregaria, flaviventris.* In this Davies and Jones agreed with Albrecht and Blackith (1957) who found that the contrast of form between the genera *Schistocerca* and *Nomadacris* lay at right angles to the "phase" and "sexual dimorphism" axes within each genus. That these axes of variation can be found almost unchanged in different species and genera is of obvious genetic interest (see Chapter 18). We are most grateful for permission to use this additional information.

The Identification of Sex and Race from Skeletal Measurements

We follow the acridological example with an anthropological one: the juxtaposition is appropriate because the two subjects have had much in common historically, turning to bivariate numerical techniques at an early date, but resisting the introduction of multivariate methods for a long period. Many of the same misunderstandings can be traced in the literature of the two subjects which are alike in requiring very fine distinctions of shape to be made at an infraspecific taxonomic level.

The sexing of skeletal human material has not been without its difficulties: when even the most experienced workers have tested themselves "blind" against material of known sex the scorings were generally in the range 75-85% correct. According to Giles and Elliot (1963), the introduction of a nine-variable discriminant function only raises this rate to some 82-89%, but it does enable this determination to be made by less experienced workers. These authors suggest that the skulls for which wrong identifications were made, either visually or as a result of the application of the discriminant function, included a fair proportion of material from individuals whose hormonal balance was tipped away from that appropriate to their "chromosomal" sex and whose skeletal conformation reflects this history.

As soon as judgements, built up on the basis of abundant contemporary material, have to be transferred to fossils, or subfossils, difficulties may arise. The application of discriminant functions may not resolve all these difficulties, but it does sometimes correct doubtful subjective judgements. Giles and Elliot quoted an example of a palaeo-Indian skeleton, previously judged subjectively to be female, which seemed to be distinctively male when measured and assessed on the multivariate basis. Moreover, Irish skeletal material from a monastery burial ground (sixth to sixteenth centuries) gave a lower proportion of females than had been assessed visually.

An extension of this approach to the medico-legal topic of race identification is also reported by Giles and Elliot (1962).

Craniometric and other anthropological studies have the honour of being the first field in which the generalized distance was applied (Mahalanobis, 1928) and have since attracted the attention of many workers interested in multivariate methods (Martin, 1936; Pons, 1955; Rao, 1961; Giles and Bleibtrau, 1961; Campbell, 1962, Defrise-Gussenhoven, 1957, to name only a few leading exponents).

Discriminant Functions and Time-series of Fossils

The unique contribution of palaeontology to morphometrics is the possibility it offers of studying morphological changes in a species over a period of time. Reyment and Naidin (1962) analysed chronologically separated occurrences of the belemnite *Actinocamax verus* (Miller) from the Senonian of the Russian platform. Although many methods of morphometric analysis were applied to these data, a discriminant analysis of the sample means, viewed as a time-series, was not tried. The dispersion matrices of the oldest (here a Santonian sample, as the Coniacian dispersion matrix was shown by Reyment and Naidin (1962) to diverge widely from the other seven dispersion matrices) and youngest samples were used to construct a discriminant function

$$y = X_1 - 11.06X_2 + 14.06X_3 - 1.59X_4, \qquad (7.5)$$

in which the seven sample means were substituted; and the result used to construct Fig. 7. The four variables analysed are morphometric characters of the rostrum of the belemnite. There is a significant grouping of the two Santonian samples and of the five Campanian samples. The Campanian samples fall in rough chronologic order, although this ordering is by no means fully certain, as it is based on unorderable field observations. It is therefore of interest to test the approximate morpho-ranking procedure we use here on data, the exact order in time of which we are sure.

Reyment (1963a) published data on the dispersion matrices of the Palaeocene ostracod *Trachyleberis teiskotensis* (Apostolescu) from the Gbekebo area of Western Nigeria. The morphological characters measured were: length, height and breadth of the carapace. The chart of discriminant scores for six samples of this ostracod, for which a

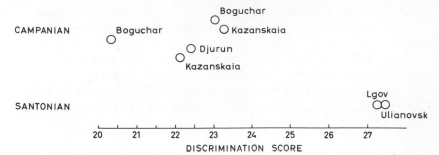

Fig. 7. Schematic diagram showing the mean discriminant scores for seven samples of Senonian belemnites from the Russian platform. The localities of the samples are given on the chart.

discriminant function was calculated between the youngest and oldest samples, is given in Fig. 8.

The interesting feature of this diagram is that there is an S-shaped curve of the multivariate size-shift over the time interval considered. There is a jump from (1) to (2), thereafter a vertical section from sample (3) to sample (5), and again a jump to (6). There may be no particular significance to the shape of the line, but it does happen to resemble the curve of logistic growth to a surprising degree. The

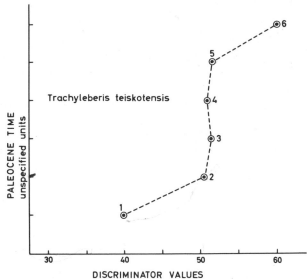

Fig. 8. Schematic diagram showing the plot of mean discriminant scores for six samples of the Paleocene ostracod *Trachyleberis teiskotensis* (Apostolescu) from a borehole at Gbekebo, Nigeria.

essential feature of the chart is that it does expose a definite shift in size and shape over time. The discriminator equation between samples (1) and (6) is:

$$Y = X_1 - 8.09X_2 + 5.11X_3. \tag{7.6}$$

The discriminator coefficients supply some further information. As already observed in an earlier part of this chapter, the sizes of the coefficients give a rough measure of the relative importance of each of the characters in the discrimination between groups. Here we see that length of carapace (of females) is of slight importance, whereas height and breadth are of approximately the same order of consequence.

Assigning specimens to either of two groups

Studies of the foregoing kind may often prompt the investigator to take up further questions. For instance, do any of the intermediate evolutionary phases in a multivariate chronocline agree sufficiently closely with either of the end members of a sequence as to be indistinguishable (statistically) with respect to the chosen variables?

We have already pointed out that this application of discriminant functions is at variance with the established statistical theory, all the more so in this case, because we actually *expect* at least some of the intermediary evolutionary phases to deviate significantly from one or other of the end groups. A suitable test for backing up such an enquiry may be constructed in the following way.

Consider a p-variate vector of values, $\mathbf{T}$, measured on a multivariate specimen (a mean vector may be inserted instead). The difference vector, $\mathbf{d}_1$, between this vector and the mean vector, $\bar{\mathbf{x}}_1$, of the first end sample is used to compute a chi-square:

$$\chi^2 = \mathbf{d}_1' \mathbf{S}^{-1} \mathbf{d}_1, \tag{7.7}$$

where $\mathbf{S}$ is the pooled dispersion matrix of the two end samples. The significance of the result is assessed for p degrees of freedom. The test should then be repeated for the second sample. The analysis of the Palaeocene ostracod material by this procedure gave the following results.

The second sample differs significantly from the earlier endpoint ($\chi^2 = 8.3$) but not from the later one. The third sample of the time series does not differ significantly from either of the arbitrary endpoints with respect to the characters studied, thus emphasizing its intermediate status in the evolutionary sequence. Sample 4 is significantly different from the first endpoint ($\chi^2 = 8.0$) but not from the last one, while sample 5 lies nearer to the second sample than the first.

Ratios of generalized distances in morphometric studies

Dempster (1964) introduced the concept of what he terms "covariance structure"' The idea is based on the comparison of ratios of generalized statistical distances and distance-like quantities. Reyment (1969f) has applied the method to fossil material as a means of tracking down the possible sources of the variability in fossil associations. Variation in a species of living, agglutinating foraminifers was used as a guide and the results obtained for that material were interpreted as being of possible genetic origin. Two species of Devonian brachiopods were also analysed in this paper. One of these was concluded not to vary significantly from sampling locality to sampling locality, while the second species could be shown to reflect sample differences of putative genetical origin.

A COMPUTER PROGRAMME FOR CALCULATING GENERALIZED STATISTICAL DISTANCES AND THE ANALYSIS OF DISPERSION MATRICES

The programme presented here is taken from a publication by Reyment, Ramdén and Wahlstedt (1969). It was designed for the calculation of generalized distances, for situations in which the dispersion matrices are not homogeneous, in addition to the normal case in which they are. It is particularly useful for the study of sexual dimorphism in fossils (see Chapter 29).

The theoretical background of the computer programme is as follows.

The Generalized Statistical Distance

Let N_1 and N_2 be the sizes of two samples drawn from two populations, each based on p-variates. The sample means (sample mean vectors) are $\bar{x}_{i1}$ and $\bar{x}_{i2}$ for the first and second samples respectively. The estimated value of the dispersion matrix is given by:

$$S = \left(\frac{1}{N_1 + N_2 - 2}\right) = \sum_{t=1}^{N_1} (x_{i1t} - \bar{x}_{i1})(x_{j1t} - \bar{x}_{j1})$$
$$+ \sum_{t=1}^{N_2} (x_{i2t} - \bar{x}_{i2})(x_{j2t} - \bar{x}_{j2})$$

The standard form of the Mahalanobis' generalized statistical distance squared between the two populations, as estimated from the sample for the p characters, is:

$$D_s^2 = \bar{d}'S^{-1}\bar{d}, \tag{7.8}$$

where S^{-1} is the inverse of matrix $S, \bar{d}$ is the vector of differences between the vectors of means of the two samples, and D_s^2 denotes the usual form of the generalized distance squared.

The test of the hypothesis specifying no difference in the p mean values for the two populations is carried out by means of the statistic

$$\frac{N_1 N_2 (N_1 + N_2 - p - 1)}{p(N_1 + N_2)(N_1 + N_2 - 2)} D_s^2, \tag{7.9}$$

which may be used as a variance ratio with p and $(N_1 + N_2 - p - 1)$ degrees of freedom. The generalized distance is connected to the Hotelling T^2 by the relationship

$$T^2 = \frac{N_1 N_2}{N_1 + N_2} D_s^2. \tag{7.10}$$

The Mahanalobis' distance has been widely used without detailed consideration of the effect of inequality of dispersion matrices. The significance level and power of T^2 may be appreciably influenced by inequality in sample sizes.

Anderson and Bahadur (1962) suggested a generalized distance for use in the case where the covariance matrices are not equal. In terms of sample quantities, the distance measure, here denoted D_h is:

$$D_h = t^{\max} \frac{2b'\bar{d}}{(b'S_1 b)^{\frac{1}{2}} + (b'S_2 b)^{\frac{1}{2}}} \tag{7.11}$$

where $b = (tS_1 + (1-t)S_2)^{-1}\bar{d}$, and $\bar{d}$ is the vector of differences of the sample means and S_1 and S_2 are the respective sample covariance matrices. When the population covariance matrices, $\Sigma_1 = \Sigma_2$, expression (7.9) becomes equal to the normal form of the distance. The method of finding D_h is therefore iterative and, without the aid of an electronic computer, must be regarded as prohibitively arduous.

D_h^2 has not yet been related to a test of significance. Therefore if it is desired to carry out an Hotelling T^2-test, and it has been ascertained that D_h^2 and D_s^2 are very different, then the significance may be checked by using the Bennet (1951) solution of the generalized Behrens-Fisher problem (cf. Anderson, 1958). The generalized distance squared statistic proposed by Reyment (1962), here referred to as D_r^2 is defined as

$$D_r^2 = 2\bar{d}'S_r^{-1}d, \tag{7.12}$$

where S_r is the sample covariance matrix of differences obtained from the random pairing of the two samples. Since $(N_1/2)D_r^2 = T^2 (N_1 < N_2)$ is an Hotelling T^2 with p and (N_1-p) degrees of freedom, the distribution of D_r^2 is known. This squared generalized distance suffers from the disadvantage that in order to obtain a usable value, it is necessary to average the results obtained from a number of permutations. This is naturally only feasible where the calculations are made on a computer.

An approximate means of producing a squared generalized distance D_a^2 for heterogeneous covariance matrices is by simply averaging the two covariance matrices.

Thus,

$$D_a^2 = \bar{d}'S_a^{-1}\bar{d},$$

where $\bar{d}$ has the same meaning as before,

$$S_a = \tfrac{1}{2}(S_1 + S_2), \tag{7.13}$$

and S_1 and S_2 are the sample dispersion matrices. The values of D_a^2 are very close to D_h^2 and D_s^2 and for most purposes offer a perfectly adequate solution to the question of a generalized distance for populations with unequal dispersion matrices.

The Equality of Dispersion Matrices

The generalized test for homogeneity of dispersion matrices (Kullback, 1959, p. 317) is:

$$2I(H_1:H_2(*)) = N_1 \log_e (\det S/\det S_1) + N_2 \log_e (\det S/\det S_2) \tag{7.14}$$

where S_1 and S_2 have the same meaning as in (7.11), N_1 and N_2 are the corresponding sample sizes, "det" denotes the determinant of a matrix and $NS = N_1 S_1 + N_2 S_2$ $(N = N_1 + N_2)$. This is the form of the test used in the computer programme employed in this study. It is approximately distributed as χ^2 with $k(k+1)/2$ degrees of freedom, where k is the number of variables. A result indicative of heterogeneity in dispersion matrices yielded by this procedure may be sufficient for many statistical purposes. Generally, however, biological studies of the kind considered in this book require further information on the nature of the heterogeneity. The first point of interest concerns the range of variability in each sample as represented in the dispersion matrices. Conceptually, this situation may be thought of in terms of inflation of the ellipsoids of scatter and is thus an expression of the degree of multivariate variability in the samples relative to each other. The second point of interest concerns the relative orientations of the ellipsoids of scatter and the connotations hereof in growth interpretation (cf. Jolicoeur and Mosimann, 1960).

An approximate test of the hypothesis that a given latent vector is the ith latent vector of a very large sample estimate of a dispersion matrix may be produced by adopting a procedure suggested by Anderson (1963) (cf. Reyment, 1969a, p. 574, and Reyment, 1969g).

One computes

$$n(\mathbf{d}_i \mathbf{b}_i' \mathbf{S}_1^{-1} \mathbf{b}_i + (1/\mathbf{d}_i)\mathbf{b}_i' \mathbf{S}_1 \mathbf{b}_i - 2) \qquad (7.15)$$

where $n + 1$ is the sample size of dispersion matrix $\mathbf{S}_1$, $\mathbf{d}_i$ is the ith sample latent root of $\mathbf{S}_1$, and the vector $\mathbf{b}_i$ is the ith latent vector of sample dispersion matrix $\mathbf{S}_2$, based on a very large sample. The criterion is distributed approximately as chi-square with $(p\text{-}1)$ degrees of freedom, where p is the number of dimensions involved.

Mainline programme ORNTDIST

Calculations are made either on the dispersion matrices directly (or on the cross-products matrices) of the two samples, A and B, and the corresponding mean vectors, or such are computed directly from the observations.

Subroutines BAHADU and HETDST: These subroutines perform the heterogeneous generalized distance calculations of formula 7.11. Convergence of the calculations done by BAHADU is checked by HETDST. For each iteration, the output from BAHADU consists of the heterogeneous generalized distance and its square. Iteration factors are given in the output from HETDST.

Subroutine DISTFN: This subroutine finds the Mahalanobis' generalized statistical distance for homogeneous dispersion matrices. The output consists of the coefficients of the discriminant function, the generalized distance, its square and Hotelling's T^2, the related variance ratio and its degrees of freedom.

Subroutine HAFCOV does the calculations of formula (7.13).

Subroutine ORIENT operates on the latent vectors of the comparative dispersion matrix in relation to the reference dispersion matrix in order to test the significance of the differences in orientations of the principal axes of the dispersion ellipsoids.

Subroutine REYDST is alternative way of finding a generalized distance for unequal dispersion matrices.

RANKRT selects cards randomly from the stored data decks. HOUSESR is a standard routine for finding latent roots and vectors. GETMAT, PUTMAT and MATOUT are subroutines for handling matrices with respect to input and output. JZUP2 substitutes the original measurements, or the logarithmically transformed measurements, into the vector of discriminant function coefficients. The values are plotted as a histogram by subroutine PLTHS1. MATINV is a standard subroutine for matrix inversion and DISTRB works out certain univariate statistical values.

The programme has proved to be particularly useful for studying the range of sexual-dimorphic differences in fossil material (see Chapter 29). It allows not only a determination of the strength of the sexual dimorphism but also provides a useful way of obtaining a quantitative appraisal of dimorphism in growth and shape patterns.

```
      PROGRAM ORNTDIST
C                    LIST OF SUBROUTINES
C     HOUSER - COMPUTES EIGENVALUES AND EIGENVECTORS
C     ORIENT - COMPARES ORIENTATIONS OF EIGENVECTORS
C     HETDST - COMPUTES ANDERSON/BAHADUR DISTANCE
C     BAHADU - PERFORMS STEPS IN THE ITERATIVE CALCULATION OF HETDST
C     REYDST - COMPUTES SIGNIFICANCE OF A HETEROGENEOUS DISTANCE
C     HAFCOV - GIVES A HETEROGENEOUS DISTANCE BY AVERAGING COVARIANCES
C     DISTFN - FINDS THE MAHANOLOBIS DISTANCE
C     RANKRT               MATINV                      JZUP2
C     PUTMAT               GETMAT                      MATOUT
C     PLTHS1 - PLOTS HISTOGRAMS
C     DISTRE - TESTS THE DISTRIBUTIONS OF EACH OF THE VARIABLES
C
C******************************************************************************
C
C     EXPLANATICN OF CONTROL CARDS
C
C     CARD 1    JOBS   = NUMBER OF JOBS BEING RUN IN SEQUENCE
C     CARD 2
C       CCL 1-3    M = NUMBER OF VARIABLES (MAXIMUM = 45)
C       CCL 4-6    NA = IS SIZE OF THE LARGER SAMPLE (MAXIMUM=500)
C       CCL 7-9    NB = SIZE OF SMALLER SAMPLE (MAXIMUM=500)
C       CCL 10-12  MATRIX = 0 FOR SERIES OF MEASUREMENTS AND 1 FOR
C                  DATA READ IN IN MATRIX FORM ,I.E. PARTIALLY PROCESSED)
C       CCL 13-15  L = 1 FOR INPUT DATA AS CROSS-PRODUCTS MATRICES
C                  L = 0 FOR COVARIANCE MATRICES AND SERIES OF
C                  MEASUREMENTS
C       CCL 16-18  LOGDEC = 1 IF LOGARITHMIC TRANSFORMATION OF RAW DATA
C                  REQUIRED AND 0 IF THE DATA ARE NOT TO BE TRANSFORMED
C       CCL 19-21  LDISC = 1 IF THE NORMAL FORM OF THE GENERALIZED
C                  DISTANCE IS TO BE COMPUTED IN ADDITION TO THE
C                  DECISION MADE BY THE PROGRAM - THIS IS A USEFUL
C                  CONTROL STEP
C       CCL 22-24, 25-27, 28-30   INVST = NUMBER OF INTERVALS FOR EACH
C                  HISTOGRAM.  IF NO SELECTION ALTERNATIVELY ZEROES
C                  PROGRAM PICKS INTERVAL OF TEN
C       CCL 31-33  KREYD = NUMBER OF TIMES IN THE REYDST-LOOP
C     CARD 3    THE TITLE CARD WITH THE HEADING IN 72 SPACES
C     CARD 4    THE VARIABLE FORMAT CARD FOR THE DATA
C     CARD 5    THE DATA IN ONE OF THE PRESELECTED FORMS
C     THE LARGER SAMPLE MUST ALWAYS BE READ FIRST
      DIMENSION E(45,45),F(45,45),AEGVAL(45),BEGVAL(45),CEGVAL(45),
     1 TDATA(30),FMT(12),TITLE(12)
      COMMON A(45,45),B(45,45),C(45,45),SA(45),SB(45),SC(45),DIFF(45),
     1 T(45),X(45),Z(45),M,NA,NB,K,MW,MATRIX,LDISCR,LOGDEC,INVST(3),IT,
     2 MATPCS,KREYD
      COMMON /UNITS/ LIN,LUT,MTA,MTB,MTMABC,MTARA,MTBRA
      DATA(TDATA= 0.0,9.21,11.34,13.28,15.09,16.81,18.48,20.09,21.67,23.
     121,24.72,26.22,27.69,29.14,30.58,32.00,33.41,34.81,36.19,37.57,38.
     293,40.29,41.64,42.98,44.31,45.64,46.96,48.28,49.59,50.89)
C-F  IV- DATA TDATA/.0,9.21,11.34,13.28,15.09,16.81,18.48,20.09,21.67,23.
C    121,24.72,26.22,27.69,29.14,30.58,32.00,33.41,34.81,36.19,37.57,38.
C    293,40.29,41.64,42.98,44.31,45.64,46.96,48.28,49.59,50.89/
C
```

```
C------- LIN IS THE STANDARD INPUT UNIT (CARD READER)
      LIN=60
C------- LUT IS THE STANDARD OUTPUT UNIT (PRINTER)
      LUT=61
C------- MTA,MTB,MTMABC,MTARA AND MTBRA ARE SCRATCH UNITS
C------- ******** SCRATCH TAPE USAGE ********
C------- TAPE MTA CONTAINS SAMPLE A
      MTA=51
C------- TAPE MTB CONTAINS SAMPLE B
      MTB=52
C------- MTMABC CONTAINS MATRICES A,B AND C IN THIS ORDER,ONE PER RECORD
      MTMABC=50
C------- RANDOM ACCES UNITS ARE USED ONLY IN REYDST AND RANKRT
C------- MTARA IS A RANDOM ACCESS UNIT WHERE SAMPLE A IS STORED
      MTARA=10
C------- MTBRA IS A RANDOM ACCESS UNIT WHERE SAMPLE B IS STORED
      MTBRA=11
      READ (LIN,380) JOBS
      KLOBS = 0
      MW=45
C------- HERE BEGINS A NEW CYCLE
 1111 CONTINUE
      REWIND MTA
      REWIND MTB
      REWIND MTMABC
      READ (LIN,3) M,NA,NB,MATRIX,L,LOGDEC,LDISCR,INVST,KREYD
      IF (M .LE. 0) CALL EXIT
      IF (M .LE. MW .AND. NA .LE. 500 .AND. NB .LE. 500) GO TO 33
      WRITE (LUT ,133) M,NA,NB
   33 DO 10101 IX = 1,3
10101 IF(INVST(IX).LE.0)INVST(IX) = 10
      DO 1620 I=1,M
      SA(I)=0.0
      SB(I)=0.0
      DO 1620 J=1,M
      A(I,J)=0.0
      B(I,J)=0.0
      C(I,J)=0.0
 1620 CONTINUE
      WRITE (LUT,207)
      IF (LOGDEC .EQ. 1) WRITE (LUT,490)
      EM=M
      ENA=NA
      ENB=NB
      READ (LIN,1) TITLE
      WRITE (LUT,333) TITLE
      READ (LIN,1) FMT
      WRITE (LUT,400) FMT
      IF(MATRIX.GT.0) GO TO 5000
      WRITE (LUT,710) NA,NB
      WRITE (LUT,5100)
      NAB=NA
      DO 6009 IREP=1,2
C------- WHEN IREP .EQ. 1  STORE SAMPLE A ON MTA
C------- WHEN IREP .EQ. 2  STORE SAMPLE B ON MTB
      LUN=MTA
```

```
      IF (IREP .EQ. 2) LUN=MTB
      DO 6008 IV=1,NAB
      READ (LIN,FMT) (X(IX),IX=1,M)
      WRITE (LUT,5020) (X(IX),IX=1,M)
      IF (LCGDEC .EQ. 0) GO TO 5051
      DO 5050 IX=1,M
5050  X(IX)=ALOG10(X(IX))
5051  WRITE (LUN) (X(IX),IX=1,M)
 625  CONTINUE
 627  IF(IREP.EG.2) GO TO 635
      DO 626 I=1,M
      SA(I)=SA(I)+X(I)
      DO 626 J=1,M
 626  A(I,J)=A(I,J)+X(I)*X(J)
      GO TO 6008
 635  DO 628 I=1,M
      SB(I)=SB(I)+X(I)
      DO 628 J=1,M
 628  B(I,J)=B(I,J)+X(I)*X(J)
6008  CONTINUE
      IF(IREP.EG.2) GO TO 6009
      NAB=NE
      WRITE (LUT,5105)
6009  CONTINUE
 655  DO 660 I=1,M
      DO 660 J=1,M
      A(I,J)=(A(I,J)-SA(I)*SA(J)/ENA)/(ENA-1.)
 660  B(I,J)=(B(I,J)-SB(I)*SB(J)/ENB)/(ENB-1.)
      DO 199 IX=1,M
      SA(IX)=SA(IX)/ENA
 199  SB(IX)=SB(IX)/ENB
      CALL DISTRB
      GO TO 1117
5000  WRITE (LUT,5100)
      DO 10 I=1,M
  10  READ (LIN,FMT) (A(I,J),J=1,M)
      CALL MATOLT (A)
      WRITE (LUT,5105)
      DO 30 I=1,M
  30  READ (LIN,FMT) (B(I,J),J=1,M)
      CALL MATOLT (B)
      READ (LIN,FMT) (SA(IX),IX=1,M)
      READ (LIN,FMT) (SB(IX),IX=1,M)
      IF(L.LE.0) GO TO1117
1116  DO 1121 I=1,M
      DO 1121 J=1,M
1121  A(I,J)=A(I,J)/(ENA-1.)
      DO 1122 I=1,M
      DO 1122 J=1,M
1122  B(I,J)=B(I,J)/(ENB-1.)
1117  DO 40 I=1,M
      DO 40 J=1,M
  40  C(I,J)=((ENA-1.)*A(I,J)+(ENB-1.)*B(I,J))/(FNA+ENB-2,)
      WRITE (LUT,1120)
      CALL MATOLT (A)
C     MATRIX A IS REFERENCE
```

```
      WRITE (LUT,1119)
      CALL MATOLT (B)
      WRITE (LUT,1118)
      CALL MATOLT (C)
      CALL FUTMAT(A)
      DO 5432 IX=1,M
 5432 SC(IX)=SB(IX)-SA(IX)
      WRITE (LUT,3333)
      WRITE (LUT,3333)
      CALL HOUSER (M,M,A,AEGVAL,1.,E)
      WRITE (LUT,716)
      WRITE (LUT,1492)
      WRITE (LUT,20) (AEGVAL(IX),IX=1,M)
      WRITE (LUT,1493)
      CALL MATOLT (E)
      DETA=0.
      DO 80 IX=1,M
      IF (AEGVAL(IX) .LE. 1.E-09) GO TO 561
   80 DETA=DETA+ALOG(AEGVAL(IX))
      CALL FUTMAT(B)
C     MATRIX  B  IS STORED
      WRITE (LUT,3333)
      CALL HOUSER (M,M,B,BEGVAL,1.,F)
      WRITE (LUT,1494)
      WRITE (LUT,20) (BEGVAL(IX),IX=1,M)
      WRITE (LUT,1495)
      CALL MATOLT (F)
      DETB=0.
      DO 100 IX=1,M
      IF (BEGVAL(IX) .LE. 1.E-09) GO TO 561
  100 DETB=DETB+ALOG(BEGVAL(IX))
      CALL FUTMAT(C)
      MATPOS=3
      WRITE (LUT,3333)
      WRITE (LUT,3333)
      CALL HOUSER (M,M,C,CEGVAL,1.,F)
      WRITE (LUT,1497)
      WRITE (LUT,20) (CEGVAL(IX),IX=1,M)
      WRITE (LUT,1498)
      CALL MATOLT (F)
      DETC=0.
      DO 120 IX=1,M
      IF (CEGVAL(IX) .LE. 1.E-09) GO TO 561
  120 DETC=DETC+ALOG(CEGVAL(IX))
      WRITE (LUT,90)
      WRITE (LUT,110)   DETA,DETB,DETC
      BSQ=(ENA-1.)*(DETC-DETA)+(ENB-1.)*(DETC-DETB)
      BETASQ=((2.*FLOAT(M)**3+3.*FLOAT(M)**2-FLOAT(M))/12.)*
     1(1./(ENA-1.)+1./(ENB-1.)-1./(ENA+ENB-2.))
      WRITE (LUT,716)
      WRITE (LUT,140) BSQ,BETASQ
      KT = M*(M+1)/2
      WRITE (LUT,151) KT
      TRACE = 0.
      DO 170 IX=1,M
  170 TRACE=TRACE+BEGVAL(IX)
```

```
      WRITE (LUT,180) TRACE
      DO 200 IX=1,M
  200 Z(IX)=(BEGVAL(IX)*100.)/TRACE
      WRITE (LUT,220) (Z(IX),IX=1,M)
      CALL GETMAT (B,2)
      DO 500 I=1,M
      DO 500 J=1,M
  500 A(I,J)=B(I,J)
C     MATRIX   A  IS FOR INVERSION
      CALL MATINV (A,M,CEGVAL,0,DETERM,MW)
      KOUNT = 1
      WRITE (LUT,160)
      DO 230 KL = 1,M
      IF(Z(KL).LE.1.) GO TO 10000
      CALL CRIENT (E,B,M,X,A,KOUNT,ENB,BEGVAL)
      KOUNT = KOUNT + 1
  230 CONTINUE
10000 CALL GETMAT(A,1)
      CALL GETMAT(B,2)
      IF(KT.GT.30) GO TO 5551
      TEST = TDATA(KT)
      WRITE (LUT,10598) TEST
      GO TO 5552
 5551 TEST=0.5*(2.*(ENA+ENB-1.)+0.96)
      WRITE (LUT,425) TEST
 5552 IF(TEST.GT.BSQ) CALL DISTFN
      IF(TEST.LE.BSQ) CALL HETDST
      GO TO 1000
  561 WRITE (LUT,562)
 1000 KLOBS = KLOBS + 1
      IF(JOBS-KLOBS) 365,365,1111
  365 WRITE (LUT,360)
      CALL EXIT
C
C
    1 FORMAT(12A6)
    3 FORMAT (25I3)
   20 FORMAT(1X,11F12.5)
   90 FORMAT(27H0          LOG DETERMINANTS /41H  REFERENCE       COMPARISO
     1N          FOOLED  )
  110 FORMAT (2X,F7.3,8X,F7.3,9X,F7.3)
  133 FORMAT (30HONUMBER OF VARIABLES TOO HIGH= ,I4,5H, NA=,I4,4H,NB=I4)
  140 FORMAT(17H     B SQUARE = ,F14.7/20H       BETA SQUARE = ,F14.7)
  151 FORMAT(27H      DEGREES OF FREEDOM = I3)
  160 FORMAT(31H1    ORIENTATION OF ELLIPSOIDS
  180 FORMAT(13H0  TRACE B = F18.7)
  207 FORMAT(50H1  ANALYSIS OF HOMOGENEITY OF COVARIANCE MATRICES)
  220 FORMAT(23H0  PERCENTAGES FOR B(I)/(X,8F14.7))
  333 FORMAT(10X12A6)
  360 FORMAT(18H1  CYCLE COMPLETED/12H0  RETURNING)
  380 FORMAT(3X,I2)
  400    FORMAT (38HODATA CARDS READ WITH VARIABLE FORMAT ,12A6)
  425 FORMAT(23H0  THE FORKING VALUE = F14.7)
  490 FORMAT(29H0  DATA LOGARITHM TRANSFORMED)
  562 FORMAT(50H0NEGATIVE EIGENVALUE, PROCEEDING TO NEXT DATA SET )
  710 FORMAT(48H0  REFERENCE SAMPLE SIZE BASED ON POPULATION OF ,I3/29H0
```

```
      1  COMPARISON SAMPLE SIZE IS ,I3)
    716 FORMAT(1H0)
   1118 FORMAT(20H0 ROW  POOLED MATRIX)
   1119 FORMAT(47H0 ROW  COVARIANCE MATRIX 2 ( COMPARISON MATRIX)  )
   1120 FORMAT(47H0 ROW  COVARIANCE MATRIX 1 ( REFERENCE MATRIX )  )
   1492 FORMAT(35H0  EIGENVALUES FOR REFERENCE MATRIX)
   1493 FORMAT(36H0  EIGENVECTORS FOR REFERENCE MATRIX)
   1494 FORMAT(36H0  EIGENVALUES FOR COMPARISON MATRIX)
   1495 FORMAT(37H0  EIGENVECTORS FOR COMPARISON MATRIX)
   1497 FORMAT(32H0  EIGENVALUES FOR POOLED MATRIX)
   1498 FORMAT(33H0  EIGENVECTORS FOR POOLED MATRIX)
   3333 FORMAT(6H0*****)
   5020 FORMAT(1X10F13.7)
   5100 FORMAT(23H0  FIRST DATA INPUT SET)
   5105 FORMAT(24H0  SECOND DATA INPUT SET)
  10598 FORMAT(10H0  TEST = F14.7)
        END



        SUBROUTINE GETMAT (STMAT,MATNR)
C------- GETMAT READS THE MATNR.TH MATRIX FROM LOGICAL UNIT MTMABC TO
C------- THE PARAMETER STMAT.
C------- MTMABC CONTAINS 3 RECORDS WHICH CONTAINS RESP MATRIX A,
C------- MATRIX B AND MATRIX C.
C------- IN THE VARIABLE MATPOS IS INDICATED WHICH MATRIX WAS READ LAST.
C------- GETMAT MUST NOT BE CALLED UNTIL PUTMAT HAS WRITTEN ALL 3
C------- MATRICES E.G. PUTMAT HAS BEEN CALLED THREE TIMES,
        DIMENSION STMAT(45,45)
        COMMON A(45,45),B(45,45),C(45,45),SA(45),SB(45),SC(45),DIFF(45),
       1 T(45),X(45),Z(45),M,NA,NB,K,MW,MATRIX,LDISCR,LOGDEC,INVST(3),IT,
       2 MATPOS,KREYD
        COMMON /UNITS/ LIN,LUT,MTA,MTB,MTMABC,MTARA,MTBRA
        L=MATNR
C------- J IS THE NUMBER OF RECORDS TO SKIP FORWARD (IF J .GT. 0) OR
C------- BACKWARD (IF J .LT. 0)
        J=L-1-MATPOS
C------- IF J .EQ. 0 MATMABC IS ALREADY POSITIONED.
        IF (J) 10,20,30
     10 J=-J
C------- BACK OVER J RECORDS
        DO 11 J1=1,J
     11 BACKSPACE MTMABC
        GO TO 20
C------- SKIP OVER J RECORDS
     30 DO 31 J1=1,J
     31 READ (MTMABC) J2
C------- READ THE MATRIX
     20 READ (MTMABC) ((STMAT(I,JO),JO=1,M),I=1,M)
C------- UPDATE MATPOS
        MATPOS=L
        RETURN
        END
```

7. FORMAL BACKGROUND OF DISCRIMINANT FUNCTION ANALYSIS   73

```
      SUBROUTINE HOUSER (N,NEV,C,EV,VEC,V)
C------ COMPUTES EIGENVALUES-EIGENVECTORS---SYMMETRIC MATRIX
      DIMENSION A(45),B(50),C(45,45),EV(45),P(45),TA(50),TB(50),W(50),
     1 Y(45),V(45,45)
      NN=N-1
      DO 16 I=1,NN
      SUM1=0
      B(I)=0
       JI=I+1
      DO 14  J =JI,N
 14   SUM1 = SUM1 + (C(I,J)*C(I,J))
      S=SQRT(SUM1)
      IF(S)15,16,15
  15 SGN=SIGN(1.,C(I,I+1))
      TEMP = SGN*(C(I, I+1))
      W(I+1)=SQRT(.5*(1.0+(TEMP/S)))
      C(I, I+1) = W(I+1)
       II=I+2
      IF(II-N)250,250,260
 250  TEMP=SGN/(2.*W(I+1)*S)
      DO 20 J=II,N
      W(J) = TEMP*C(I,J)
 20   C(I,J) = W(J)
 260  B(I) = -SGN*S
      DO 22  L = JI,N
      SUM2 = 0.0
      DO 21  M = JI,N
 21   SUM2 = SUM2 + (C(L,M) * W(M))
 22   P(L) = SUM2
      XKAP = 0.0
      DO 23  K = JI,N
 23   XKAP = XKAP + (W(K) * P(K))
      DO 24  L = JI,N
 24   P(L) = P(L) - (XKAP * W(L))
      DO 26  J = JI,N
      DO 25  K = J,N
      C(J,K) = C(K,J) - (2.0 * ((P(J) * W(K)) + (P(K) * W(J))))
 25   C(K,J)=C(J,K)
 26   CONTINUE
 16   CONTINUE
 17   DO 18  K = 1,N
 18   A(K) = C(K,K)
      B(N-1)=-B(N-1)
      B(N) = 0.0
      DO 500 I = 1,N
 500  W(I) = B(I)
C------ STURM METHOD (MODIFIED FROM SAND HOUSESR)
  29 U=ABS(A(1))+ABS(B(1))
      DO 30 I=2,N
      BD=ABS(A(I))+ABS(B(I))+ABS(B(I-1))
  30 IF (BD .GT. U) U=BD
      U2=U**2
      DO 32 I=1,N
      B(I)=B(I)**2/U2
      A(I)=A(I)/U
      EV(I)=-1.
```

```
32    CONTINUE
      BD=L
      U=1.
      DO 1500 K=1,NEV
      EL=EV(K)
38    ELAM=.5*(U+EL)
      IF (ABS(ELAM-U) .LT. 1.E-20) GO TO 34
  33  IF (ABS(ELAM-EL) .LT. 1.E-20) GO TO 34
  35  P0=1.
      P1=A(1)-ELAM
      ZNSIG2=1.
      B2=0.
      B1=B(1)
      IF(P1)1051,1052,1052
1051  ZNSIG1=-1.
      IAG=0
      GO TO 1053
1052  ZNSIG1=1.
      IAG=1
1053  DO 100 I=2,N
      ALPH=A(I)-ELAM
      IF(B1)115,111,115
 111  P2=ZNSIG1*ALPH
      GO TO 114
115   IF(B2)116,117,116
 116  IF (ABS(P1)+ABS(P0) .GE. 1.E-20) GO TO 152
 151  P1=1.E20*P1
      P0=1.E20*P0
 152  P2=ALPH*P1-B1*P0
      GO TO 114
117   P2=ALPH*P1-ZNSIG2*B1
 114  P0=P1
      B2=B1
      B1=B(I)
      P1=P2
      ZNSIG2=ZNSIG1
      IF(P2) 121,125,122
 121  ZNSIG1=-1.
      GO TO 123
122   ZNSIG1=1.
 123  IF (ZNSIG1*ZNSIG2 .LE. 0) GO TO 100
125   IAG=IAG+1
100   CONTINUE
      IF (IAG .GE. K) GO TO 40
  42  U=ELAM
      GO TO 38
40    M=K+1
      DO 41 MM=M,IAG
41    EV(MM)=ELAM
      EL=ELAM
      GO TO 38
34    EV(K)=ELAM
1500  CONTINUE
      DO 1060 I=1,N
      A(I)=A(I)*BD
      EV(I)=EV(I)*BD
```

```
1060  CONTINUE
43    IF(VEC) 44,700,44
44    L = NEV - 1
      DO 502 K = 1,L
      IF (EV(K) .GT. EV(K+1)) GO TO 502
501   EV(K+1)=.999999999*EV(K)
502   CONTINUE
      DO 600 I = 1,NEV
      II = 0
      DO 503 J=1,N
503   Y(J)=1.
601   DO 504 K=1,N
      P(K) = 0.0
      TB(K) = W(K)
504   TA(K) = A(K) - EV(I)
      L = N-1
      DO 505 J = 1,L
      IF (ABS(TA(J))-ABS(W(J))) 507,506,506
506   F = W(J) / TA(J)
      GO TO 509
507   F = TA(J)/W(J)
      TA(J) = W(J)
      T = TA(J+1)
      TA(J+1) = TB(J)
      TB(J) = T
      P(J) = TB(J+1)
      TB(J+1)=0.0
      IF(II-1)  509,508,509
508   T = Y(J)
      Y(J) = Y(J+1)
      Y(J+1) = T
509   TB(J+1) = TB(J+1) -F*P(J)
      TA(J+1) = TA(J+1)-F*TB(J)
      IF(II-1)  505,510,505
510   Y(J+1) = Y(J+1) - F*Y(J)
505   CONTINUE
      IF(TA(N)) 511,512,511
512   TA(N) = 10.E-30
511   IF(TA(N-1)) 513,514,513
514   TA(N-1) = 10.E-30
513   Y(N) = Y(N)/TA(N)
      Y(N-1) = (Y(N-1)-Y(N)*TB(N-1))/TA(N-1)
      L = N-2
      DO 515 J = 1,L
      K = N-J-1
      IF(TA(K))  515,516,515
516   TA(K) = 10.E-30
515   Y(K) = (Y(K) - Y(K+1)*TB(K)-Y(K+2)*P(K))/TA(K)
      IF(II) 517,518,517
518   II = 1
      GO TO 601
517   DO 521 J=1,L
      T = 0.0
      K=N-J-1
      M = K+1
      DO 519 KK=M,N
```

```
519   T = T+C(K,KK)*Y(KK)
      DO 520 KK=M,N
520   Y(KK) = Y(KK) - 2.*T*C(K,KK)
521   CONTINUE
      T=0.0
      DO 523 J = 1,N
523   T = T + Y(J)**2
      XNORM=SQRT(T)
      DO 524 J = 1,N
524   V(J,I) = Y(J) / XNORM
600   CONTINUE
700   RETURN
      END


      SUBROUTINE PUTMAT(STMAT)
C------- PUTMAT WRITES THE MATRIX STMAT ON LOGICAL UNIT MTMABC.
C------- PUTMAT EXPECTS THAT MTMABC IS REWOUND BEFORE FIRST MATRIX IS
C------- STORED.
C------- FIRST CALL IS FOR STORING MATRIX A, SECOND FOR MATRIX B AND
C------- THIRD FOR MATRIX C.
C------- EACH MATRIX WILL OCCUPY ONE RECORD.
      DIMENSION STMAT(45,45)
      COMMON A(45,45),B(45,45),C(45,45),SA(45),SB(45),SC(45),DIFF(45),
     1 T(45),X(45),Z(45),M,NA,NB,K,MW,MATRIX,LDISCR,LOGDEC,INVST(3),IT,
     2 MATPOS,KREYD
      COMMON /UNITS/ LIN,LUT,MTA,MTB,MTMABC,MTARA,MTBRA
      WRITE (MTMABC)((STMAT(I,J),J=1,M),I=1,M)
      RETURN
      END


      SUBROUTINE ORIENT(E,H,M,X,O,KOUNT,ENB,EV)
      DIMENSION E(45,45),H(45,45),O(45,45),X(45),EV(45)
      COMMON /UNITS/ LIN,LUT,MTA,MTB,MTMABC,MTARA,MTBRA
      DO 170 IX=1,M
170   X(IX)=0.
      WRITE (LUT,100)
      DO 200 I=1,M
      DO 200 J=1,M
200   X(I) = X(I) + H(I,J)*E(J,KOUNT)
      AQ=0.
      DO 230 I=1,M
230   AQ = AQ + X(I)*E(I,KOUNT)
      DO 260 I=1,M
      X(I)=0.
      DO 260 J=1,M
260   X(I) = X(I) + O(I,J)*E(J,KOUNT)
      AP=0.
      DO 271 I=1,M
271   AP = AP +E(I,KOUNT)*X(I)
      HLAMB = EV(KOUNT)
      CHISQ=ENB*(AP*HLAMB+AQ/HLAMB-2.)
      CHISQ = ABS(CHISQ)
      K=M-1
      WRITE (LUT,280) KOUNT,CHISQ,K
      RETURN
C
C
100   FORMAT (33H0   VECTOR      CHISQUARE    DF  )
280   FORMAT (6X,I2, 9X,F9.3,4X,I3)
      END
```

```
      SUBROUTINE HETDST
      COMMON A(45,45),B(45,45),C(45,45),SA(45),SB(45),SC(45),DIFF(45),
     1 T(45),X(45),Z(45),M,NA,NB,K,MW,MATRIX,LDISCR,LOGDEC,INVST(3),IT,
     2 MATPCS,KREYD
      COMMON /UNITS/ LIN,LUT,MTA,MTB,MTMABC,MTARA,MTBRA
C     PROGRAM HETDST
C     ANDERSON AND BAHADURS GENERALIZED DISTANCE (R.REYMENT,JULY,1966)
C     THE PROGRAM COMPUTES THE GENERALIZED DISTANCE FOR UNEQUAL
C     COVARIANCE MATRICES ACCORDING TO ANDERSON/BAHADUR(1962).
C
C
C     INPUT = ON FIRST CARD...NO. VARIABLES, NO.SPECIMENS IN SAMPLE A,
C     AND NO. SPECIMENS,IN,SAMPLE B...INDIVIDUALS OF A AND INDIVIDUALS
C
C     OUTPUT = COMPUTED COVARIANCE MATRICES,SAMPLE SIZES,FOR EACH
C     ITERATION,THE NUMBER OF ITERATION,VALUE OF ITERATION FACTOR T,
C     COMPUTED Z DERIVING FROM THIS,INVERSE MATRIX AND ITS DETERMINANT,
C     THE ESTIMATION VECTORS,INTERMEDIATE VALUES FOR DISTANCE,DISTANCE,
C     SQUARE OF DISTANCE.
C
C     NOTE...INFORMATION ON HOMOGENEITY OF COVARIANCE MATRICES OBTAINED
C     FROM PROGRAM HOMOGEN (REYMENT JUNE 1966).
C
C     SUBROUTINES = MATINV(COOLEY AND LOHNES,1962),BAHADU,REYMENT
C     JULY,1966).
C
      WRITE (LUT,1)
      ENA=NA
      ENB=NB
      WRITE (LUT,18) NA,NB
C     REQUIRES COVARIANCE MATRICES,A,B,A VECTOR,SB(I) TO BE DETERMINED.
      IT=0
      K=1
      T(K)=0.5
      NV=M
C     VECTOR SC CONTAINS THE DIFFERENCES OF MEANS
      WRITE (LUT,993)
      DO 997 IX=1,NV
  997 WRITE (LUT,304) SC(IX)
      WRITE (LUT,733)
      DO 69 KK=1,4
      CALL BAHADU
      IF(KK.EQ.4) GO TO 720
      I=1.+(Z(K)*10.)
      IF (I .EQ. 1) GO TO 161
      K=K+1
      IF (I .GT. 1) GO TO (162,2001,2003),KK
      IF (KK-2) 30,31,32
   30 T(K)=0.75
      GO TO 69
   31 T(K)=0.875
      GO TO 69
   32 T(K)=1.0
      GO TO 69
 2003 T(K)=.8125
   69 CONTINUE
```

```
2001 T(K)=.625
     DO 2004 KK=1,2
     CALL EAHADU
     IF(KK.EQ.2) GO TO 720
     I=1.+(Z(K)*10.)
     IF (I .EQ. 1) GO TO 161
     K=K+1
     IF (I .GT. 1) GO TO 2005
     T(K)=.6875
     GO TO 2004
2005 T(K)=.5625
2004 CONTINUE
 162 T(K)=.25
     DO 2169 KK=1,3
     CALL EAHADU
     IF(KK.EQ.3) GO TO 720
     I=1.+(Z(K)*10.)
     IF (I .EQ. 1) GO TO 161
     K=K+1
     IF (I .GT. 1) GO TO (736,2007),KK
     T(K)=.375
     IF(KK.EQ.2) T(K)=.4375
     GO TO 2169
2007 T(K)=.3125
2169 CONTINUE
 736 T(K)=.125
     DO 2008 KK=1,2
     CALL EAHADU
     IF(KK.EQ.2) GO TO 720
     I=1.+(Z(K)*10.)
     IF (I .EQ. 1) GO TO 161
     K=K+1
     IF (I .GT. 1) GO TO 2009
     T(K)=.1875
     GO TO 2008
2009 T(K)=.0625
2008 CONTINUE
 720 IT = K-1
     I=1.+(Z(K)*10.)
     IF (I-1) 2710,161,2711
2710 SIGNTM=1.0
     GO TO 2703
2711 SIGNTM=-1.0
2703 IT=IT+1
     K=K+1
     DIFF(K)=ABS(T(K-1)-T(K-2))
     IF (DIFF(K) .LE. DIFF(K-1))GO TO  2705
     T(K)=T(K-1)+SIGNTM*DIFF(K)/FLOAT(IT-2*IT/3)
2706 IF (DIFF(K) .LT. 0.00001) GO TO 161
     CALL EAHADU
     IF(IT-19)720,720,169
2705 T(K)=T(K-1)+SIGNTM*DIFF(K)*C.5
     GO TO 2706
 169 WRITE (LUT,184)
C
 161 WRITE (LUT,174)
```

```
      DO 735 I=1,NV
  735 WRITE (LUT,734) SB(I)
      IF(MATRIX.EQ.1) CALL EXIT
      CALL JZLP2
      CALL HAFCOV
      CALL HEYDST
      IF(LDISCR.EQ.1) CALL DISTFN
C     HEYDST COMPUTES T SQUARE AND VARIANCE RATIO FOR HET MATRICES
      RETURN
C
C
    1 FORMAT(81H1 STATISTICAL DISTANCE FOR HETEROGENEOUS COVARIANCE MATR
     1ICES                                                       HETDST
   18 FORMAT(8H0    NA =,I3,7H    NB =,I3/38H0  ITERATIONS FOR DISTANCE CA
     1LCULATICN)
  174 FORMAT(33H0ESTIMATE OF DISCRIMINANT VECTOR
  184 FORMAT(19H0WOULD NOT CONVERGE)
  304 FORMAT(1X,7F12.5)
  733 FORMAT(48H0ITERATION    Z         D         DSQ       DIFF T
  734 FORMAT(2X,F10.5)
  993 FORMAT(31H  SAMPLE DIFFERENCE MEAN VECTOR)
      END


      SUBROUTINE HAFCOV
      COMMON A(45,45),B(45,45),C(45,45),SA(45),SB(45),SC(45),DIFF(45),
     1 T(45),X(45),Z(45),M,NA,NB,K,MW,MATRIX,LDISCR,LOGDEC,INVST(3),IT,
     2 MATPCS,KREYD
      COMMON /UNITS/ LIN,LUT,MTA,MTB,MTMABC,MTARA,MTBRA
      WRITE (LUT,407)
      CALL GETMAT(A,1)
      CALL GETMAT(B,2)
      DO 408 I=1,M
      DO 408 J=1,M
  408 C(I,J)=(A(I,J)+B(I,J))/2.
      WRITE (LUT,200)
      CALL MATOLT (C)
      CALL MATINV(C,M,Z,0,DETERM,MW)
      DO 410 I=1,M
      Z(I)=0.
      DO 410 J=1,M
  410 Z(I) = Z(I) + C(I,J)*SC(J)
      DSQ=0.
      DO 411 I=1,M
  411 DSQ = DSQ + SC(I)*Z(I)
      DSQ=ABS(DSQ)
      DS=SQRT(DSQ)
      WRITE (LUT,222) DS,DSQ
      RETURN
C
C
  200 FORMAT (24H0 ROW  COVARIANCE MATRIX )
  222 FORMAT( 33H0     DISTANCE, DISTANCE SQUARED   /7X,F7.3,6X,F9.3/)
  407 FORMAT(61H1    HETEROGENEOUS DISTANCE BY AVERAGING COVARIANCE MATR
     1ICES          )
      END
```

```
      SUBROUTINE BAHADU
      DIMENSICN ASC(45)
      COMMON A(45,45),B(45,45),C(45,45),SA(45),SB(45),SC(45),DIFF(45),
     1 T(45),X(45),Z(45),M,NA,NB,K,MW,MATRIX,LDISCR,LOGDEC,INVST(3),IT,
     2 MATPCS,KREYD
      COMMON /UNITS/ LIN,LUT,MTA,MTB,MTMABC,MTARA,MTBRA
C     SC IS DIFFERENCE OF MEANS
      NV=M
      TNEW=T(K)
      DO 151 I=1,NV
      DO 151 J=1,NV
  151 C(I,J)=TNEW*A(I,J)+(1.-TNEW)*B(I,J)
      CALL MATINV (C,NV,ASC,0,DETERM,MW)
C     SOLUTION CF EQUATION FOR VECTOR SB(I) THRU MATINV AND MEAN VECTOR
      DO 153 I=1,NV
      SB(I)=0.
      DO 153 J=1,NV
  153 SB(I)=C(I,J)*SC(J)+SB(I)
C     THIS GIVES THE VECTOR ESTIMATION DESIRED
C     DISCRIMINANT VECTOR IS SB(I)
      DO 154 I=1,NV
      DO 154 J=1,NV
  154 C(I,J)=(A(I,J)*TNEW**2)-((1.-TNEW)**2)*B(I,J)
      DO 155 I=1,NV
      X(I)=0.
      DO 155 J=1,NV
  155 X(I)=X(I)+C(I,J)*SB(J)
      ZNEW=0.
      DO 156 I=1,NV
  156 ZNEW=ZNEW+X(I)*SB(I)
C     EXPANSION OF THE QUADRATIC FORM
      Z(K)=ZNEW
      TNEW=0.
      Q=0.0
      DO 167 I=1,NV
      X(I)=0.0
  167 Q=Q+SC(I)*SB(I)
      Q=2.0*Q
      DO 168 I=1,NV
      ASC(I) = 0.
      DO 168 J=1,NV
  168 ASC(I) = ASC(I) + A(I,J)*SB(J)
      R=0.0
      DO 170 I=1,NV
  170 R = R + ASC(I)*SB(I)
      R=SQRT(ABS(R))
      DO 171 I=1,NV
      ASC(I) = 0.
      DO 171 J=1,NV
  171 ASC(I) = ASC(I) + B(I,J)*SB(J)
      AS=0.0
      DO 172 I=1,NV
  172 AS = AS + ASC(I)*SB(I)
      AS=SQRT(ABS(AS))
      D = Q/(R+AS)
      DSQ = D**2
```

```
      WRITE (LUT,159) IT,7(K),D,DSQ,DIFF(K)
      RETURN
C
C
  159 FORMAT(2X,I2,7X,F8.5,2X,F6.3,3X,F6.3,3X,F10.7)
      END

      SUBROUTINE REYDST
      DIMENSION XA(45),XB(45),SUM(45),DISC(45),XC(45)
      COMMON A(45,45),B(45,45),C(45,45),SA(45),SB(45),SC(45),DIFF(45),
     1 T(45),X(45),Z(45),M,NA,NB,K,MW,MATRIX,LDISCR,LOGDEC,INVST(3),IT,
     2 MATPCS,KREYD
      COMMON /UNITS/ LIN,LUT,MTA,MTB,MTMABC,MTARA,MTBRA
C------- RETURN IF KREYD IS LE 0
      IF (KREYD .LE. 0) RETURN
      WRITE (LUT,4165)
      DSUM = 0.
      ENB = NB
      ENA = NA
      NAB = NA
      TK = M
      AN = SQRT(ENB/ENA)
      WRITE (LUT,3) AN
C     REMEMBER THAT A IS THE LARGER SAMPLE AND NA IS GRT TH NB
C     REMEMBER THAT VECTOR SC CONTAINS DIFFERENCE OF MEANS
C****************************************************************
C------- THE STATEMENTS FROM  DO 6008 UNTIL 200 CONTINUE ARE MACHINE
C------- DEPENDENT. THEY USE CD 3600-FORTRAN STATEMENTS TO STORE SAMPLE
C------- A AND SAMPLE B ON A RANDOM ACCESS DRUM STORAGE UNIT.
C------- IF YOU HAVE NO RANDOM ACCESS UNIT YOU CAN SKIP THIS STATEMENT
C------- AND MAKE SOME CHANGES IN ROUTINE  RANKRT.
      REWIND MTARA
      REWIND MTA
      DO 6008 IV = 1,NAB
      READ (MTA) (XA(IX),IX=1,M)
      BUFFER OUT (MTARA,1) (XA(1),XA(M))
 6011 IF (UNIT,MTARA) 6011,6008
 6008 CONTINUE
C------- SAMPLE A STORED NOW ON DRUM IN RANDOM ACCESS
      REWIND MTBRA
      REWIND MTB
      DO 200 IV=1,NB
      READ (MTB) (XB(IX),IX=1,M)
      BUFFER OUT (MTBRA,1) (XB(1),XB(M))
  205 IF (UNIT,MTBRA) 205,200
  200 CONTINUE
C------- SAMPLE B IS NOW STORED ON DRUM IN RANDOM ACCESS
C****************************************************************
 6009 NAB = NB
      DO 500 KK=1,KREYD
      DO 6050 I=1,M
      XC(I) = 0.
 6050 X(I) = 0.
      DO 6007 I=1,M
      SUM(I) = 0.
      DO 6007 J=1,M
 6007 A(I,J) = 0.
      CALL CLEAR
      DO 40 IV=1,NAB
      CALL RANKRT(XA,0)
      CALL RANKRT(XB,1)
      DO 20 I=1,M
   20 XC(I) = XA(I)*AN
```

```
      DO 22 I=1,M
   22 X(I) = ABS(XB(I)-XC(I))
C     NEW OBSERVATIONS STORED IN X (I)
      DO 23 I=1,M
   23 SUM(I) = SUM(I) + X(I)
      DO 25 I=1,M
      DO 25 J=1,M
   25 A(I,J) = A(I,J) + X(I)*X(J)
   40 CONTINUE
      DO 45 I=1,M
      DO 45 J=1,M
   45 A(I,J)=(A(I,J) - SUM(I)*SUM(J)/ENB)/(ENB-1.0)
C     COVARIANCE MATRIX OF THE NEW VARIABLES
      CALL MATINV(A,M,B,0,DETERM,MW)
      DO 60 I=1,M
      DISC(I)=0.
      DO 60 J=1,M
   60 DISC(I) = DISC(I) + A(I,J)*SC(J)
      DS = 0.
      DO 70 I=1,M
   70 DS = DS + DISC(I)*SC(I)
      DS = DS*2.0
      DSUM = DSUM + DS
      TEST = DSUM/FLOAT(KK)
      WRITE (LUT,76) KK,DS
  500 CONTINUE
      WRITE (LUT,80)
      DS=DSUM/FLOAT(KREYD+1)
      TT = DS*ENB/2.
      WRITE (LUT,85) DS,TT
      F = (TT*(ENB-FLOAT(M)))/((ENB-1.)*FLOAT(M))
      NAA = NB-1
      WRITE (LUT,95) F,NAA,M
      RETURN
C
C
    3 FORMAT(24H0  MULTIPLICATION FACTOR /8H0  AN = ,F14.7/32H0   ITERATI
     1ON  REYMENTS DSQ   TEST  )
   76 FORMAT(6X,I2,8X,F14.4,8X,F14.4)
   80 FORMAT(28H0  SIGNIFICANCE COMPUTATIONS /1H0)
   85 FORMAT (25H0          AVE.DSQ      TSQ /2(3X,F10.3))
   95 FORMAT(23H0  VARIANCE RATIO IS = F14.2,4H FORI4,5H AND I2,19H DEGR
     1EES OF FREEDOM     )
 4165 FORMAT(30H1  HETEROGENEOUS SIGNIFICANCES             )
      END


      SUBROUTINE CLEAR
C------- CLEAR IS A ROUTINE WHICH MUST BE CALLED BEFORE FIRST CALLRANKRT
C------- IT CLEARS THE INDICATOR ARRAY KAT.
      COMMON  /RANKO/ KAT(500)
      DO 40 I=1,500
   40 KAT(I)=0
      RETURN
C********************************************************************
C------- IF YOU HAVE NO RANDOM ACCESS UNIT YOU MUST ADD THE FOLLOWING
C------- CODE.
C     COMMON /UNITS/ .....
C     REWIND MTA
C     REWIND MTB
C********************************************************************
      END
```

```
      SUBROUTINE RANKRT(XD,KABSW)
C------- RANKRT PICKS RANDOMLY ONE RECORD FROM SAMPLE A (IF KABSW=0 )
C------- OR FROM SAMPLE B ( IF KABSW=1 ). THIS VERSION ASSUMES THAT
C------- SAMPLE A AND SAMPLE B IS PLACED EARLIER ON RANDOM ACCESS UNITS
C------- MTARA RESP MTBRA. BEFORE FIRST CALL RANKRT THE INDICATOR ARRAY
C------- KAT MUST BE CLEARED BY CALL CLEAR.
C------- VERSION 2, WHICH CAN HANDLE TWO PARALLEL RANDOM FILES
      DIMENSION XD(45)
      COMMON A(45,45),B(45,45),C(45,45),SA(45),SB(45),SC(45),DIFF(45),
     1 T(45),X(45),Z(45),M,NA,NB,K,MW,MATRIX,LDISCR,LOGDEC,INVST(3),IT,
     2 MATPCS,KREYD
      COMMON /UNITS/ LIN,LUT,MTA,MTB,MTMABC,MTARA,MTBRA
      COMMON /RANKO/ KAT(500)
      KAB=KABSW
      IF (KAB) 45,45,47
C------- PICK UP A CARD FROM UNIT MTARA
   45 NT=NA
      LUN=MTARA
      GO TO 48
C------- PICK UP A CARD FROM UNIT MTBRA
   47 NT=NB
      LUN=MTBRA
   48 CONTINUE
C------- GET A RANDOM NUMBER FROM 1 TO NT+1
C------- X=RANF(-1) WILL GIVE A NEW FLOATING POINT RANDOM NUMBER IN THE
C------- INTERVAL 0.0 UNTIL 1.0 EVERY TIME RANF IS CALLED.
   20 I=RANF(-1)*FLOAT(NT)+1.0
C------- DON'T ACCEPT I .GT. NT
      IF (I .GT. NT ) GO TO 20
C------- GET INDICATOR FOR UNIT MTBRA
      IK=KAT(I)/10
C------- GET INDICATOR FOR UNIT MTARA
      IR=KAT(I)-IK*10
C------- TEST IF WE JUST NOW ARE USING MTARA
      IF (KAB .EQ. 0) IK=IR
C------- IF THIS CARD ALREADY USED TRY TO GET A NEW RANDOM NUMBER
      IF (IK .NE. 0) GO TO 20
C------- POSITION THE DRUM
      CALL LOCATE (LUN,(I-1)*M)
      BUFFER IN (LUN,1) (XD(1),XD(M))
   11 IF (UNIT,LUN) 11,12
C------- INDICATE THAT THIS CARD IS USED
   12 KAT(I)=KAT(I)+KAB*10
      RETURN
C*********************************************************************
C------- IF YOU HAVE NO RANDOM ACCES UNIT YOU CAN USE THE FOLLOWING CODE
C      COMMON  ......
C      DIMENSION  ........
C      DIMENSION LUNPOS(2)
C      DATA (LUNPOS=0,0)      OR      DATA LUNPOS/0,0/
C      KAB=KABSW
C      IF (KAB) 45,45,47
C   45 NT=NA
C      LUN=MTA
C      GO TO 48
C   47 NT=NB
```

```
C       LUN=MTB
C   48 CONTINUE
C   20 I=RANF(-1)*FLOAT(NT)+1.0
C       IF (I .GT. NT) GO TO 20
C       IK=KAT(I)/10
C       IR=KAT(I)-IK*10
C       IF (KAB.EG. 0) IK=IR
C       IF (IK .NE. 0) GO TO 20
C       J=I-1-LUNFOS(KAB+1)
C       IF (J) 50,60,70
C   50 J=-J
C       DO 51 J1=1,J
C   51 BACKSPACE LUN
C       GO TO 60
C   70 DO 71 J1=1,J
C   71 READ(LUN) J2
C   60 LUNFOS(KAB+1)=I
C       READ (LUN) (XD(I)X),IX=1,M)
C   12 KAT(I)=KAT(I)+KAB*10
C       RETURN
C*****************************************************************
      END


      SUBROUTINE DISTFN
      COMMON A(45,45),B(45,45),C(45,45),SA(45),SB(45),SC(45),DIFF(45),
     1 T(45),X(45),Z(45),M,NA,NB,K,MW,MATRIX,LDISCR,LOGDEC,INVST(3),IT,
     2 MATPCS,KREYD
      COMMON /UNITS/ LIN,LUT,MTA,MTB,MTMABC,MTARA,MTBRA
C     ALL INPUT OF DATA IS DONE IN CALLING PROGRAM
      WRITE (LUT,300)
      ENA=NA
      ENB=NB
      NDF=NA+NB-2
      WRITE (LUT,200) NA,NB,NDF
      DO 13 I=1,M
   13 WRITE (LUT,272) I,SA(I),SB(I),SC(I)
 4448 CONTINUE
      CALL GETMAT (C,3)
      CALL MATINV(C,M,Z,0,DETERM,MW)
      DO 17 I=1,M
      Z(I)=0.
      DO 17 J=1,M
   17 Z(I) = Z(I) + C(I,J)*SC(J)
      WRITE (LUT,309)
      WRITE (LUT,222) (Z(I),I=1,M)
      DO 1064 I=1,M
 1064 SB(I) = Z(I)
      DSQ=0.
      DO 18 I=1,M
   18 DSQ = DSQ + SC(I)*Z(I)
      DSQ=AES(DSQ)
      WRITE (LUT,310)
      EM=M
      DS=SQRT(DSQ)
      WRITE (LUT,223) DS,DSQ
      WRITE (LUT,312)
      DSQ=(ENA*ENB/(ENA+ENB))*DSQ
      WRITE (LUT,313) DSQ
      DSQ=DSQ*((ENA+ENB-EM-1.)/(EM*(ENA+ENB-2.)))
      WRITE (LUT,314) DSQ
      NP = NA+NB-M-1
      WRITE (LUT,315) M,NP
      CALL CZLP2
      RETURN
```

```
C
C
  200 FORMAT(13H            N1 =,I3,12H            N2 =I3,12H        DF =,I3/
     151H  VAR    MEAN VECTOR 1  MEAN VECTOR 2  DIFF.VECTOR  )
  222 FORMAT(2X,12F10.5)
  223 FORMAT (2X,F10.5,11X,F10.5)
  272 FORMAT(I4,F14.3,2F15.3)
  300 FORMAT (32H1            STATISTICAL DISTANCE ,//1H0)
  309 FORMAT(28H0 DISCRIMINATOR COEFFICIENTS)
  310 FORMAT(33H0 RESULTS FOR MAHALANOBIS  DSQUARE /33H              D
     1         D SQUARE )
  312 FORMAT(40H0 SIGNIFICANCE FOR D SQUARE AND T SQUARE/)
  313 FORMAT(11H0T SQUARE =F13.3)
  314 FORMAT( 4H0F =F13.2)
  315 FORMAT(6H0DF1 =  I3,7H  DF2 = I3  )

      END


      SUBROUTINE MATINV(A,N,B,M,DETERM,JW)
      DIMENSION IPIVOT(45),A(JW,JW),B(JW,1),INDEX(45,2),PIVOT(45)
      EQUIVALENCE(IROW,JROW),(ICOLUM,JCOLUM),(AMAX,T,SWAP)
      DETERM=1.
      DO 20 J=1,N
   20 IPIVOT(J)=0
      DO 550 I=1,N
      AMAX=0.
      DO 105 J=1,N
      IF(IPIVOT(J)-1)60,105,60
   60 DO 100 K=1,N
      IF(IPIVOT(K)-1)80,100,740
   80 IF(ABS(AMAX)-ABS(A(J,K)))85,100,100
   85 IROW=J
      ICOLUM=K
      AMAX=A(J,K)
  100 CONTINUE
  105 CONTINUE
      IPIVOT(ICOLUM)=IPIVOT(ICOLUM)+1
C     INTERCHANGE ROWS TO PUT PIVOT ELEMENT ON DIAGONAL
      IF(IROW-ICOLUM)140,260,140
  140 DETERM=-DETERM
      DO 200 L=1,N
      SWAP=A(IROW,L)
      A(IROW,L)=A(ICOLUM,L)
  200 A(ICOLUM,L)=SWAP
      IF(M)260,260,210
  210 DO 250 L=1,N
      SWAP=B(IROW,L)
      B(IROW,L)=B(ICOLUM,L)
  250 B(ICOLUM,L)=SWAP
  260 INDEX(I,1)=IROW
      INDEX(I,2)=ICOLUM
      PIVOT(I)=A(ICOLUM,ICOLUM)
      DETERM=DETERM*PIVOT(I)
C     DIVIDE PIVOT ROW BY PIVOT ELEMENT
      A(ICOLUM,ICOLUM)=1.
      DO 350 L=1,N
  350 A(ICOLUM,L)=A(ICOLUM,L)/PIVOT(I)
      IF(M)380,380,360
  360 DO 370 L=1,M
  370 B(ICOLUM,L)=B(ICOLUM,L)/PIVOT(I)
  380 DO 550 L1=1,N
      IF(L1-ICOLUM)400,550,400
  400 T=A(L1,ICOLUM)
      A(L1,ICOLUM)=0.
      DO 450 L=1,N
  450 A(L1,L)=A(L1,L)-A(ICOLUM,L)*T
```

```
      IF(M)550,550,460
  460 DO 500 L=1,M
  500 B(L1,L)=B(L1,L)-B(ICOLUM,L)*T
  550 CONTINUE
      DO 710 I=1,N
      L=N+1-I
      IF(INDEX(L,1)-INDEX(L,2))630,710,630
  630 JROW=INDEX(L,1)

      JCOLUM=INDEX(L,2)
      DO 705 K=1,N
      SWAP=A(K,JROW)
      A(K,JROW)=A(K,JCOLUM)
      A(K,JCOLUM)=SWAP
  705 CONTINUE
  710 CONTINUE
  740 RETURN
      END


      SUBROUTINE PLTHS1(XX,III,START,ENDX,NUM)
C     PLOTS A FREQUENCE DISTRIBUTION IN NUM INCREMENTS FOR THE VALUES IN
C     THE ARRAY XX BETWEEN START AND ENDX IN VSLUE. THE FIRST III
C     VALUES IN THE XXX ARRAY ARE CONSIDERED.
C
C     ----------------------------------------------------------------
      DIMENSION XX(100),KONT(100),LINE(101)
      DATA (LB=1H ),(LAST=1H*),(LDASH=1H-),(LI=1HI)
C-F IV- DATA LB,LAST,LDASH,LI /1H ,1H*,1H-,1HI/
C     ----------------------------------------------------------------
C
      COMMON /UNITS/ LIN,LUT,MTA,MTB,MTMABC,MTARA,MTBRA
      DO 1 I=1,NUM
    1 KONT(I) = 0
C     SET NUM COUNTERS TO ZERO.
      XNUM = NUM
      XINT = (ENDX-START)/XNUM
C     THE INTERVAL SIZE FOR THE NUM INTERVALS IS XINT.
      NUM2 = 100/NUM
      NUM3 = NUM2 - 1
      NUM4 = NUM2*NUM+1
      DO 5 I=1,101
    5 LINE(I) = LB
C     LINE IS SET EQUAL TO BLANKS.
      LINE(NUM4) = LI
C     THE LAST PLACE TO BE USED IN LINE IS SET EQUAL TO AN I.
      DO 10 I=1,III
C     LOOK AT III PINTS IN THE XX ARRAY.
      IF(XX(I).LT.START.OR.XX(I).GT.ENDX) GO TO 10
C     IF XX(I) IS OUTSIDE THE RANGE IGNORE IT.
      KNT = (XX(I)-START)/XINT
      KNT = KNT + 1
C     KNT IS THE INTERVAL THAT XX(I) FALLS INTO.
      IF(KNT.GT.NUM) KNT = NUM
C     IF XX(I) = ENDX KNT COULD BE GREATER THAN NUM.
      KONT(KNT) = KONT(KNT) + 1
C     INCREASE THE KNT COUNTER BY 1.
   10 CONTINUE
C
      KMAX = KONT(1)
      DO 20 I=2,NUM
      IF(KONT(I).GT.KMAX) KMAX = KONT(I)
C     FIND THE VALUE OF MAXIMUM COUNTER.
   20 CONTINUE
C
```

```
      KSCALE = KMAX/50
C     KSCALE IS THE VERTICAL SCALE FACTOR.
      IF(KSCALE*50.NE.KMAX) KSCALE = KSCALE + 1
      WRITE (LUT,100) XINT,START,ENDX,NUM
      NTOP = KMAX + KSCALE
      DO 21 I=1,NUM
      IF(KONT(I).EQ.0) KONT(I) = -KSCALE
   21 CONTINUE
   30 NTOP =NTOP-KSCALE
      IF(NTOP.LE.(-KSCALE)) GO TO 40
      IF(NTOP.LT.0) NTOP = 0

      ICONT = 0
      DO 32 I=1,NUM
      IF(KONT(I).LT.NTOP) GO TO 32
C     IF THE COUNTER FOR THE INTERVAL IS GREATER THAN OR EQUAL TO NTOP
C     PUT ASTERISKS INTO THE PART OF LINE ALOTTED TO THIS INTERVAL.
      DO 33 II=1,NUM2
C     THERE ARE NUM2 SPACES IN LINE FOR EACH INTERVAL.
      INX = II + ICONT
   33 LINE(INX) = LAST
   32 ICONT = ICONT + NUM2
      IF(NTOP.EG.0) GO TO 40
      WRITE (LUT,101) NTOP,LINE
      GO TO 30
C
   40 IF(NUM3.EG.0) GO TO 50
C     THIS SEGMENT FROM HERE TO RETURN PRINTS THE BASE LINE WITH I AT
C     THE END OF EACH INTERVAL AND MINUSES BETWEEN.
      ICONT = 0
      DO 41 I=1,NUM
      DO 42 II=1,NUM3
      INX = II + ICONT
   42 LINE(INX) = LDASH
      LINE(INX+1) = LI
   41 ICONT = ICONT + NUM2
      WRITE (LUT,102) LINE
      RETURN
C
   50 DO 51 I=1,NUM
   51 LINE(I) = LI
      WRITE (LUT,102) LINE
      RETURN
C
C
  100 FORMAT(34H1         HISTOGRAM FOR DISCRIMINANTS /14H0   INTERVAL = ,
     1F10.4,16H DATA STARTS AT ,F10.4/11H0   ENDS AT ,F10.4,19H CLASS INT
     2ERVALS = ,I4)
  101 FORMAT(I5,2X,1HI,101A1)
  102 FORMAT(7X,1HI,101A1)
      END
```

8
Canonical Variates

Formal Background

We have already touched upon some of the properties of a multivariate normally distributed universe, and shall return to this topic in Chapter 12. One of the more interesting morphometric applications of multivariate statistics concerns the method usually known as canonical variate analysis. In one way, this method may be regarded as a generalization of component analysis, for it treats two or more samples simultaneously. In another, it is a generalization of the construction of generalized distance charts in which the underlying axes of variation are first constructed and then used to plot the positions of the various groups. The taxonomic implications of this procedure are interesting, some are almost self-evident, others much more subtle.

We may think about canonical variates in the following way. We have k samples, each drawn from a universe (with the theoretical requirement that the dispersion matrices of the k universes are homogeneous). Each of the universes forms a swarm of points in a p-variate Cartesian space: these swarms will be ellipsoidal in shape if the variables are multivariate normally distributed.

The calculations for canonical variates are similar to those for principal components and here, again, transformed axes are produced. The first axis is inclined in the direction of greatest variability between the means of the k samples. The second axis is perpendicular to the first and inclined in the direction of next greatest variability, and so on for the subsequent axes. In principal components analysis, the latent vectors are orthogonal to each other. In canonical variates this is generally not so.

We have already pointed out the discriminant properties of the latent vectors or canonical variates in the foregoing chapter. The procedure of canonical variates can also be interpreted in the light of the multivariate analysis of variance, single classification.

The canonical variates are found from the latent roots (z_i) and vectors of the determinantal equation:

$$|B - zW| = 0, \tag{8.1}$$

where B is the matrix known as the "among matrix of sums of squares and cross products" and W is the "within sums of squares and cross products matrix".

If there are more variables than groups, and in particular, if $p > k - 1$, there will only be $k - 1$ non-zero roots to equation (8.1). Conversely, if $p < k - 1$, there will be p latent roots.

The latent vectors corresponding to the determinantal equation (8.1) and thus corresponding to the latent roots z_i, are found from

$$(B - zW)t = 0, \tag{8.2}$$

and are the p-component vectors, t. These vectors may be standardized to give the canonical axes, u:

$$\left(t' \frac{W}{n-k} t \right)^{-\frac{1}{2}} = u. \tag{8.3}$$

These calculations, as well as tests of significance and multivariate analysis of variance computations, are done in the programme CANVR2, appended to this chapter. This programme is mainly based on Reyment and Ramdén (1970).

For morphometric problems, significance testing and the multivariate analysis of variance applications are definitely of subordinate value. The main use of canonical variates in morphometric studies is for disclosing relationships between groups.

Methods of Representing the Results of Canonical Variate Analyses

The end product of an analysis along canonical variates consist of (i) the vectors which represent the axes themselves, (ii) the positions of each individual entity or group of entities measured along each of these axes of variation, and, possibly, (iii) various ancillary significance tests, should the nature of the problem call for them. Although from a strictly statistical point of view the analysis has been substantially completed at this stage, from the biologists' or geologists' point of view, the task of interpreting the results is just beginning.

A first task is to find out how many of the canonical variates correspond to biologically or geologically meaningful sources of variation, that is, how many such variates so order the data that the analyst can learn something of consequence from the way in which the data are ordered. For instance, where the measurements are the dimensions of animals and there is sexual dimorphism in the species being measured, at least one of the canonical variates should order the material so that males have distinctly higher or lower scores than the females, although there may be some overlap if the sexual

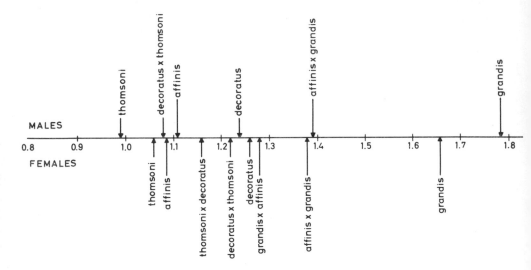

FIRST CANONICAL VARIATE

(size variation)

Fig. 9. Size relationships between four species of the bug *Scolopostethus* and some of their hybrids: based on ten morphometric characters compounded into the first canonical variate. Redrawn from Eyles and Blackith (1965).

dimorphism is slight in comparison with the intrasex variation. Sometimes, the first canonical variate (that corresponding to the largest root of the determinantal equation) takes up such a large fraction of the total amount of variation in the material that merely plotting the positions of the various entities along this first axis gives good insight as to the nature of the axis. Eyles and Blackith (1965) performed a canonical analysis of ten measurements made on several species of lygaeid bugs and their hybrids. The first canonical variate represented nearly all the variation in size, displaying the hybrids as intermediate in size between the parent species (Fig. 9).

There is, at this point, a clear distinction to be drawn between the statistical and the biological importance of an axis of variation. Statistically, the importance is measured by the size of the canonical root which generates the vector, and roots which do not contribute significant amounts to the total sum of squares in the analysis of dispersion are to be disregarded. From the biologist's standpoint, such an assessment of "importance" is as misleading for what it does not do as for what it does. The largest of the canonical roots generates a vector (the first canonical variate) which often represents some quantity which is numerically large but of little consequence to the biologist. In many morphometric analyses, size variation will occupy one of the variates corresponding to the largest canonical roots, but size variation is seldom an object of study in itself. There is an analogy here between the analysis of dispersion, which is a form of the multivariate analysis of variance, on the one hand, and the widely used univariate analysis of variance in agricultural field trials on the other. The analysis of a field trial often brings out a substantial sum of squares corresponding to differences between the experimental blocks (in a randomized block or similar experiment) but this sum of squares has been segregated precisely because it contributes nothing of interest and would inflate the residual variance to no good purpose unless it was removed. No analyst would point to a large "between-blocks" variance and claim that it was, on that account, of agricultural interest.

Conversely, it does not follow that because a canonical variate has been generated by a root which is not statistically significant, it is not capable of ordering the material in a meaningful way. Such a situation is liable to happen when only a few measurements have been made, limiting the number of axes of variation which can be generated. The last of these axes that comes out of the analysis, corresponding to the smallest root, can never be shown to have statistical significance by the usual tests, but may well be of service in ordering the data. A good example occurs in the work of Albrecht and Blackith (1957) who analysed simultaneously three measurements made on the sexes and phases of two species of locusts. They found that three axes of variation could be discerned in a generalized distance chart, each one with a clearly defined biological meaning, namely, sexual dimorphism, phase variation, and specific differences of form.

Although only two of these axes could be shown to have statistical significance using multivariate theory, because only three characters were measured, the third axis is just as interesting biologically as the first two. Non-parametric tests based on the capacity of the third vector to ordinate the organisms could usefully be employed at this stage.

It is often useful to plot the data arising from the analysis on graphs whose axes are the canonical variates taken in pairs. In the paper cited, Eyles and Blackith, after disposing of the less interesting size variation in their bugs, plotted the second axis of variation against the third (Fig. 10). The second axis proved to order the bugs

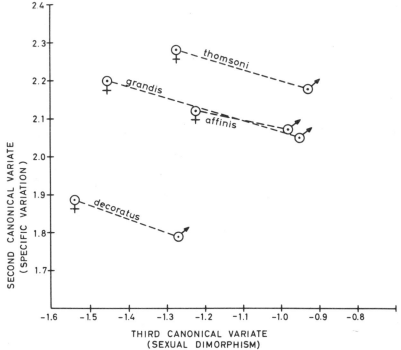

Fig. 10. Chart showing the shapes of four species of *Scolopostethus* along the second, and sexual dimorphism along the third, of three canonical variates based on ten morphometric characters. Adapted from Eyles and Blackith (1965).

into species, again with the hybrid forms intermediate, and the third axis ordered them into sexes.

When only a limited amount of data has to be plotted, as was usually the case in early multivariate studies because of the difficulty of processing the data by hand, graphs of this kind are often sufficient to elucidate all the important features of the investigation. If there are p dimensions measured, there will be $p(p-1)/2$ graphs to be drawn, but a few quick sketches will usually disclose whether all these graphs need to be drawn out in full, since some display the data more clearly than others.

When there are large amounts of data to be plotted, the problem can become quite tedious and difficult. Many computers carry data-plotting equipment, usually off-line, but each of the plotted points may have to be identified and this identification is almost as tedious as the drawing of the original graphs. Sneath (1966) remarks that "it is surprising how difficult it is to imitate on a computer a process which is swiftly accomplished by the eye whenever the data can be presented graphically in a suitable form". He is referring to the attempted curve-seeking programme designed to pick out trends from scattered data. The remark could apply with equal force to the problems of using computers to present data in a form suitable for interpretation. So great are these problems that it is sometimes well worth while spending quite a lot of time in exploring rearrangements of the computer output so as to present the result of the experiment to best advantage. The programmes published in this book contain limited facilities for machine identification of the points on the plots.

CASE HISTORIES INVOLVING THE USE OF CANONICAL VARIATES

The Ecology of Insects Living on Broom Plants

Eight attributes of some mirid bugs living on broom plants were measured by Waloff (1966) for populations living in South East England; California, U.S.A. and British Columbia, Canada. The first three canonical variates were found to be of interest and the first two are plotted in Fig. 11. Waloff has drawn on her chart the underlying dimensions of sexual dimorphism and specific differences (between the measured two species of the genus *Orthotylus*) as reasonably inferred from the fact that all the male insects are located by these two canonical variates at the top left-hand of the chart, with the females down in the bottom right-hand portion, whereas all *O. virescens* fall into the top right-hand portion and all *O. concolor* in the bottom left-hand.

Although the method of canonical variates is not usually thought of as a clustering method, because it operates on measurements made on groups of organisms (groups which are known prior to the analysis), it does afford an effective method of clustering these

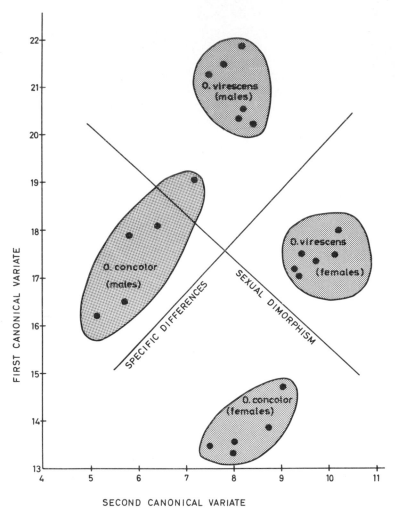

Fig. 11. Canonical variate chart showing the specific and sexual variation of the mirid bug *Orthotylus* on broom plants in England and Western N. America. Based on eight morphometric characters. Redrawn from Waloff (1966).

groups into biologically meaningful entities. The four entities here are determined by sex and species. The axes drawn inside the chart are purely inferential and do not require computation, they are in fact simply a guide to the reader's eye to tell him what to look for in his interpretation.

Within each of the four main clusters there is a visible structure. In each one of the four, the sample from England has the highest score on the first canonical axis, whereas the samples from Canada have the lowest scores on the same axis, with the remaining groups intermediate. Geographical variation is thus symbatic with component 1, and so far as geographically determined changes of form are concerned there is much in common between them and the sexual dimorphism.

This interpretation, at least in its main features as outlined here, is quite straightforward and presents few problems. Reference to the original publication will show that there is more to be extracted from the data than has been discussed above, but the paper does show that the introduction of *Orthotylus* to British Columbia has changed the shape of this British bug in both species.

The Nature of Variation in a "Difficult" Grasshopper Genus

It is a commonplace that, amongst any group of animals, some genera or other subdivisions are particularly difficult to classify. The reasons vary; among parasitic animals the great importance of behaviour patterns integrated with those of the host ensures that species are differentiated, all too readily from the taxonomist's point of view, by behavioural traits which may leave little record in the anatomy. Even in non-parasitic groups, however, "difficult" genera are to be found. One such is the pyrgomorphid genus *Chrotogonus,* with a vast range from Egypt to Soviet Central Asia and down to Ceylon. As Blackith and Kevan (1967) comment, the genus shows a morphological plasticity, even within individual species, which is bewildering. These authors measured seven morphological characters (length and width of the pronotum; length and width of the hind femur; width of the mesosternal interspace; elytron length; and head width) on 1093 insects of both sexes. This mass of information related to 34 different predetermined taxa according to the species, subspecies, sex, and degree of alary polymorphism, since *Chrotogonus* species may include individuals with well-developed wings and elytra or with these organs rudimentary.

A canonical variate analysis of the data showed that there were effectively only two basic dimensions of variation, the first accounting for 94.6% of the total variation and the second for 4.1%. These two dimensions are represented by the vectors

$$(-0.147, +0.297, +0.359, -0.832, +0.108, -0.001, +1.000)$$
and
$$(-0.220, +0.018, +0.006, +1.000, -0.921, +0.350, -0.826)$$

which, although orthogonal in the seven dimensional Riemannian space in which the groups are separated by their proper generalized distances, make an angle of some 40° in three-dimensional Euclidean space.

The interpretation of the chart constructed from these vectors, using them as ordinate and abscissa respectively and plotting the positions of each group along the axes, is unusual. Each canonical vector subsumes two biologically distinct patterns of variation.

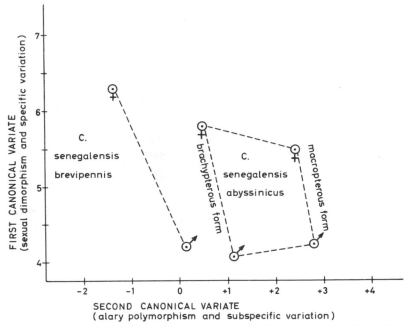

Fig. 12. Canonical variate chart for the grasshopper *Chrotogonus senegalensis* showing that subspecific variation is symbatic with alary polymorphism. Adapted from Blackith and Kevan (1967).

In Fig. 12 the subspecific distinction between *C. senegalensis brevipennis* and *C. s. abyssinicus* occupies the same (second) variate as the alary polymorphism contrasting the long-winged and short-winged forms. In Fig. 13 the first variate subsumes both sexual dimorphism and some of the differences between species, since *C. homalodemus* always has higher scores along this variate than the various elements of *C. hemipterus*. Many other examples can be seen in the original paper.

We thus have a disquieting situation for the taxonomist; sexual variation is, at least on the basis of these seven characters, indistinguishable from the specific contrasts of form corresponding to the division between species with slender ovipositors and species with stouter ovipositors. Moreover, alary polymorphism is inextricably bound up with many of the differences between subspecies. Patterns of growth which are biologically quite distinct in their causation but phenotypically identical in their expressions have been called "symbatic" by Blackith and Albrecht (1959) who found

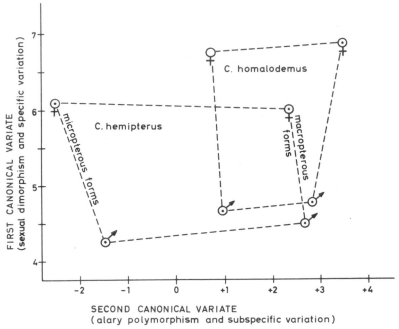

Fig. 13. Canonical variate chart showing that specific differences between *Chrotogonus hemipterus* and *C. homalodemus* are symbatic with sexual dimorphism. Based on seven morphometric characters. Adapted from Blackith and Kevan (1967).

another example in the polymorphism of the red locust. It seems as if the genus *Chrotogonus* is incapable of exhibiting more than these two patterns of contrasting forms, so that all the ecological and environmental variation has to be expressed along these two variates. As a consequence, the range of variation of any one species or subspecies along them is enormous; within one species individuals can be found whose shape and size correspond with any of the other species (sex for sex) within the genus. Little wonder that the subtle appreciation of shape which is normally the basis of a skilled taxonomist's work should have led earlier workers further into difficulties. It is, in fact, virtually impossible to be sure of an identification of any individual *Chrotogonus* unless the provenance is known, since the forms which are most alike often have different geographical origins.

Reyment (1969c) has revealed that a somewhat similar situation arises in the ammonites of the Jurassic genus *Promicroceras* where five putative species can be splayed out on principal coordinates charts (in this context functionally similar to canonical variates) without any apparent grouping into species, at least on the basis of the seven measured characters.

The Simultaneous Analysis of Many Different Groups of Grasshoppers

When large numbers of groups of organisms are to be analysed, problems can arise that are rare when few groups are involved, Blackith and Blackith (1969) analysed 196 groups belonging to the Morabinae, a subfamily of the eumastacid grasshoppers, by means of a canonical variate analysis based on ten measurements; 1450 insects were measured altogether. Each axis of variation reflects a biologically meaningful contrast of form; however, some of these contrasts will not affect some of the groups. For instance, sexual dimorphism is a relevant contrast of form for all except the single parthenogenetic species of the subfamily. The fifth of the canonical variates contrasts the two divisions of the genus *Keyacris,* for instance, and is thus irrelevant to members of other genera. All these other groups will, however, have scores along the fifth axis and it becomes necessary to pick out the meaningful contrasts of form from the apparently meaningless ones.

An extreme example of this necessity is revealed by consideration of the seventh canonical variate which contrasts the single parthenogenetic species with all the remainder. This contrast of *virgo* and a few related, but bisexual, species with the remainder of the subfamily is very clear-cut, there being no overlap between the two groups, yet the canonical variate takes up only a small fraction of the total variation just because so few species are involved in the outlying group. This situation should be a clear warning not to consider the biological importance of a canonical variate too closely in terms of its capacity to take up a large fraction of the total variance.

Any rules comparable with those used by some factor analysts, by which the process of extracting factors (or, in this case, canonical variates) is brought to an end, are hardly likely to be worth applying in canonical variate analysis. For the same reason noted above, that a variate may be biologically important but statistically trivial, no rule, based as it must be on the statistical importance, can be relied upon not to exclude the very canonical variate which matters most to the experimenter.

A Psychometric Example Using Dichotomous Variables

Maxwell (1961) has pointed out that one can quite correctly employ canonical variate analysis even when the data are dichotomous. Indeed, as Claringbould (1958) has shown, most of the standard techniques of multivariate analysis can be worked successfully with data of the presence-absence type. In Maxwell's example, 224 schizophrenics, 279 manic depressives, and 117 patients with various anxiety states were scored for the presence or absence of the following four symptoms: (a) anxiety, (b) suspicion, (c) schizophrenic type of thought disorder, (d) delusions of guilt.

When these scores were converted into a matrix, and the canonical variates extracted, a graph was constructed on which the positions of the three groups of patients could be plotted (Fig. 14). The manic depressives and patients with anxiety states form one pole of the first canonical variate, accounting for 83% of the total variance, and the schizophrenics form the other pole. The manic depressives are separated from the patients with anxiety states by the second of the canonical variates, accounting for 17% of the total variation. It should be noted that since there are only three groups under comparison, only two variates can be extracted from the data. The weights of the characters in these two variates, which are respectively

$$(-0.49, +0.46, +0.70, -0.06)$$

and

$$(+0.29, +0.03, +0.06, -0.34),$$

show that the distinction between schizophrenics and the rest depends mainly on the first three of the characters, whereas that between manic depressives and anxious patients depends mainly on the first and last characters.

Naturally, this kind of experiment can be performed with continuously distributed data, instead of dichotomized data; an example is given by Rao and Slater (1949).

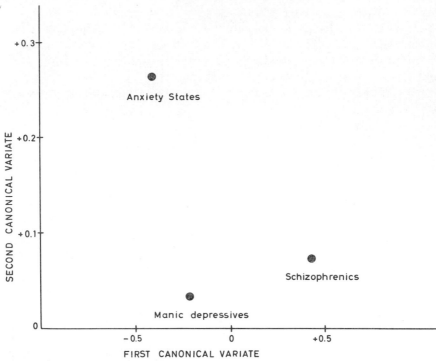

Fig. 14. The psychometric relationships between three sets of mentally disturbed patients, assessed along two canonical variates. Based on four dichotomous characters. Redrawn from Maxwell (1961).

The Primate Shoulder in Locomotion

Nine measurements of the scapulae of a number of Anthropoidea and Prosimii were compounded into nine canonical variates by Ashton *et al.*, 1965; Oxnard, 1967, 1968. The first three of these variates were found to have a readily interpretable biological significance. The separation afforded by the first two of the canonical variates is shown in Fig. 15a.

Although differentiation of the material at the highest (subordinal) level was poor, good discrimination was obtained within the Hominoidea, separating the genus *Homo* from the four genera of apes in the Pongidae. These apes are all brachiators. Again, within the Cercopithecoidea, the semibrachiators of the Colobinae are distinct from the quadrupedal Cercopithecinae. The basic method of locomotion adopted by a species influences, as might be expected, the shape of its scapula. Among the Prosimii, a similar contrast of the

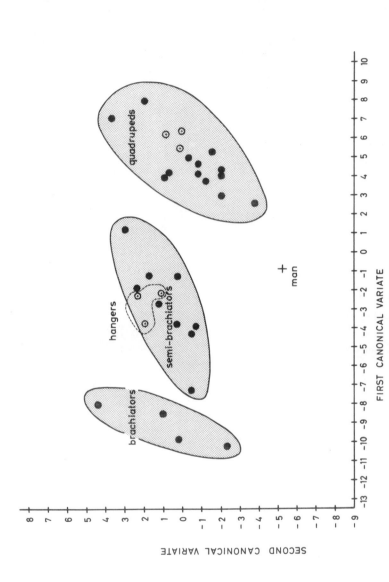

Fig. 15a. Canonical variate chart showing the mutual relations of the shoulder girdles of some primates. Based on nine morphometric characters. Adapted from Ashton, Healy, Oxnard and Spence (1965). Open circles ○, prosimiids; closed circles ●, anthropoids.

forms of the scapulae is shown by the separation of the quadrupedal Galaginae from two genera of Lorisinae which are hangers. This distinction is pervasive, and turns up again in the lemurs so that *Lemur,* a quadrupedal genus, is clearly separated from the genus *Propithecus,* which is a hanger. Most of these distinctions are made by the first variate, which reflects primarily those anatomical changes which have occurred as a response to the amount of use of the forelimbs for suspension. The second variate, however, seems mainly to differentiate ground-dwellers from arboreal dwellers, a distinction which involves changes of the scapula apart from those involved in the use of the forelimb in hanging, for there are important aspects of locomotion other than the suspensory ones. The almost exclusively arboreal forms *Pongo, Hylobates, Brachyteles,* and *Rhinopithecus* cluster at one extreme of this ordination, whereas the more terrestrial members cluster at the opposite extreme.

The third of the canonical variates does not distinguish between the subhuman primate genera, but it does clearly separate man from them. *Homo* is thus unique in being the only genus to stand out from the plane formed by the first two canonical variates into a third dimension of variation. In some respects, man has a shoulder girdle like that of the quadrupeds, in other respects it is like the brachiators, but it is a mosaic of these trends and not, as are the semibrachiators, intermediate between the two extremes in respect of separate features. This analysis is an excellent illustration of the use of multivariate analyses in sorting out what has been termed mosaic evolutionary trends, where advances (or retrogressions) have proceeded simultaneously along several quite distinct axes of variation. The success of the technique is all the more striking because considerable efforts had previously been made to recover the information outlined above from a study of the individual measurements and their ratios, without conspicuous success. It may be noted that the canonical variates are often referred to in these papers as discriminant functions, a correct usage in that they are linear functions which are being used for discrimination, but perhaps apt to mislead the uninitiated into thinking that each is a discriminant function between a pair of groups. This question was discussed in the introduction to Chapter 7.

These papers are also of interest in that they carry, round each of the plotted points, a measure of the precision with which each point is determined, in the form of a cross whose arms are equal to twice the standard error of the mean, together with a circle including 90%

of the individuals comprising that mean. For detailed interpretation these additions to the charts are of considerable value, although they do make the initial inspection more difficult because of the relatively "cluttered" appearance of the chart with these additions.

Functional relationships of the kind revealed by these studies can be traced in the evolution of the shoulder girdle in marsupials, edentates, rodents and carnivores (Oxnard, 1968) and in the teeth of hominoids (Ashton *et al.,* 1957). No doubt they form an important controlling influence on the integration of shape and size in the mammalian skeleton as discussed by Olson and Miller (1958). Oxnard, in fact, goes on to suggest that these influences on the form of the shoulder girdle may have become established so early in the evolution of the mammalia as to restrict their development to a limited number of pathways.

It seems that the multivariate analysis of mammalian skeletal structures is able to throw light on the evolution of the group in a way which is quite distinctive and which has escaped the more classical approaches to the subject. The multiplicity of forms in the shoulder girdle that have evolved can be reduced, in fact, to the results of the interplay of only a very few, essentially two, uncorrelated patterns of development. These do not appear to differ appreciably even in such widely separated groups as the marsupials, the edentates, and the primates. The adaptation to a gliding habit gives a third pattern of variation. Some classical workers had suggested that these functionally adaptive features of the skeletal joints were superimposed on the basic pattern of development of the animals: the facts seem to be exactly the contrary; the functional adaptations, as reflected in the interplay of the two patterns of development, measurable along the first two canonical variates illustrated by Oxnard (1968), and repeated here for a small part of his total findings, *are* the basic patterns of development whose degree of interlocking is controlled by the evolutionary process. That the evolution of form depends essentially on the degree of interlocking of a few distinctive patterns of growth has already been suggested for insects by Blackith (1965). Thus multivariate studies of growth and adaptation can make a contribution to functional anatomical studies which more classically oriented methods have not seemed capable of making.

Oxnard has shown (Fig. 15b) that the adaptation to actual gliding is distinct from the adaptations that go with living arboreally. Moreover, some of the bats which have a butterfly-like flight are

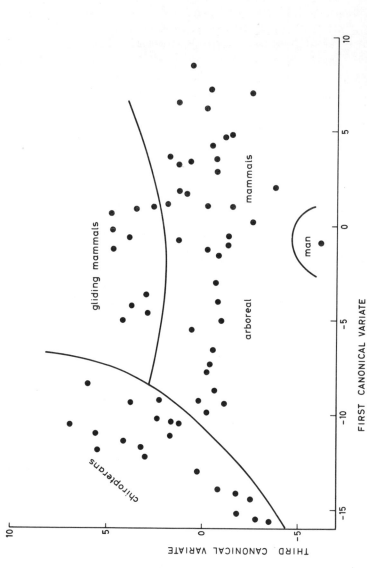

Fig. 15b. Canonical variate analysis of the shapes of the shoulder girdle of some mammals associated with trees, based on nine morphometric measurements. Scale units are one standard error along each axis. Redrawn and adapted from Oxnard (1968).

separated along axis 3 from those which have either powerful long-distance flight or swift straight flight. This axis also separates man from the other primates.

Within the arboreal carnivores, almost all the useful separation was along axis 1: this axis seems to separate those forms in which the forelimbs are designed to withstand pulling from those whose forelimbs are designed to take thrusts, this is the essential distinction between the terrestrial and arboreal ways of life.

As often happens when very extensive bodies of data are analysed, some of the later canonical variates, in Oxnard's work up to the ninth, afforded useful contrasts of form with evident biological interpretations.

The Skeletal Form of Some African Shrews

In the papers of the late Dr. T. P. Burnaby, deposited in the Palaeontological Institute, Uppsala, were two analyses of the skeletons of some Pleistocene African shrews, based on measurements made by Professor P. M. Butler and Mrs. M. Greenwood, of Royal Holloway College, Surrey, England. We are most grateful to Professor Butler for allowing us to publish these analyses, which have the special interest that one was carried out using the measurements and the other using the rankings of the various species of shrew in respect of the five characters measured. As will be discussed later (p. 300, Chapter 26) tests of the robustness of multivariate analyses by using rankings, and thus rendering the analyses non-parametric, are as valuable as they are rare, and we take this opportunity of paying tribute to the imaginative scope of Dr. Burnaby's contributions to statistics.

Figure 16a shows the results in the form of a plot of the first two canonical variates based on measurements on which the positions of the 19 species of shrew are marked. The *deserti-hindei-fulvaster-marrensis-katherina* group form a natural and distinctive subdivision of the genus *Crocidura,* other subdivisions of which are made up of the *fumosa-turba-luna-monax* group, and the *russula-suaveolens-cyanea* groups. The species *gracilipes,* although forming part of the genus *Crocidura,* has some characters much like those of the genus *Sylvisorex,* made up of *megalura-granti-johnstoni* and *lunaris.* The species *etruscus* and *lixus* fall into the genus *Suncus.*

Figure 16b shows the same species plotted on the first two canonical variates based solely on the rankings. At least from the point of view of a biologist seeking to interpret the charts to discover

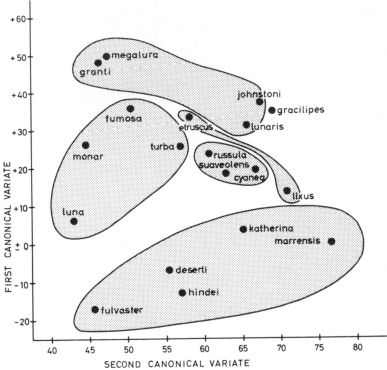

Fig. 16a. Nineteen species of African shrews clustered along two canonical variates, based on five cranial characters. Data due to Prof. P. M. Butler and Mrs. M. Goodman. Analysis by Dr. T. P. Burnaby.

the affinities of the various groups, almost no loss of information has taken place, suggesting that the canonical variate analysis is quite robust, at least in the sense of the term used here. In both cases not only the generic groups but various infrageneric groupings are evident. The extent to which canonical variate analysis can be used to give information about infraspecific and geographical relationships is shown by Delany and Whittaker (1969), who studied variations of field-mice with the aid of nine skull characters.

Subspeciation in South American Anteaters; a Confused Taxonomy

The development of the taxonomy of the South American anteaters of the genus *Tamandua* seems to have been unusually confused, authorities disagreeing flatly about the number and nature of the subspecies of *T. tetradactyla*. Appreciating that much of this

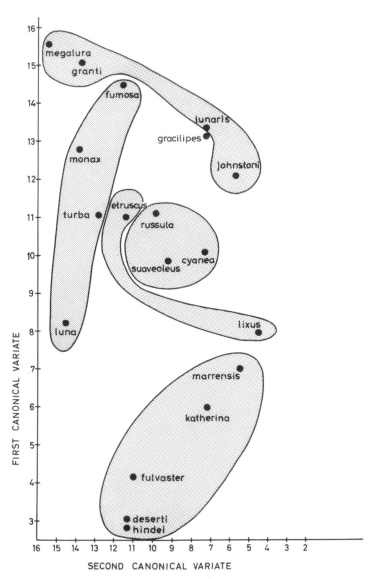

Fig. 16b. The same data and analysis as for Fig. 16a, but based on the rankings of the 19 species in respect of the five characters instead of the absolute measurements.

confusion arose from the failure to realize that individuals of different sizes will often have distinct morphometric ratios of attributes, Reeve (1941) quite rightly stressed the importance of considering the taxonomy of the genus in the light of what was at that time known about allometric growth patterns, and making a suitable allowance for the size of the animal.

Later, Seal (1964) reanalysed the data collected by Reeve, involving the measurements of the basal length of the skull excluding the premaxilla; the occipitonasal length; and the greatest length of the nasals. Seal found that the two canonical variates in which the three attributes could be linked to describe variation in the group consisted of

| I | 108.918 | −33.823 | −35.497 |
| II | −46.608 | 59.952 | 22.821 |

Seal notes that the picture which emerged from the canonical analysis was substantially different from the conclusions drawn by Reeve, who felt that the subspecies *chapadensis* and *mexicana* were the only ones with a claim to distinct entities on the basis of the skull measurements. However, Seal's analysis showed that *chiriquensis* and *instabilis* were distinct from one another and from the first two subspecies. Moreover Reeve's conclusion that *instabilis* seemed to be only a small edition of *mexicana* and *chiriquensis* is not upheld by the later analysis, but it is only fair to mention that Reeve was here speaking solely in terms of skull measurements, and considered that on the basis of tail-length, and other body characters, "this group clearly deserves to be placed in a separate subspecies".

The almost total discrepancy between the interpretation of Seal's analysis and that of Reeve based on the same data is disconcerting, and it is worth while trying to find out how such a situation could arise, because, although quantitative multivariate analyses often supplement and clarify more conventional ones, they rarely run flatly counter to them. We have here a situation where Reeve's analysis also disagreed with those of earlier workers and Reeve is probably correct in considering that it is the earlier workers who were at fault in forgetting that a morphometric ratio can be misleading when taken on animals of different sizes, because of allometric growth; see, for example, Christensen (1964). On the other hand, examination of the nature of Seal's canonical variates shows that neither is representative of "size", since the vector (1,1,1) evidently makes a large angle with each of the canonical variates (see p. 41). There seem to be two variates, neither strongly size-sensitive,

which are capable of ordering the putative sub-species of
T. tetradactyla; allometric studies such as those of Reeve are
particularly at risk when more than one underlying pattern of
development has to be assessed.

Anthropometry of some Nigerians

A very extensive collection of cranial measurements of various
subdivisions of the people of Eastern Nigeria was made by Talbot,
and analysed after his death by Mulhall in Talbot and Mulhall, 1962,
who used generalized distances to make an excellent arrangement of
the material based on eight characters. Subsequently, Reyment and
Ramdén (1970) have reanalysed the data, using canonical variates to
construct a discriminatory topology, which on general theoretical
grounds, should resemble Mulhall's analysis closely. This material is
highly suitable for a practical test of the concordance of the two
approaches, because Mulhall had originally examined the homogeneity of the basic dispersion matrices for each group and found
that, as is common in anthropometric work, an encouraging level of
homogeneity was present.

In the event, all Mulhall's interpretations were confirmed. The
discriminatory topology required at least three dimensions for its
proper representation, but the first two canonical variates,
accounting for 68% of the total variation, gave a chart which seemed
to include all the essential features of the relationships between the
groups. As is almost always the case with anthropometric data, which
deals with material differing at much less than subspecific level even
by the generous standards employed in hominoid taxonomy for the
erection of categories, there is little apparent "structure" to the
analysis. The first canonical variate contrasted facial height against
head length and bigonial breadth. The second canonical variate
contrasted head length and facial height against minimum frontal
breadth and bizygomatic breadth. The third canonical variate
contrasted head length against all the remaining characters and
appears to have many of the features of a "size" vector. It singled
out the Etche and Abakaliki peoples as being, in this sense, small
headed.

The Form of Some Brachiopods from the Devonian

Figure 17 shows the canonical variate analysis based on three
morphometric measurements for four samples of the brachiopod
Martinia inflata (Schnur) from the Devonian of Bergisch-Gladbach,

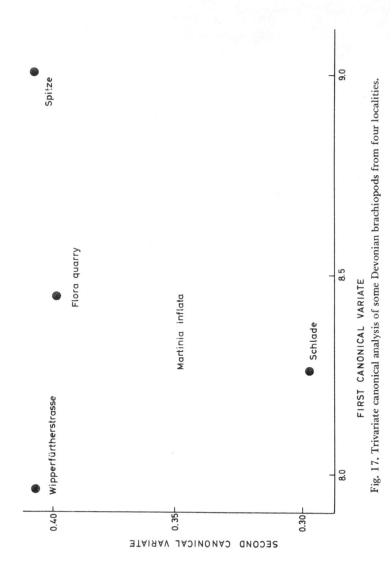

FIRST CANONICAL VARIATE

Fig. 17. Trivariate canonical analysis of some Devonian brachiopods from four localities.

Germany (Jux and Strauch, 1965, 1966). Large numbers of these brachiopods were measured, 212 from the Wipperfürtherstrasse exposure at Flora, 214 from the Spitze exposure 8 from Schlade, and 103 from the quarry at Flora. The variables measured were length, height and breadth of the shell.

The first canonical variate, accounting for about 84% of the total variation, contrasts the first character against the remaining pair, whereas the second canonical variate takes up the remaining 16% of the variation and contrasts the second character against the first and third.

This variation seems to be of two distinct kinds, that along the first canonical variate, which separates the first, second and fourth localities clearly, and that along the second variate which distinguishes the third, Schlade, sample from all the others. It is the Schlade sample which is the "odd man out" of the four. The three grouped samples come from localities fairly close to each other, whereas the small sample from the Schlade occurrence is not from the immediate vicinity. A further point of some palaeoecological significance is worth mentioning. The three more or less linearly located points, Wipperfürtherstrasse, Flora quarry (located close to each other) and Spitze, seem to reflect some kind of homogeneous development of the brachiopods at these three localities. We wish to thank Professor Jux for making this material available.

A COMPUTER PROGRAMME FOR CALCULATING CANONICAL VARIATES

The following programme listing is based on a publication by Reyment and Ramdén (1970); it has been very slightly modified with respect to the output in order to make it more pliable for morphometric work. It is written in general FORTRAN IV and should work on all computers with a compiler for this dialect, providing attention is paid to a few notes interspersed in the listing. The programme is made up of the following major parts.

The basic statistics for each sample are computed and the dispersion matrices tested for homogeneity. The generalized determinantal equation is solved and the canonical roots tested for significance. The mean vectors are tested in a one-way multivariate analysis of variance for significant unlikeness. The means of the first three canonical variates are plotted as are also the corresponding transformed observations. Where great amounts of data are involved, the plots tend to become cluttered. For this reason it is possible to select the number of transformed observations.

The basic elements of canonical variate analysis have already been presented in the introductory part of this chapter. We shall here only review briefly aspects of the programme, not already discussed in the text.

The programme carries out a one-way multivariate analysis of variance (also called analysis of dispersion; a commonly used abbreviation is "manova") for the sample means and may therefore be used to test differences in the means of several groups. As we have already pointed out on earlier occasions, testing of group means is not of primary interest in morphometrics.

However, some situations can and do arise in which it is enlightening to carry out such a test.

Manova is based on a natural breakdown of the variability into "between" and "within" components—in the present case, the "between sums of squares and cross products" matrix and the "within sums of squares and cross products" matrix. The manova breakdown is given in Table 2. S_T is the total dispersion matrix and

Table 2

Breakdown for manova

Matrix	Meaning	Degrees of freedom
$\mathbf{B} = \mathbf{T} - \mathbf{W}$	The among s.s.c.p. matrix	$k-1$
$\mathbf{W} = (N-k)\mathbf{S}_W$	The within s.s.c.p. matrix	$N-k$
$\mathbf{T} = (N-1)\mathbf{S}_T$	The total s.s.c.p. matrix	$N-1$

S_W is the within-groups dispersion matrix, k is the number of groups and N the total number of observations.

The canonical variates are found from the latent roots and vectors of the determinantal equation:

$$|\mathbf{B} - z\mathbf{W}| = 0.$$

The latent vectors corresponding to the z_i are required for finding the canonical variates. These are found from

$$(\mathbf{B} - z\mathbf{W})\mathbf{t} = \mathbf{0}$$

and are the p-component vectors $\mathbf{t}$.

Significance of the Latent Roots

Although in this programme we only make use of the first two canonical variates for Canvar, it is important to test the latent roots for significance, as this gives an indication of whether much information lies with the remaining roots. The well-known test of Bartlett used in this programme is

$$\{(N-1)-(p+k)/2\}\log_e\left[\prod_{j=f+1}^{m}(1+z_j)\right], \qquad (8.4)$$

which is approximately a chi-square variable with $(p-f)(k-f-1)$ degrees of freedom. In this formula N is the total sample size, p the number of dimensions, k the number of groups, f the number of latent roots and the z_j the latent roots themselves. The idea behind this procedure is that the roots are tested to see whether they can be given zero values.

Multivariate Analysis of Variance

Referring back to Table 2, the manova is made in terms of the matrices $\mathbf{W}$ and $\mathbf{T}$ by means of what is known as the Wilks' Λ (lambda) criterion. This is the determinantal ratio:

$$\Lambda = \frac{|\mathbf{W}|}{|\mathbf{T}|} \qquad (8.5)$$

which can vary between nought and one. If it is one then there is no between-groups variance–covariance. The smaller the value of the ratio the greater is the between-groups variance–covariance. An approximate test of significance is made by a chi-square test.

Homogeneity of Dispersion Matrices

Canonical variates analysis requires the assumption of homogeneity of dispersion matrices. There is a body of empirical evidence available that suggests that the method may be moderately robust (resilient) to departures from homogeneity. It is nevertheless advisable to test the sample dispersion matrices in this respect. In this programme the test used is the following well known one, which can be found in any text on multivariate statistics. The version used here is from Kullback (1959, p. 318). The formula required is

$$\sum_{i=1}^{k}\frac{n_i}{2}\log_e\frac{|\mathbf{S}|}{|\mathbf{S}_i|}, \qquad (8.6)$$

where n_i is the sample size, less one, S the pooled dispersion matrix and S_i the dispersion matrix of the ith sample. This is distributed asymptotically as chi-square with $(k-1)p(p+1)/2$ degrees of freedom.

PROGRAMME DESCRIPTION

Operating instructions

Card 1 Column 3: Number of jobs
Card 2 Title of work in columns 1-72
Card 3 Columns 2-3: M = number of variables
        Columns 4-5: MG = number of groups
        Column 6: For matrix data put L = 0
        For raw data put L = 1.

MATRIX DATA

Card 4 --- Format card
Card 5 --- columns 2-6: NI, = the total number of observations
Card 6 and following contain the input matrices in the following order:

    (1) C(I,J) = M × MG matrix of mean vectors
    (2) A(I,J) = "Among groups" sums of squares and cross products matrix
    (3) B(I,J) = "within groups" sums of squares and cross products matrix.

RAW DATA

The further input information needed for raw data is that of the subroutine CORREL, listed here for convenience.

Card 4 Columns 1-5: Size of the sample
        Column 6: For the transformation of the data to base 10 logarithms put KLØG = 1, otherwise KLØG is put = 0.
        Column 7: MØUT = 1 if output from CORREL for this sample is to be printed, otherwise MØUT = 0.
Card 5 Title for this sample in columns 1-72.
Card 6 Variable format for this sample
Card 7 and following, contain the observations.

LIST OF SUBROUTINES

Mainline programme CANVAR

The mainline programme reads and prints on the formalized units LIN (read) and LUT (write). Matrix data are read in here and raw data by calling subroutine CORREL, whereafter the various sums of squares and cross products are accumulated. The generalized latent roots returned by subroutine EIGGEN are tested for significance by a chi-square procedure (8.4). The transformed canonical variate means are printed out from this section.

Subroutine CORREL

This subroutine computes means, standard deviations, sums of squares and cross products, covariances and variances and correlations from the raw data. If required, these calculations can be made on the logarithmically transformed data (base 10).

Subroutine EIGGEN

This computes the generalized latent roots and vectors for the generalized determinantal equation obtained from the "between groups" and "within groups" sums of squares and cross products matrices as in equations (1) and (2).

A negative latent root indicates a divergence from the theoretical requirements of these matrices for canonical variates analysis and should be watched for.

Subroutine GETMAT

This subroutine is a tape-rationalizing procedure. It fetches a labelled stored matrix from a place in the core memory back into the course of action. It may be easily modified to allow recalling of a particular matrix from an external medium such as a tape or drum. GETMAT works in conjunction with PUTMAT; these must be exactly compatible.

Subroutine HDIAG

This computes the latent roots and vectors of a square symmetric matrix. The latter part of the subroutine contains some modifications for the CDC3600, which may be easily changed to suit an IBM FORTRAN compiler.

Subroutine HOMMAT

This carries out an intermediate step for the test of homogeneity of covariance matrices. A byproduct of the calculations is an abridged principal component analysis for each sample. HOMMAT is only called into play for raw data.

Subroutine HOMO

This subroutine carries out a test for homogeneity of the dispersion matrices (8.6) derived from the samples. It is only called where raw data are used. The result is expressed in terms of chi-square.

Subroutine MATOUT

Prints out MG x M matrices with a title and a maximum of ten numbered columns to a page. In its present form it is adapted for the CDC3600 computer, which has eight characters in each cell (NCHPW set to eight). On other computers NCHPW will have to be changed to the pertinent cell size—this involves a corresponding change in format statement 103.

Subroutine PLT 3

This subroutine plots a scatter diagram, used for the transformed means and the transformed observations, on the printer. It takes account of each sample in the latter case, plotting for each an identifying letter. The number of transformed observations to be plotted is to be specified.

Subroutine PUTMAT

This stores an M x N matrix and labels it. The subroutine complements GETMAT.

Subroutine STAND

The subroutine standardizes the elements of the matrix of the generalized latent vectors as an intermediate step in producing the coordinates for the plotting section of the programme. The calculations of equation (8.3) are done here.

Subroutine TRANSF

This transforms the original observations to a canonical variate set and prepares the first two elements of each vector for plotting by PLT 3.

Subroutine WILKS

This subroutine reports on the one-way multivariate analysis of the samples; i.e. it is a test of equality of mean vectors. The test of significance is based on the Wilks' lambda (8.5) criterion. The result is given as chi-square.

```
      PROGRAM CANVR2
C-----------PROGRAM FOR COMPUTING CANONICAL VARIATES
C----------------------------OUTPUT-----------------------------
C----------BASIC STATISTICS FOR EACH SAMPLE (OPTIONAL)
C----------MATRIX OF MEAN VECTORS
C----------MATRIX OF CANONICAL MEAN VECTORS
C----------EIGENVALUES AND EIGENVECTORS OF THE GENERALIZED
C-------------------------DETERMINANTAL EQUATION
C----------TESTS OF SIGNIFICANCE OF THESE EIGENVALUES
C----------TEST   OF HOMOGENEITY OF COVARIANCE MATRICES
C----------WILK,S TEST OF EQUALITY OF THE MEAN VECTORS (GENERALIZED
C-------------ONE-WAY CLASSIFICATION ANALYSIS OF VARIANCE)
C----------GRAPH OF FIRST THREE TRANSFORMED MEANS-----------------
C----------GRAPH OF FIRST THREE TRANSFORMED OBSERVATIONS----------
C----------TO AVOID OVERCROWDED PLOTS,NUMBER OF POINTS TO BE SPECIFIED---
C----------N.B. IF THERE ARE LESS THAN 4 GROUPS THE PLOTS MAY BE
C----------------ADVERSELY AFFECTED ----------------------------
C-----------------------------INPUT----------------------------
C----------IN THE CASE OF PROCESSED DATA
C----------MEAN VECTORS, WITHIN SUMS OF SQUARES AND CROSS PRODUCTS
C----------MATRIX AND BETWEEN SUMS OF SQUARES AND CROSS PRODUCTS MATRIX
C----------IN THE CASE OF UNPROCESSED (RAW) DATA
C-------------INFORMATION INPUTTED AS OBSERVATION VECTORS
C---- MATRIX A IS AMONG GROUPS SSP-MATRIX
C---- MATRIX B IS WITHIN GROUPS SSP-MATRIX**TOTAL=WITHIN + AMONG
C-------------------ORDER OF THE CONTROL CARDS----------------
C------FIRST CARD********** NUMBER OF JOBS IN COL 3
C------SECOND CARD***** TITLE OF WORK IN COLUMNS 1 THROUGH 72
C------THIRD CARD M= NUMBER OF VARIABLES IN COLS 2-3
C---------------MG = NUMBER OF GROUPS IN COLS 4-5
C----------------L IN COL  6 IS PUT = 0 FOR MATRIX INPUT AND = 1
C----------------FOR RAW DATA
C------------------THE CASE OF MATRIX DATA------------------
C-------CARD 4 THE VARIABLE FORMAT CARD
C-------CARD 5  COLUMNS 2-6  NI = THE TOTAL NUMBER OF OBSERVATIONS
C-------CARD 6 AND FOLLOWING CARDS CONTAIN THE INPUT MATRICES IN THE
C-----------------------FOLLOWING ORDER.......
C----------(1) C(I,J) = M X MG MATRIX OF MEAN VECTORS
C----------(2) A(I,J) = ,,AMONG GROUPS,, SUMS OF SQUARES AND CROSS
C--------------------PRODUCTS MATRIX
C----------(3) B(I,J) = ,,WITHIN GROUPS,, SUMS OF SQUARES AND CROSS
C-----------------PRODUCTS MATRIX
C------------------THE CASE OF RAW DATA--------------------
C-------THE FURTHER INPUT INFORMATION NEEDED FOR RAW DATA IS THAT
C-------OF THE SUBROUTINE   CORREL   . LISTING HERE FOR CONVENIENCE.
C----------CARD 4  COLUMNS 1-5  SIZE OF THE SAMPLE
C----------COLUMN 6  FOR THE BASE 10 LOG-TRANSFORMATION OF THE DATA
C------------------PUT KLOG = 1, OTHERWISE KLOG = 0.
C----------COLUMN 7  MOUT = 1 GIVES FULL OUTPUT FROM CORREL. IF NO
C----------------OUTPUT REQUIRED MOUT = 0.
C----------CARD 5  TITLE FOR THIS SAMPLE IN COLUMNS 1-72.
C----------CARD 6  VARIABLE FORMAT FOR THIS SAMPLE
C----------CARD 7, AND FOLLOWING CARDS, CONTAIN THE OBSERVATIONS
C------------------------------------------------------------
      COMMON A(30,30),B(30,30),C(60,30),DET,DETM(30),ENI,ENUMB(30),
     1 L,M,MG,NI,NG,SS(30,30),SSD(30,30),SX(30,30),X(30,30),
```

```
      2 XD(30),XL(30),XM(30),Z(30,30)
        COMMON /DIMENS/ MDIM,MGDIM
        COMMON /MATRIX/ MATNR,MTMAT
        COMMON /PLOT/ IVPLOT(60),CHAR(25),KGROUP
        COMMON /SCRCOM/ TS(30,30),DUMMY(2700)
        COMMON /UNITS/ LIN,LUT,MTDATA
        DATA (CHAR=1HA,1HB,1HC,1HD,1HE,1HF,1HG,1HH,1HJ,1HK,1HL,1HM,1HN,1HO
      1,1HP,1HQ,1HR,1HS,1HT,1HU,1HV,1HW,1HX,1HY,1HZ)
        DIMENSION FMT(12),SUMM(30),XME(30),TITLE(12)
        MDIM=30
        MGDIM=60
C------- LIN=STANDARD INPUT UNIT (CARD-READER)
        LIN=60
C------- LUT=STANDARD OUTPUT UNIT  (PRINTER)
        LUT=61
C------- MTDATA IS A SCRATCH UNIT TO STORE SPECIAL VALUES
        MTDATA=1
        READ(LIN,7025) JOBS
        DO 7606 KLOBS=1,JOBS
        REWIND MTDATA
        MATNR=0
        WRITE(LUT,1709)
        READ(LIN,3) TITLE
        WRITE(LUT,3)TITLE
        READ(LIN,1) M,MG,L
        WRITE(LUT,8) MG,M
        MT=M
        IF(M.GE.MG) WRITE(LUT,505)
        IF (M .GT. MDIM .OR. MG .GT. MGDIM) GO TO 9900
        IF(L)100,100,200
C------------------------INPUT FOR MATRIX DATA-----------------------
  100 READ(LIN,3) FMT
        READ(LIN,40) NI
        DO 5 I=1,MG
    5 READ(LIN,FMT)(C(I,J),J=1,M)
C---- MEANS READ IN AS C(I,J)
        CALL PUTMAT (C,MG,M,MGDIM,3)
        CALL MATOUT (C,MG,M,MGDIM,19,19HO     TABLE OF MEANS )
        DO 20 I=1,M
   20 READ(LIN,FMT)(A(I,J),J=1,M)
C------MATRIX A IS THE AMONG-GROUPS MATRIX
        CALL MATOUT (A,M,M,MDIM,34,34HO   AMONG GROUPS SSCP MATRIX,A(I,J) )
        DO 21 I=1,M
   21 READ(LIN,FMT)(B(I,J),J=1,M)
        CALL PUTMAT (B,M,M,MDIM,1)
        CALL MATOUT (B,M,M,MDIM,28,28HO   WITHIN GROUPS SSCP MATRIX )
        DO 23 I=1,M
        DO 23 J=1,M
   23 X(I,J) = A(I,J) + B(I,J)
        CALL PUTMAT (X,M,M,MDIM,2)
        GO TO 270
C------STARTING POINT FOR UNPROCESSED DATA
  200 N=0
        KM=0
        DO 205 I=1,M
        SUMM(I)=0.
```

```
        XME(I)=0.
        DO 205 J=1,M
        A(I,J)=0.
        B(I,J)=0.
        X(I,J)=0.
  205   TS(I,J)=0.
        KGROUP=0
        DO 2206 I=1,MG
        DO 2206 J=1,M
 2206   C(I,J)=0.
  300   KM=KM+1
        CALL CORREL
        N=N+NG
        DO 210 I=1,M
        SUMM(I)=SUMM(I)+SX(I)
        DO 210 J=1,M
        B(I,J)=B(I,J)+SSD(I,J)
  210   TS(I,J)=TS(I,J)+SS(I,J)
        DETM(KM) = DET
        ENUMR(KM) = ENI
        WRITE(LUT,7061) KM
        I=KM
        DO 211 J=1,M
  211   C(I,J)=XM(J)
        KGROUP=KGROUP+1
        IVPLOT(KGROUP)=N
        IF(KGROUP-MG)300,215,215
  215   EN = N
        DO 220 I=1,M
        XME(I)=SUMM(I)/EN
        DO 220 J=1,M
  220   X(I,J)=TS(I,J)-(SUMM(I)*SUMM(J)/EN)
        CALL MATOUT (B,M,M,MDIM,31,31H1    POOLED WITHIN MATRIX B(I,J) )
        CALL PUTMAT (B,M,M,MDIM,1)
C------- THE WITHIN MATRIX IS NOW SAVED
        DO 255 I=1,M
        DO 255 J=1,M
  255   A(I,J)=X(I,J)-B(I,J)
        CALL MATOUT (A,M,M,MDIM,24,24H0    AMONG GROUPS MATRIX    )
        CALL MATOUT (X,M,M,MDIM,16,16H0   TOTAL MATRIX )
        CALL PUTMAT (X,M,M,MDIM,2)
C------MATRIX OF TOTAL SUMS IS NOW SAVED
        WRITE(LUT,245)N
        WRITE(LUT,250)
        WRITE(LUT,231)(XME(I),I=1,M)
        NI = N
        CALL MATOUT (C,MG,M,MGDIM,19,19H0    TABLE OF MEANS )
        CALL PUTMAT (C,MG,M,MGDIM,3)
C------MATRIX OF MEANS NOW SAVED
        WRITE(LUT,1060)
C------------------------------------------------------------------------
  270   CALL EIGGEN
        WRITE(LUT,1058)
        K=0
        KR=0
  320   K=K+1
```

```
      PROD=1.
      DO 1300 I=K,M
 1300 PROD = PROD*(1. + XL(I))
      PROD = ALOG(PROD)
      WRITE(LUT,305) K
      WRITE(LUT,310) PROD
      CHISQ = -(FLOAT(NI-1) - .5*(FLOAT(M+MG)))*PROD
      CHISQ = ABS(CHISQ)
      WRITE(LUT,315)CHISQ
      KDF = (M-KR)*(MG-KR-1)
      WRITE(LUT,1088)KDF
      KR = KR + 1
      IF(MT-K)325,325,320
  325 CONTINUE
      CALL WILKS
      IF(L.EQ.0) GO TO 7606
      CALL HOMO (DETM,M,ENUMB,NI,MG)
 7606 CONTINUE
 7027 CALL EXIT
 9900 WRITE (LUT,9901)
      GO TO 7027
C-------------------------- FORMATS --------------------------------
    1 FORMAT(1X,I2,I2,I1)
    3 FORMAT(12A6)
    8 FORMAT(42H0   CANONICAL VARIATES STUDY FOR GROUPS = I2/29H0          N
     1UMBER OF VARIABLES = I2)
   40 FORMAT(1X,I5)
  231 FORMAT(1X,10F13.5)
  245 FORMAT(24H0   TOTAL SAMPLE SIZE = I5)
  250 FORMAT(21H0   TOTAL MEAN VECTOR)
  305 FORMAT(12H0 LAMBDA NO I1)
  310 FORMAT(11X,F14.7)
  315 FORMAT(15H0   CHISQUARE = F14.7)
  505 FORMAT(30H0   LESS GROUPS THAN DIMENSIONS)
 1058 FORMAT(45H1   TESTS OF SIGNIFICANCE FOR THE EIGENVALUES       )
 1060 FORMAT(27H1   OUTPUT FROM EIGENROUTINE)
 1088 FORMAT(10H0   D.F. = I3)
 1709 FORMAT(1H1)
 7025 FORMAT(1X,I2)
 7061 FORMAT(17H-   END OF SAMPLE I2)
 9901 FORMAT (25H0*** TOO BIG A PROBLEM***          )
      END


      SUBROUTINE CORREL
C-------SUMMARY OF INPUT INSTRUCTIONS
C-------FIRST CONTROL CARD FOR EACH GROUP DEFINED AS
C-------COLUMNS 1-5 SAMPLE SIZE
C------- COLUMN 6  KLOG = 0 FOR RAW DATA AND = 1 FOR LOGARITHMS
C-------MOUT = 0 FOR NO OUTPUT AND = 1 FOR ALL OUTPUT FROM SUBROUTINE
C-------SECOND CARD BEARS GROUP TITLE IN COLUMNS 1 - 72
C-------THIRD CARD BEARS VARIABLE INPUT FORMAT
      COMMON A(30,30),B(30,30),C(60,30),DET,DETM(30),ENI,ENUMB(30),
     1 L,M,MG,NI,NG,SS(30,30),SSD(30,30),SX(30,30),X(30,30),
     2 XD(30),XL(30),XM(30),Z(30,30)
      COMMON /DIMENS/ MDIM,MGDIM
      COMMON /SCRCOM/ DUMMY1(1800),D(30,30),RX(30,30)
      COMMON /UNITS/ LIN,LUT,MTDATA
```

```
      DIMENSION IFMT(12),XX(30),SD(30),TITLE(12)
      READ(LIN,21) NG,KLOG,MOUT
      READ(LIN,22) TITLE
      ENG=NG
      READ(LIN,22) IFMT
      DO 230 I=1,M
      SX(I)=0.
      DO 230 J=1,M
  230 SS(I,J)=0.
      CASES=ENG
  240 READ(LIN,IFMT)(XX(I),I=1,M)
      IF(KLOG) 2240,2240,2241
 2240 GC TO 2244
 2241 DO 2243 I=1,M
 2243 XX(I) = ALOG10(XX(I))
 2244 CONTINUE
      WRITE(MTDATA)(XX(I),I=1,M)
      DO 260 I=1,M
      SX(I) = SX(I) + XX(I)
      DO 260 J=I,M
      SS(I,J) = SS(I,J) + XX(I)*XX(J)
  260 SS(J,I)=SS(I,J)
C---- RAW SUMS OF SQUARES AND CROSS PRODUCTS
      CASES=CASES-1.
      IF(CASES)280,280,240
  280 DO 286 I=1,M
      DO 286 J=1,M
      SSD(I,J)=SS(I,J)-SX(I)*SX(J)/ENG
  286 SSD(J,I)=SSD(I,J)
      DO 295 I=1,M
      XM(I)=SX(I)/ENG
  295 SD(I)=SQRT(SSD(I,I)/(ENG-1.))
C     MEANS AND STANDARD DEVIATION
      IF(MOUT) 435,435,690
  690 WRITE (LUT,31) TITLE,M,NG,(XM(I),I=1,M)
      WRITE (LUT,32)(SD(I),I=1,M)
      CALL MATOUT (SS,M,M,MDIM,39,39H1RAW SUMS OF SQUARES AND CROSS PROD
     1UCTS )
      IF(MOUT.GT.0.AND.M.GT.10) WRITE(6,909)
      CALL MATOUT (SSD,M,M,MDIM,45,45H0DEVIATION SUMS OF SQUARES AND CRO
     1SS PRODUCTS)
  435 DO 441 I=1,M

      DO 441 J=I,M
      D(I,J)=SSD(I,J)/(ENG-1.)
  441 D(J,I)=D(I,J)
C---- DISPERSION MATRIX
      IF (MOUT .GT. 0) CALL MATOUT (D,M,M,MDIM,18,18H1COVARIANCE MATRIX)
  670 DO 486 I=1,M
      DO 486 J=1,M
      RX(I,J)=D(I,J)/(SD(I)*SD(J))
  486 RX(J,I)=RX(I,J)
      IF(MOUT.GT.0.AND.M.GT.10) WRITE(6,909)
C     CORRELATION MATRIX
      IF (MOUT .GT. 0)CALL MATOUT(RX,M,M,MDIM,19,19H0CORRELATION MATRIX)
```

```
      CALL HOMMAT(M,D,DET)
C----------
      ENI = ENG
  700 RETURN
C------------------------------- FORMATS ---------------------------
   21 FORMAT(I5,2I1)
   22 FORMAT(12A6)
   31 FORMAT (26H1CORRELATION ANALYSIS FOR ,12A6/23H0NUMBER OF VARIABLES
     1 = ,I3/26H0NUMBER OF OBSERVATIONS = ,I4/13H0MEAN VECTOR /(9F14.5))
   32 FORMAT (35H0STANDARD DEVIATIONS OF THIS SAMPLE /(9F14.5))
  909 FORMAT(1H1)
      END


      SUBROUTINE EIGGEN
C-----------THIS SUBROUTINE COMPUTES THE GENERALIZED EIGENVALUES
C--------AND EIGENVECTORS OF A NON-SYMMETRIC MATRIX OF THE FORM
C     B-INVERSE-*A VECTOR XL CONTAINS THE CANONICAL VARIATE
C-----COEFFICIENTS AS COLUMNS
      COMMON A(30,30),B(30,30),C(60,30),DET,DETM(30),ENI,ENUMB(30),
     1 L,M,MG,NI,NG,SS(30,30),SSD(30,30),SX(30,30),X(30,30),
     2 XD(30),XL(30),XM(30),Z(30,30)
      COMMON /DIMENS/ MDIM,MGDIM
      COMMON /UNITS/ LIN,LUT,MTDATA
      WRITE(LUT,1060)
      DO 1 I=1,M
      XL(I)=0.
      DO 1 J=1,M
    1 X(I,J)=0.
      CALL HDIAG(B,M,0,X,NR)
      DO 10 I=1,M
   10 XL(I)=1./SQRT(ABS(B(I,I)))
      DO 20 I=1,M
      DO 20 J=1,M
   20 B(I,J)=X(I,J)*XL(J)
      DO 35 I=1,M
      DO 35 J=1,M
      X(I,J)=0.
      DO 35 K=1,M
   35 X(I,J)=X(I,J)+B(K,I)*A(K,J)
      DO 40 I=1,M
      DO 40 J=1,M
      A(I,J)=0.
      DO 40 K=1,M
   40 A(I,J)=A(I,J)+X(I,K)*B(K,J)
      CALL HDIAG(A,M,0,X,NR)
      WRITE(LUT,1066)(A(I,I),I=1,M)
      DO 55 I=1,M
   55 XL(I)=A(I,I)
      SUMMA=0.
      DO 60 I=1,M
      SUMMA=SUMMA+XL(I)
      IF(XL(I).LT.0.) WRITE(LUT,432)
   60 WRITE(LUT,65)I,XL(I)
      DO 70 I=1,M
      DO 70 J=1,M
      A(I,J)=0.
      Z(I,J)=0.
      DO 70 K=1,M
```

```
   70 A(I,J)=A(I,J)+B(I,K)*X(K,J)
      CALL MATOUT (A,M,M,MDIM,15,15H0  EIGENVECTORS)
      DO 71 I=1,M
      DO 71 J=1,M
   71 Z(I,J)=A(I,J)
      DO 75 J=1,M
      SUM=0.
      DO 80 I=1,M
   80 SUM=SUM+(A(I,J)**2)
      DEN=SQRT(SUM)
      DO 85 I=1,M

   85 X(I,J)=A(I,J)/DEN
   75 CONTINUE
      CALL MATOUT (X,M,M,MDIM,26,26H0  NORMALIZED EIGENVECTORS)
      CALL STAND
      RETURN
C-----------------------------  FORMATS  ----------------------------
   65 FORMAT(16H0  EIGENVALUE  I2,F14,7)
  432 FORMAT(33H0  NEGATIVE EIGENVALUE OBTAINED                  )
 1060 FORMAT(49H1  SOLUTION OF GENERALIZED DETERMINANTAL EQUATION)
 1066 FORMAT(1X,12F10,4)
      END

      SUBROUTINE GETMAT (STMAT,M,N,MDIM,MATNO)
C------- GETMAT FETCHES THE STORED MATRIX NO MATNO.
C------- THIS MATRIX IS PLACED IN THE PARAMETER STMAT,
C------- DIMENSIONED IN THE CALLING PROGRAM TO (MDIM,SOMETHING)
C------- THE ACTUAL MATRIX IS OF SIZE M*N.
C------- THIS VERSION OF GETMAT FETCHES THE MATRIX FROM A PLACE IN
C------- THE CORE MEMORY (COMMON/SAVMAT/ ) BUT THE ROUTINE CAN BE EASILY
C------- REWRITTEN TO FETCH THE MATRIX FROM AN EXTERNAL MEDIUM SUCH AS A
C------- MAGNETIC TAPE OR DRUM.
C------- NOTE THAT THE MATRICES MUST FIRST BE STORED BY ROUTINE PUTMAT
C------- WHICH MUST WORK IN THE SAME MANNER AS GETMAT,
      DIMENSION STMAT (MDIM,25)
      COMMON /MATRIX/ MATNR,MTMAT
      COMMON /SAVMAT/ IX(3),SAVE(3600)
      DATA (IX=1,901,1801)
      JX=IX(MATNO)
      DO 1 I=1,M
      DO 1 J=1,N
      STMAT(I,J)=SAVE(JX)
    1 JX=JX+1
      RETURN
      END

      SUBROUTINE HDIAG (H,N,IEGEN,U,NR)
C------- THIS SUBROUTINE CALCULATES THE EIGENVALUES AND EIGENVECTORS
C------- OF A SQUARE SYMMETRIC MATRIX
C------------      --------      --------      --------      --------
C-------****NOTE****NOTE****
C------- THE LATTER PART OF THE SUBROUTINE MUST BE REWRITTEN FOR
C------- AN IBM FORTRAN COMPILER
      DIMENSION H(30,30),U(30,30),X(30),IQ(30)
      IF(IEGEN)15,10,15
   10 DO 14 I=1,N
      DO 14 J=1,N
      IF(I-J)12,11,12
   11 U(I,J)=1.
      GO TO 14
   12 U(I,J)=0.
   14 CONTINUE
```

```
15      NR=0
        IF(N-1)1000,1000,17
17      NMI1=N-1
        DO 30 I=1,NMI1
        X(I)=0.
        IPL1=I+1
        DO 30 J=IPL1,N
        IF(X(I)-ABS (H(I,J)))20,20,30
20      X(I)=ABS (H(I,J))
        IQ(I)=J
30      CONTINUE
        RAP=7.450580596E-9
        HDTEST=1.0E38
40      DO 70 I=1,NMI1
        IF(I-1)60,60,45
45      IF(XMAX-X(I))60,70,70
60      XMAX=X(I)
        IPIV=I
        JPIV=IQ(I)
70      CONTINUE
        IF(XMAX)1000,1000,80
80      IF(HDTEST)90,90,85
85      IF(XMAX-HDTEST)90,90,148
90      HDIMIN=ABS (H(1,1))
        DO 110 I=2,N
        IF(HDIMIN-ABS (H(I,I)))110,110,100
100     HDIMIN=ABS (H(I,I))
110     CONTINUE
        HDTEST=HDIMIN*RAP
C       RETURN IF MAX. H(I,J) LESS THAN (2**-27)ABSF(H(K,K)*MIN)
        IF(HDTEST-XMAX)148,1000,1000
148     NR=NR+1
C       COMPUTE TANGENT, SINE AND COSINE , H(I,I),H(J,J)
150     IF(H(IPIV,IPIV)-H(JPIV,JPIV))91,92,92
91      A=-2.0
        GO TO 93
92      A=2.0
93      TANG=A*H(IPIV,JPIV)/(ABS (H(IPIV,IPIV)-H(JPIV,JPIV))+SQRT ((H(IPIV
        1,IPIV)-H(JPIV,JPIV))**2+4.0*H(IPIV,JPIV)**2))
        COSINE=1.0/SQRT (1.0+TANG**2)
        SINE=TANG*COSINE
        HII=H(IPIV,IPIV)
        H(IPIV,IPIV)=COSINE**2*(HII+TANG*(2.*H(IPIV,JPIV)+TANG*H(JPIV,JPIV
        1)))
        H(JPIV,JPIV)=COSINE**2*(H(JPIV,JPIV)-TANG*(2.*H(IPIV,JPIV)-TANG*HI
        1I))
        H(IPIV,JPIV)=0.
        IF(H(IPIV,IPIV)-H(JPIV,JPIV))152,153,153
152     HTEMP=H(IPIV,IPIV)
        H(IPIV,IPIV)=H(JPIV,JPIV)
        H(JPIV,JPIV)=HTEMP
        IF(SINE)94,95,95
94      HTEMP=COSINE
        GO TO 96
95      HTEMP=-COSINE
96      COSINE=ABS (SINE)
        SINE=HTEMP
153     CONTINUE
        DO 350 I=1,NMI1
        IF(I-IPIV)210,350,200
200     IF(I-JPIV)210,350,210
210     IF(IQ(I)-IPIV)230,240,230
230     IF(IQ(I)-JPIV)350,240,350
240     K=IQ(I)
250     HTEMP=H(I,K)
```

```
      H(I,K)=0.
      IPL1=I+1
      X(I)=0.
      DO 320 J=IPL1,N
      IF(X(I)-ABS (H(I,J)))300,300,320
300   X(I)=ABS (H(I,J))
      IQ(I)=J
320   CONTINUE
      H(I,K)=HTEMP
350   CONTINUE
      X(IPIV)=0.
      X(JPIV)=0.
      DO 530 I=1,N
      IF(I-IPIV)370,530,420
370   HTEMP=H(I,IPIV)
      H(I,IPIV)=COSINE*HTEMP+SINE*H(I,JPIV)
      IF(X(I)-ABS (H(I,IPIV)))380,390,390
380   X(I)=ABS (H(I,IPIV))
      IQ(I)=IPIV
390   H(I,JPIV)=-SINE*HTEMP+COSINE*H(I,JPIV)
      IF(X(I)-ABS (H(I,JPIV)))400,530,530
400   X(I)=ABS (H(I,JPIV))
      IQ(I)=JPIV
      GO TO 530
420   IF(I-JPIV)430,530,480
430   HTEMP=H(IPIV,I)
      H(IPIV,I)=COSINE*HTEMP+SINE*H(I,JPIV)
      IF(X(IPIV)-ABS (H(IPIV,I)))440,450,450
440   X(IPIV)=ABS (H(IPIV,I))
      IQ(IPIV)=I
450   H(I,JPIV)=-SINE*HTEMP+COSINE*H(I,JPIV)
      IF(X(I)-ABS (H(I,JPIV)))400,530,530
480   HTEMP=H(IPIV,I)
      H(IPIV,I)=COSINE*HTEMP+SINE*H(JPIV,I)
      IF(X(IPIV)-ABS (H(IPIV,I)))490,500,500
490   X(IPIV)=ABS (H(IPIV,I))
      IQ(IPIV)=I
500   H(JPIV,I)=-SINE*HTEMP+COSINE*H(JPIV,I)
      IF(X(JPIV)-ABS (H(JPIV,I)))510,530,530
510   X(JPIV)=ABS(H(JPIV,I))
      IQ(JPIV)=I
530   CONTINUE
      IF(IEGEN)40,540,40
540   DO 550 I=1,N
      HTEMP=U(I,IPIV)
      U(I,IPIV)=COSINE*HTEMP+SINE*U(I,JPIV)
550   U(I,JPIV)=-SINE*HTEMP+COSINE*U(I,JPIV)
      GO TO 40
1000  RETURN
      END
```

```
      SUBROUTINE HOMMAT(M,D,DET)
C---------INTERMEDIATE STEP IN THE TEST FOR HOMOGENEITY OF
C---------COVARIANCE MATRICES****
C---------AS A SIDELINE IT GIVES AN ABRIDGED PRINCIPAL COMPONENTS
C---------BREAKDOWN***
      COMMON /DIMENS/ MDIM,MGDIM
      COMMON /SCRCOM/ DUMMY1(900),H(30,30),DUMMY2(1800)
      COMMON /UNITS/ LIN,LUT,MTDATA
      DIMENSION D(30,30)
      WRITE(LUT,2)
      CALL HDIAG(D,M,0,H,NR)
      DET=1.0
      DO 1 I=1,M
    1 DET = DET*D(I,I)
C---------
      WRITE(LUT,15) DET
C---------
      WRITE (LUT,3) (D(I,I),I=1,M)
      SUM = 0.0
      DO 4 I=1,M
    4 SUM = SUM + D(I,I)
      DO 5 I=1,M
      PRO = (D(I,I)/SUM)*100.
      WRITE(LUT,6) I,PRO
    5 CONTINUE
      CALL MATOUT (H,M,M,MDIM,35,35H0  ELEMENTS OF PRINCIPAL COMPONENTS)
      RETURN
C------------------------- FORMATS ---------------------------
    2 FORMAT(20H1 OUTPUT FROM HOMMAT  )
    3 FORMAT (33H0      PRINCIPAL COMPONENTS SUMMARY/20H0   VARIANCES FOR P
     1CA/(X,10F13.6))
    6 FORMAT(18H0  PERCENTAGE FOR I2,3H = F14.4)
   15 FORMAT(9H0  DET = E18.11)
      END

      SUBROUTINE HOMO (DETM,M,ENUMB,NI,MG)
      COMMON /DIMENS/ MDIM,MGDIM
      COMMON /SCRCOM/ B(30,30),X(30,30),DUMMY(1800)
      COMMON /UNITS/ LIN,LUT,MTDATA
      DIMENSION DETM(30),ENUMB(30)
C------THIS SUBRT COMPUTES THE TEST FOR HOMOGENEITY OF COVARIANCE
C------MATRICES...IT IS ONLY CALLED WHEN RAW DATA ARE USED
      WRITE(LUT,30)
      CALL GETMAT (B,M,M,MDIM,1)
      DO 2 I=1,M
      ENUMB(I) = ENUMB(I)-1.0
      DO 2 J=1,M
    2 B(I,J) = B(I,J)/FLOAT(NI-MG)
      CALL HDIAG(B,M,1,X,NR)
      DET2=1.0
      DO 5 I=1,M
    5 DET2 = DET2*B(I,I)
      RES = 0.0
      DO 6 I=1,MG
    6 RES = RES + ENUMB(I)*ALOG(DET2/DETM(I))
      WRITE(LUT,9) RES
      KDF = (MG-1)*M*(M+1)/2
      WRITE(LUT,8) KDF
      RETURN
C------------------------- FORMATS ---------------------------
    8 FORMAT(8H0   DF = I3)
    9 FORMAT(9H0 CHSQ = F14.6)
   23 FORMAT(12F10.6)
   30 FORMAT(41H1    HOMOGENEITY OF COVARIANCE MATRICES          )
      END
```

```
      SUBROUTINE MATOUT (STMAT,MG,M,NDIM,NCHAR,TITLE)
C------- MATOUT PRINTS OUT THE MG*M MATRIX STMAT.
C------- STMAT IS SUPPOSED TO BE DIMENSIONED TO (NDIM,SOMETHING)
C------- BEFORE WRITING THE MATRIX A HEAD TITLE IS WRITTEN
C------- THE TITLE LENGTH IS NCHAR CHARACTERS WHICH ARE WRITTEN IN
C------- COLUMNS 1 TO NCHAR.
C------- IN CD 3600 EACH CELL CONTAINS 8 CHARACTER SO NCHPW IS SET TO 8
C------- IF NCHPW IS CHANGED , DO NOT FORGET TO CHANGE FORMAT 103 TOO.
      DIMENSION STMAT (NDIM,25),TITLE(1)
      COMMON /UNITS/ LIN,LUT,MTDATA
      DATA(NCHPW=8)
      IW=(NCHAR+NCHPW-1)/NCHPW
      WRITE (LUT,103) (TITLE(J),J=1,IW)
      MEND = 0
    1 MBEG = MEND + 1
      MEND = MBEG + 9
      IF(MEND.GT.M)MEND = M
      WRITE(LUT,101)(J,J=MBEG,MEND)
      DO 2 I=1,MG
    2 WRITE(LUT,102) I,(STMAT(I,J),J=MBEG,MEND)
      IF(MEND.EQ.M) RETURN
      GO TO 1
C------------------------------ FORMATS ------------------------------------
  101 FORMAT(1H0,10I12)
  102 FORMAT(2X,I2,2X,10F12.5)
  103 FORMAT (17A8)
      END


      SUBROUTINE PUTMAT (STMAT,M,N,MDIM,MATNO)
C------- PUTMAT STORES MATRIX STMAT AND NOTES IT AS MATRIX NO. MATNO.
C------- STMAT IS SUPPOSED TO BE DIMENSIONED TO (MDIM,SOMETHING).
C------- THE ACTUAL SIZE OF STMAT IS M*N ELEMENTS.
C------- PUTMAT CORRESPONDS TO SUBROUTINE GETMAT AND THIS VERSION
C------- STORES THE MATRICES IN MEMORY AT COMMON/SAVMAT/. IT IS EASY
C------- TO REWRITE PUTMAT AND GETMAT TO STORE THE MATRICES ON AN
C------- EXTERNAL MEDIUM LIKE MAGNETIC TAPE OR DRUM.
      DIMENSION STMAT (MDIM,25)
      COMMON /MATRIX/ MATNR,MTMAT
      COMMON /SAVMAT/ IX(3),SAVE(3600)
      DATA (IX=1,901,1801)
      JX=IX(MATNO)
      DO 1 I=1,M
      DO 1 J=1,N
      SAVE(JX)=STMAT(I,J)
    1 JX=JX+1
      RETURN
      END
```

```
      SUBROUTINE PLT3 (X1,Y1,N1,IVPLOT,CHAR,MG)
C------- PLT3 PLOTS THE N1 COORDINATES IN X1,Y1 ON THE PRINTER.
C------- IVPLOT IS AN ARRAY THAT CONTAINS INDEX LIMITS FOR DIFFERENT
C------- GROUPS OF DATA. EACH GROUP IS WRITTEN WITH A SPECIAL SYMBOL.
C------- MG IS THE NUMBER OF GROUPS. IF MG IS NEGATIVE EACH X1,Y1 PAIR
C-------  IS PLOTTED WITH A NEW SYMBOL.
C------- CHAR CONTAINS THE SYMBOLS THAT ARE TO BE USED. ONE CHARACTER IN
C-------  EACH CELL IN A1-FORMAT.
      COMMON /UNITS/ LIN,LUT,MTDATA
      DIMENSION X1(10),Y1(10),IVPLOT(10),CHAR(10),ALINE(101)
      DATA(BLANK=1H ),(DASH=1H-),(UP=1HI),(DOT=1H.),(AST=1H*)
      XMAX = X1(1)
      XMIN = XMAX
      YMAX = Y1(1)
      YMIN = YMAX
      DO 1 I=2,N1
      IF(X1(I).GT.XMAX) XMAX = X1(I)
      IF(X1(I).LT.XMIN) XMIN = X1(I)
      IF(Y1(I).GT.YMAX) YMAX = Y1(I)
      IF(Y1(I).LT.YMIN) YMIN = Y1(I)
    1 CONTINUE
      KEY = 1
      ZMAX = XMAX
      ZMIN = XMIN
    4 RANGE = ZMAX - ZMIN
      SCALE = 1.E-9
    5 SCALE = 10.*SCALE
      IF(SCALE.LT.RANGE) GO TO 5
      ZMIN = ZMIN/SCALE
      ZMAX = ZMAX/SCALE
      MIN = 20.0*(ZMIN - 0.025)
      MAX = 20.0*(ZMAX + 0.025)
      BOTTOM = 0.05*FLOAT(MIN)
      TOP = 0.05*FLOAT(MAX)
      RANGE = 0.1*(TOP-BOTTOM)
    6 IF(BOTTOM.LE.ZMIN) GO TO 7
      BOTTOM = BOTTOM - RANGE
      GO TO 6
    7 IF(TOP.GE.ZMAX) GO TO 8
      TOP = TOP + RANGE
      GO TO 7
    8 YINC = 0.01*(TOP -BOTTOM)
      IF(KEY.EQ.2) GO TO 11
      KEY = 2
      ZMAX = YMAX
      ZMIN = YMIN
      XSCALE = SCALE
      XTOP = TOP
      XINC = YINC*SCALE
      XBOT = BOTTOM*SCALE
      GO TO 4
C
   11 YLOW = TOP + YINC
      YINC = 2.0*YINC
C
      KEY = 5
```

```
      WRITE(LUT,1000) XSCALE,SCALE
   15 KKEY = 10
      IF(KEY.NE.5) GO TO 19
      DO 18 I=1,101
      IF(KKEY.NE.10) GO TO 16
      ALINE(I) = DOT
      GO TO 17
   16 ALINE(I) = DASH
   17 KKEY = KKEY - 1
      IF(KKEY.EQ.0) KKEY = 10
   18 CONTINUE
      GO TO 23
   19 DO 22 I=1,101
      IF(KKEY.NE.10) GO TO 20
      ALINE(I) = UP
      GO TO 21
   20 ALINE(I) = BLANK
   21 KKEY = KKEY - 1
      IF(KKEY.EQ.0) KKEY = 10
   22 CONTINUE
   23 CONTINUE
C     PUT POINTS ON ALINE
      YHIGH = YLOW
      YLOW = YHIGH -YINC
      YHS = SCALE*YHIGH
      YLS = SCALE*YLOW
      DO 24 I=1,N1
      IF(Y1(I).GT.YHS.OR.Y1(I).LE.YLS) GO TO 24
      INDEX = (X1(I) - XBOT)/XINC
      INDEX = INDEX + 1
      IF(INDEX.GT.101) INDEX = 101
      MGA=IABS(MG)
      IF (MG .GT. 0) GO TO 102
      IF (MGA .EQ. 0) MGA=1
      L=I-1
      L=L-L/MGA*MGA+1
      GO TO 100
  102 DO 10 L=1,MG
   10 IF (I .LE. IVPLOT(L)) GO TO 103
      L=MG
  103 IF (L .LE. 25) GO TO 100
      L=L-1
      L=L-L/25*25+1
  100 IF (ALINE (INDEX) .EQ. UP .OR. ALINE(INDEX) .EQ. DASH .OR.
     1 ALINE (INDEX) .EQ. DOT .OR. ALINE(INDEX) .EQ. BLANK) GO TO 101
      IF (ALINE (INDEX) .EQ. CHAR(L) ) GO TO 24
      ALINE (INDEX) =AST
      GO TO 24
  101 ALINE(INDEX)=CHAR(L)
   24 CONTINUE
C     WRITE ALINE OUT
      IF(KEY.NE.5) GO TO 28
      WRITE(LUT,1001) TOP,ALINE
      GO TO 29
   28 WRITE(LUT,1002) ALINE
   29 CONTINUE
```

MM—5

```
C
      KEY = KEY - 1
      IF(KEY.EQ.0) KEY = 5
      TOP = TOP - YINC
      IF(TOP.GE.BOTTOM) GO TO 15
      IF(KEY.NE.4) GO TO 15
C
      XINC = 10.*XINC/XSCALE
      ALINE(1) = XBOT/XSCALE
      DO 30 I=2,11
   30 ALINE(I) = ALINE(I-1) +XINC
      WRITE(LUT,1003) (ALINE(I),I=1,11)
      RETURN
C-------------------------------- FORMATS ------------------------------
 1000 FORMAT(22H0SCALE FACTOR ON X IS ,  E9.2,4X,21HSCALE FACTOR ON Y IS
     1 ,E9.2/1H )
 1001 FORMAT(F10.3,1X,101A1)
 1002 FORMAT(11X,101A1)
 1003 FORMAT(5X,11F10.3)
      END


      SUBROUTINE STAND
C----------STANDARDIZES THE COORDINATES THROUGH THE MATRIX OF
C----------EIGENVECTORS
      COMMON A(30,30),B(30,30),C(60,30),DET,DETM(30),ENI,ENUMB(30),
     1 L,M,MG,NI,NG,SS(30,30),SSD(30,30),SX(30,30),X(30,30),
     2 XD(30),XL(30),XM(30),Z(30,30)
      COMMON /DIMENS/ MDIM,MGDIM
      COMMON /PLOT/ IVPLOT(60),CHAR(25),KGROUP
      COMMON /UNITS/ LIN,LUT,MTDATA
      DIMENSION F(30)
      DO 998 I=1,M
      SX(I)=0.0
  998 XM(I)=0.0
      CALL GETMAT (B,M,M,MDIM,1)
      DO 210 I=1,M
      DO 210 J=1,M
  210 B(I,J) = B(I,J)/FLOAT(NI-MG)
      K=1
  555 CONTINUE
  570 DO 556 I=1,M
  556 XD(I) = Z(I,K)
      DO 560 I=1,M
      DO 560 J=1,M
  560 XM(I) = XM(I) + B(I,J)*XD(J)
      DO 565 I=1,M
  565 DEN = DEN + XM(I)*XD(I)
      SX(K) = DEN
      K = K+1
      IF(K-M)570,570,571
  571 DO 572 J=1,M
      DO 572 I=1,M
  572 Z (I,J) =  Z(I,J)/ SQRT(SX(J))
      CALL MATOUT  (Z,M,M,MDIM,32,32H0  ADJUSTED CANONICAL VARIATES  )
      CALL GETMAT (C,MG,M,MGDIM,3)
```

```
      DO 51 I1=1,MG
      DO 50 I2=1,M
      F(I2)=0
      DO 50 I3=1,M
   50 F(I2)=F(I2)+C(I1,I3)*Z(I3,I2)
      DO 51 I2=1,M
   51 C(I1,I2)=F(I2)
      DO 275 K=1,MG
      WRITE(LUT,280) K
      WRITE(LUT,231)(C(K,J),J=1,M)
  275 CONTINUE
      WRITE(LUT,350)
      CALL PLT3 (C,C(1,2),MG,MG,CHAR,-25)
      WRITE(LUT,3500)
      CALL PLT3 (C,C(1,3),MG,MG,CHAR,-25)
      WRITE(LUT,3501)
      CALL PLT3(C,C(2,3),MG,MG,CHAR,-25)
      IF(L.EQ.0) GO TO 666
      CALL TRANSF ( Z,M,NI)
  666 RETURN
C---------------------------- FORMATS ---------------------------------
  231 FORMAT(1X,10F13.5)

  280 FORMAT(30H0    CANONICAL VARIATE MEAN NO.I2)
  350 FORMAT(21H1CANMEANS FOR 1 AND 2          )
 3500 FORMAT(21H1CANMEANS FOR 1 AND 3        )
 3501 FORMAT(21H1CANMEANS FOR 2 AND 3         )
      END


      SUBROUTINE TRANSF (A,M,N)
      COMMON /DIMENS/ MDIM,MGDIM
      COMMON /PLOT/ IVPLOT(60),CHAR(25),KGROUP
      COMMON/SCRCOM/ B(1300,2),DUMMY(1000)
      COMMON /UNITS/ LIN,LUT,MTDATA
      DIMENSION A(30,30),X(30),Y(30)
C------THE EIGENVECTORS ARE PLACED IN MATRIX A
C------THE TOTAL DATA IS IN MATRIX B WITH A MAXIMUM OF 1300 OBSERVATIONS
      KZ = 1
      KY = 2
      DO 555 II=1,3
      REWIND MTDATA
      DO 20 KOUNT=1,N
      DO 14 I=1,M
   14 Y(I) = 0.0
      READ(MTDATA)(X(I),I=1,M)
      DO 1 J=1,M
      DO 1 I=1,M
    1 Y(J) = Y(J) + X(I)*A(I,J)
      B(KOUNT,1) = Y(KZ)
      B(KOUNT,2) = Y(KY)
   20 CONTINUE
      IF(II.EQ.1) WRITE(LUT,600)
      IF(II.EQ.2) WRITE(LUT,3500)
      IF(II.EQ.3) WRITE(LUT,3501)
```

```
      DO 5 I=1,N
      DO 4 J=1,KGROUP
   4  IF (I .LE. IVPLOT(J)) GO TO 40
      J=KGROUP
  40  IF (J .LE.25) GO TO 5
      J=J-J/25*25+1
   5  WRITE(LUT,10)I,B(I,1),B(I,2),CHAR(J)
      WRITE(LUT,3505)
      CALL PLT3 (B,B(1,2),N,IVPLOT,CHAR,KGROUP)
      IF(II.EQ.1) KY = 3
      IF(II.EQ.2)KZ = 2
 555  CONTINUE
      RETURN
C------------------------------ FORMATS ------------------------------*
  10  FORMAT(15X,I3,6X,F15.6,10X,F15.6,5X,A1)
 600  FORMAT(50H1  PLOT OF TRANSFORMED OBSERVATIONS FOR 1 AND 2      )
3500  FORMAT(50H1  PLOT OF TRANSFORMED OBSERVATIONS FOR 1 AND 3     )
3501  FORMAT(50H1  PLOT OF TRANSFORMED OBSERVATIONS FOR 2 AND 3      )
3505  FORMAT(1H1)
      END


      SUBROUTINE WILKS
      COMMON A(30,30),B(30,30),C(60,30),DET,DETM(30),ENI,ENUMB(30),
     1 L,M,MG,NI,NG,SS(30,30),SSD(30,30),SX(30,30),X(30,30),
     2 XD(30),XL(30),XM(30),Z(30,30)
      COMMON /DIMENS/ MDIM,MGDIM
      COMMON /UNITS/ LIN,LUT,MTDATA
      WRITE(LUT,150)
      KDF = M*(MG-1)
      CALL GETMAT (B,M,M,MDIM,1)
      CALL GETMAT (X,M,M,MDIM,2)
      CALL HDIAG (B,M,0,SS,NR)
      CALL HDIAG (X,M,0,Z,NR)
      DO 5 I=1,M
      XL(I) = B(I,1)
      IF(B(I,I).LE.0.)GO TO 90
      XD(I) = X(I,I)
      IF(X(I,I).LE.0.) GO TO 95
   5  CONTINUE
      PROD = 1.
      PRODD = 1.
      DO 10 I=1,M
      PROD = PROD*XL(I)
  10  PRODD = PRODD*XD(I)
      WLAMDA=PROD/PRODD
      CHISQ = -(FLOAT(NI-M) - (FLOAT(M) + FLOAT(MG))*0.5)*ALOG(WLAMDA)
      WRITE (LUT,15) PROD,PRODD,WLAMDA,CHISQ,KDF
      RETURN
  90  MX=1
      GO TO 99
  95  MX=2
  99  WRITE (LUT,100) MX
      RETURN
C------------------------------ FORMATS ------------------------------*
  15  FORMAT (10H0  DETB = E18.9,9H  DETX = E18.9/18H0  WILKS LAMBDA =
     1F10.5/15H0  CHISQUARE = F12.3,7H  DF = I2 )
 100  FORMAT (8H0 MATRIX,I2,22H NOT POSITIVE DEFINITE)
 150  FORMAT(28H1  TEST OF EQUALITY OF MEANS )
      END
```

9
Canonical Correlations

In the course of morphometric studies, and also of a wide range of ecological investigations, we may wish to decide whether one set of variables, taken as a whole, varies with another set of variables, and, if the answer is positive, to discover the nature of this joint variation.

In ecological investigations, there may be special interest attached to the changes of the elements of the fauna, such as the collembolan fauna of a peat bog, as the bog is drained, a process which in itself will entrain a large number of other changes such as the depth of the anaerobic layers, of root systems, etc. In strictly morphometric work, there are occasions when we need a linear compound of the measurements of an organism which are, as far as possible, uncorrelated with environmentally induced changes of shape, which in turn will be assessed in terms of another linear compound of measurements. To watch the mutual fluctuations of two linear compounds in this way is to take, in effect, a synoptic view of the processes at work; not every worker approves of the synoptic approach to biological problems, probably because in the past this approach has been associated with very generalized statements which, in practice, have proved essentially non-operational in that the hypotheses generated by synoptic approaches have been untestable in any sufficiently critical quantitative terms. Canonical variates, and particularly canonical correlations, do afford a means of testing hypotheses; if we set up the hypothesis that, say, the collembola of a bog vary together with the nematodes of the bog, as the latter is drained, we can estimate these two groups of animals in the bog at different stages of draining, and determine whether there is or is not an association between the two sets of estimates. The price paid for this general approach is a certain loss of detail, but we can agree that the synoptic view is inadequate for special purposes without denying that such a view is in other respects rewarding. Potential uses in geology have been suggested by Lee (1969) and in meteorology by Glahn (1968).

Formal Framework of Canonical Correlation

The method of canonical correlation is in effect a generalization of the concept of multiple regression. The method was developed by Hotelling (1935, 1936) for analysing the interrelationship between two sets of measurements. The method of canonical correlations may be described as a way of finding the maximum correlation between linear functions of the two sets of variables. In many studies it will be discovered that several linear combinations of the two sets of variables are frequently possible. In other words, this correlation technique provides a means of studying the dependencies between the two sets.

For our purposes, it is easiest to talk about the subject of canonical correlations in terms of a morphometric experiment in ecology. The one set of variables (say, p in number) may be thought of as being composed of measurements on some species. The other set of variables (say, q in number) are a set of ecological measurements.

Thus, each observation vector (X'_1, X'_2) is made up of p morphometrical elements and q ecological elements. The dispersion matrix of the sample of these variables, S, will therefore be divisible (or partitionable) into four blocks (N.B. the subject may be equally as well presented in terms of the correlation matrix).

$$S = \begin{bmatrix} S_{11} & S_{12} \\ S'_{12} & S_{22} \end{bmatrix} \qquad (9.1)$$

The submatrices S_{11} and S_{22} refer to set 1 and set 2, respectively, while submatrices S'_{12} (read, the transpose of S_{12}) and S_{12} represent the interaction between all the variables of both sets.

The required canonical correlations are obtained by finding the roots of the determinantal equations:

$$|S_{12}S_{22}^{-1}S'_{12} - \lambda S_{11}| = 0, \qquad (9.2)$$

or,

$$|S'_{12}S_{11}^{-1}S_{12} - \lambda S_{22}| = 0. \qquad (9.3)$$

The square roots of the latent roots, λ, are the canonical correlations. The next step is to find the linear equations corresponding to these roots. The coefficients, a_i and b_i, of the i-th pair of equations are obtained from,

$$(S_{12}S_{22}^{-1}S'_{12} - \lambda_i S_{11})a_i = 0, \qquad (9.4)$$

and,

$$(S'_{12}S_{11}^{-1}S_{12} - \lambda_i S_{22})b_i = 0. \qquad (9.5)$$

If λ_i is a distinct root, the vectors $\mathbf{a}_i$ and $\mathbf{b}_i$ will be unique and their linear compounds will be uncorrelated with the other canonical variates. For the case in which $q = 1$ and $p > 1$, the canonical correlation structure reduces to one of multiple regression.

Where the correlation matrix has been employed to find the (same) canonical correlations as have just been obtained using the dispersion matrix, the evaluation of the linear equation for producing transformed observations should be made in terms of the "standard scores":

$$y_{ij} = \frac{x_{ij} - \bar{x}_{ij}}{s_{ij}},$$

of the original observations, because the canonical variates derived from the correlation matrix are dimensionless. The formula says that the standard score is found by subtracting the mean from each observation and dividing the result by the standard deviation of the variable. Anderson (1958, p. 303) gives a good practical account of the steps required for the calculation of canonical variates.

We note, that if q is greater than p, there are q possible roots of the determinantal equation, $q\text{-}p$ of which are equal to nought. In order to save computational time it is expeditious to define the left and right sets so that q is less than p.

A suitable FORTRAN II listing for doing the calculations is given in Cooley and Lohnes (1962), and most computer centres carry a similar programme. These computer programmes include a test of significance of the latent roots. We point out that whilst it is certainly valuable to know how many of the latent roots are statistically significantly different from zero (the null hypothesis is that the p variates are uncorrelated with the q variates), lack of statistical significance does not categorically rule out the possibility of biologically meaningful combinations.

CASE HISTORIES INVOLVING THE USE OF CANONICAL CORRELATIONS

Construction of a "Phase" Vector for Locusts

We offer as a simple example of the technique a study of locust shapes, in which it was desired to create a linear compound of attributes sensitive to the effects of crowding (which leads to "phase" changes preparatory to swarming) but insensitive to differences of size, which may arise from other ecological and environmental conditions irrelevant to this study. This example was used by Blackith and Albrecht (1959) and although, with further knowledge of the biology of locusts one might now wish to modify the basis of the calculation slightly it serves quite adequately as a worked example.

On the basis of earlier morphometric investigations on locusts, six measurements were chosen, three (the weight, Z; the elytron length, E; and the compound eye width, Oh) as indicators of the size of the locust, and the other three (the hind femoral length, F; the head width, C; and the pronotal height, H) as indicators of the "phase" differences consequent upon crowding or isolation of the insects and their parents.

Naturally, these two sets of attributes will be correlated, and we wish to choose a pair of linear compounds which will be as nearly uncorrelated as possible. Note that the problem of trying to obtain the most highly correlated linear compounds is the same, both the highest and the lowest correlation coefficients emerging from the same computation. The theory and practice of the calculations has been given by Dwyer (1951), Kendall (1957) and Quenouille (1952), for example.

If we denote the canonical correlation coefficient by λ, we require to solve the determinantal equation below, in which the a_{ij} are the entries in the dispersion matrix (the variances and covariances of the attributes) and $a_{ij} = a_{ji}$

$$\begin{vmatrix} -a_{11}\lambda & -a_{12}\lambda & -a_{13}\lambda & a_{14} & a_{15} & a_{16} \\ -a_{12}\lambda & -a_{22}\lambda & -a_{23}\lambda & a_{24} & a_{25} & a_{26} \\ -a_{13}\lambda & -a_{23}\lambda & -a_{33}\lambda & a_{34} & a_{35} & a_{36} \\ a_{14} & a_{24} & a_{34} & -a_{44}\lambda & -a_{45}\lambda & -a_{46}\lambda \\ a_{15} & a_{25} & a_{35} & -a_{45}\lambda & -a_{55}\lambda & -a_{56}\lambda \\ a_{16} & a_{26} & a_{36} & -a_{46}\lambda & -a_{56}\lambda & -a_{66}\lambda \end{vmatrix} = 0.$$

This pooled dispersion matrix is as follows:

	Z	E	Oh	F	C	H
Z	0.102715	0.289420	0.009692	0.169622	0.040382	0.059732
E	0.289420	2.776243	0.064040	0.991217	0.222717	0.251552
Oh	0.009692	0.064040	0.008092	0.044762	0.010532	0.010198
F	0.169622	0.991217	0.044762	0.771332	0.132497	0.144837
C	0.040382	0.222717	0.010532	0.132497	0.041860	0.036699
H	0.059732	0.251552	0.010198	0.144837	0.036699	0.049418

The three solutions of positive sign are:

0.8177
0.2132
0.0022

representing the squares of the three canonical correlations:

0.9042
0.4617
0.0468

As the approximate standard errors of the highest and lowest coefficients are 0.023 and 0.120 respectively, the lowest correlation coefficient is clearly not significantly different from zero.

Having obtained the appropriate solution for λ, we now have to substitute this in the determinantal equation, to generate the vectors representing "size" uncorrelated with "phase". We need to solve the set of six equations consisting of the rows of the determinant. As there are six equations and six unknown coefficients, we make the equations determinate by standardizing H, arbitrarily, as 1.0000 and proceed to evaluate the remainder, the last equation being linearly dependent on the remaining five and hence contributing no additional information.

The solutions are

$$
\begin{aligned}
Z &= + \ \ 1.76 \\
E &= - \ \ 1.15 \\
Oh &= + \ 18.12 \\
F &= - \ \ 2.79 \\
C &= + \ 13.20 \\
H &= + \ \ 1.00
\end{aligned}
$$

which may be standardized by dividing through by the square root of the sum of squares of the values (22.71). The required variate representing "size" independent of "phase" is then

$$0.0775Z - 0.0504E + 0.7980Oh$$

and that for "phase" independent of "size" becomes

$$-0.1225F + 0.5811C + 0.0440H.$$

Influence of the Superincumbent Water on the Pore Water of a Sediment

The chemical composition of interstitial water in the pores of a sediment, at least in the top layers of sediment, will be influenced by a variety of factors including the amount of exchange with the superincumbent water and its composition. Reyment (1968) measured the pH, Eh, free oxygen, carbonate and water contents of the sediment, and the pH, Eh, and free oxygen of the superincumbent water.

The linear compound for the sediment which was most closely correlated with the equivalent compound for the superincumbent water was 0.12; −0.51; 0.53; 0.35; −0.56 and that for the superincumbent water was 0.12; −0.65; 0.75. Evidently pH plays only a minor part in this canonical correlation, which borders on significance for its upper value (0.50).

Nepionic Variation in Upper Cretaceous *Orbitoides*

During a study of growth in *Orbitoides* the question arose as to whether the dimensions of the principal chambers, the deuteroconch and the protoconch, depended on the number of chambers in the megalospheric test. The material, mainly from the French Upper Cretaceous, was measured by Dr. J. E. van Hinte of the University of Utrecht, and analysed by R. A. Reyment in an unpublished manuscript.

Two sets of variables were measured; the first comprised:

x_1... Outer maximum diameter of the initial four chambers, measured across the principal auxiliary chambers.

x_2... Inner maximum diameter, measured across the principal auxiliary chambers.

x_3... Outer maximum diameter, measured across the protoconch and deuteroconch.

x_4... Inner maximum diameter, measured across the protoconch and deuteroconch.

The second set of variables comprised:

x_5... budding number of youngest pseudonepionic closing chambers.

x_6... budding number of youngest periembryonic closing chamber.

x_7... Number of periembryonic chambers, excluding the epiauxiliary and closing chambers.

The objective of the analysis is to determine whether there is any significant canonical correlation between the two sets of variables and to determine the biological nature of the correlation should one be demonstrated to exist.

The pooled samples contained 314 specimens, and the correlation matrix for the seven variables was computed. A highly significant canonical correlation coefficient of 0.953 was found, generating the two vectors (0.48, 0.29, 0.76, 0.31) and (0.92, 0.02, 0.38). Thus the linear dimensions of the early part of the test, especially that of the outer maximum diameter across the protoconch and deuteroconch, depend fairly strongly on the numbers of chambers, especially the number of the youngest pseudonepionic closing chambers.

The pooled samples do not tell the whole story, however. When nine of the samples, from successively younger horizons, were analysed separately, the first three were dominated by x_4 but in the last six samples x_1 predominates either alone or with characters other than x_4. Perhaps this represents an evolutionary trend.

10
Variation in the Apparently Homogeneous Sample

Earlier in this book we have been considering the problems arising from the comparison of different groups of organisms in as many ways as are biologically or geologically meaningful. Now we turn to the much more nebulous problem of discovering "structure", not in the relations between different groups, but within any one group of undifferentiated material. This step is often treated as a major logical one, and a persistent criticism of discriminatory topology has been that it does not provide for the initial classification into the groups whose mutual relationships may afterwards be studied. For a variety of reasons, this criticism lacks validity and one of these reasons is that the distinction between discrete groups and homogeneous groups represents the extreme forms of a continuum, the intermediate members of which we have to deal with in real life. A large part of the difficulty arises from the need to abstract from the real-life situation such entities as the homogeneous groups, in which the variation is represented by neat spherical or ellipsoidal clouds round the mean, as a basis for a mathematical model. Useful as such models are, they should not usurp the properties of the samples of material we have to deal with.

We are accustomed to talk about random variation about a mean to such an extent that constant preoccupation with the possible meanings of the word random becomes tiresome. In any case departures from randomness are commonly thought of in terms of departures from normality of sections through the mean parallel with the ordinate representing the frequency, and the ideal situation is pictured by Seal (1964; 101). However, Pearce (1959) has raised the substantial point that variation, even if random in certain senses, is not necessarily uniform in the multidimensional space, but is in effect canalized along certain patterns of growth whose existence can be demonstrated objectively in multivariate studies. The variation, though possibly random in magnitude, is certainly not random in direction.

In other words, we should have, in any supposedly homogeneous sample of organisms, incipient variation which bears a relationship to the contrasts of form between closely affine discrete groups. Looked at in this light, there is a continuum between those situations in which we can easily distinguish the groupings (appropriate for analyses by, say, canonical variates) to those situations where we cannot, or at least have not, distinguished them (appropriate for analyses by principal components or principal coordinates). To a considerable extent, and certainly far more so than a perusal of statistical texts would suggest, the difference between discrete groups and the continuous, homogeneous, case is subjective rather than objective, for as the contrasts of form between the groups become less and less marked, more intensive and perhaps more sophisticated techniques will be required to detect the heterogeneity of the continuum. Indeed, a homogeneous group has been cynically defined as one in whose heterogeneity the experimenter has not yet become interested.

Evidently, regarded in this way, the canalized variation within the supposedly homogeneous group should merge with the axes representing variation between discrete groups, assuming that these groups are affine. We can see that this is so if we consider a group of animals in which there is some sexual dimorphism in terms of size, but in which the primary sexual characters are difficult to see, so that an inexperienced worker might fail to spot the distinction. If the same batch of animals were subjected to a morphometric analysis by two workers independently, the one able to distinguish sexes, the other not, in all probability two types of analysis would emerge. One analyst would consider the variation of the group as a whole, and would extract a series of latent vectors, the most considerable of which would describe the size variation; the other analyst would spot the difference between the sexes, perform an analysis, using discriminant functions or their equivalent, and find that one such function described the size/sex variation.

Because this dual approach to multivariate problems is always possible, Blackith (1960) considered the vectors describing variation within a group as "signposts" to possible variation between that group and other groups. Depending on one's views as to the relationship between polymorphism and speciation, one can even describe them as signposts to evolution in the species, though this imagery is less well founded than that of the vectors within a group merging with those between groups created out of the original one. When testing such hypotheses, one needs to be careful to examine the parallelism of two vectors in the same space, and not merely in spaces of the same dimensionality; it is easy to forget this when testing the concordance of vectors derived from different analyses.

Great care, however, is needed in the interpretation of vectors describing variation within groups, since objective tests of the nature of this variation are less easy to perform. Moreover, there is considerable disagreement as to the relative importance of various potential sources of trouble. To some authors, departures from multivariate normality may distort the tests. To others, the lack of scale-invariance in principal component analyses is a serious handicap (Gower, 1967a), although for many this has never seemed to be important in practice (Stuart, 1964). As always, there is a gulf between the attitudes of statisticians, with their concern for the theoretical niceties, and biometricians, who are interested in the useful application of statistical methods, which it is hard for any one person to bridge.

One of the sources of strength in multivariate morphometric procedures is that the vectors which the procedures uncover often correspond to features of the organism which have long been appreciated in an intuitive way; for instance, the concepts of "vigour" and "precocity" in the growth of apple-trees, were found by Moore (1965, 1968) to be adequately represented by principal components; so were "taper" in tree-growth (Holland, 1968a; Fries and Matérn, 1966). Holland has emphasized that even where the components themselves have no obvious biological interpretation the space which, as axes, they define does.

Similarly, in plant and animal breeding experiments, a quantitative assessment of the body conformation at which the breeders, rightly or wrongly, are aiming is invaluable; Rouvier and Ricard (1965) have shown how component analysis can help to provide this assessment.

11
Du Praw's Non-Linnean Taxonomy

One of the most promising developments in recent years has been Du Praw's exploitation of the continuum between the formally continuous and formally discontinuous cases in morphometric analysis, in order to create a method of classifying organisms (or other objects) free from the restraints of Linnean taxonomy and also free from the unsatisfactory theoretical basis of numerical taxonomy, which has shown signs of becoming somewhat inward-looking and reluctant to tackle its theoretical weaknesses.

What Du Praw (1964; 1965a) does is to make a reasonable number of measurements (in the event, on the wing venation of honey-bees) on a large and at least potentially heterogeneous sample of individuals. He then uses canonical variate analyses, as set out in Rao's (1952) book, to extract vectors linking provisional groups of individuals taken from the whole. These groups need not be taxa in the Linnean sense, that is, they may well overlap, or constitute truncated samples, or be based on provenance. Du Praw extracts only the first two canonical vectors (which he described as discriminant functions, quite correctly in the sense that they are functions which discriminate, but perhaps a little misleadingly since, in general, they will not have been computed as we described on p. 46). There does not seem to be any reason why, apart from some inconvenience in presentation of the results, more than two such vectors should not be used if the nature of the variation in the material studied calls for it.

The two vectors are used as the axes of a chart on which the individuals, both those which belong to the provisional defining groups and the remainder, are plotted by computing their scores along each axis of variation. The clustering of the data is then inspected, and if there are large groups within which some further discrete variation is suspected, that part of the chart may be further examined, as if it were at "higher magnification", by repeating the analysis using new characters and less material. Although Du Praw does not make the point specifically, the process is essentially an iterative one, in which the provisional groupings lead to provisional canonical vectors, which in turn form the basis for what ought to prove a rapidly convergent sequence.

Already, in Du Praw's work on the honey bee, this method has shown a resolving power, to borrow another expression from microscopy, much greater than that of standard Linnean methods. This improvement showed in particular in the ability to discriminate certain geographical variants of the honey bee. As Du Praw (1965b) comments, the system he advocates merges with the application of morphometric analyses to specific and subspecific entities that have already been described in this book. Both Rao (1948) and Jolicoeur (1959) and the present authors have drawn attention to the taxonomic promise of such methods.

The advantages of this technique in enabling computer recovery of information about taxa to be made rapidly are also mentioned by Du Praw; there is, however, a difficulty which arises when taxonomic categories higher than the species are being dealt with, namely, the increasing proportion of those distinctions which are qualitative and the reduced importance of quantitative differences between the taxa. However, with the introduction of programmes which can cope with both kinds of data this difficulty should become less daunting than it has been in the past, when it was one of the prime reasons for the growth of the school of numerical taxonomists handling, in effect, only qualitative characters.

The fact that non-Linnean taxonomy, in the sense in which Du Praw uses the term, is possible illustrates an important feature of morphometric studies. The initial variates on which the individual organisms are plotted are not necessarily optimal in any useful sense of the word. They could be, but so far as one can tell from Du Praw's publications are not, refined by iteration. However, the mere fact that the individuals have been splayed out on some axes is enough to permit the eye to sort them into clusters, a process which no doubt could also be entrusted to a computer programme. What matters is the initial decision to use some reasonable vectors as the axes of a chart on which the individuals are plotted. Gower (1969) has also commended this approach to classification by a combined ordination-clustering technique. There is here a clue to the fact that useful results have been recovered from various kinds of factor analysis in which all sorts of mathematically unwarranted, and perhaps biologically undesirable, rotations and other processes have been inflicted on the axes of variation. Admittedly, in principle (a principle more honoured, perhaps, in the breach than the observance) the use of a factor technique presupposes that the experimenter knows enough of the structure of the material in question to justify the imposition of such a structure on the analysis. In real life we must acknowledge that the use of factor-analytic models is often not so much the result of *a priori* convictions as to the structure of the material but the fact that factor analyses have been widely publicized and are often the only multivariate techniques that come to hand, when an inexperienced worker feels the need to employ multivariate methods. In such circumstances the use of factor analyses, which have been extensively "explained" in relatively simple phrases understandable to the biologist or geologist, represents accessibility rather than suitability. Because of the great power of the human eye to detect clusters, and the feed-back by which "structure" can be imposed on the charts in terms of the experimenter's past experience, optimal vectors are merely desirable, not essential; useful results can be had from markedly suboptimal vectors.

This conclusion is gratifying, not only because it helps to explain the persistence of markedly suboptimal techniques in morphometrics, but also because it may encourage us to look with a less jaundiced eye at the flood of poor quality "data processing", which we may expect to pour from computers as programmes and access to computers become generally available. If analyses were once sometimes chosen because the experimenter knew enough of only one analysis to practise it, the same, essentially, is now sometimes true if for analysis we read programme. Yates and Healy (1964) commented some years ago that "there has been plenty of statistical nonsense produced on desk machines, but this will be nothing compared with the flood that will emerge from computers if they are not wisely used and firmly controlled".

12
Principal Components and Principal Coordinates Analyses

FORMAL PRESENTATION OF THE METHOD OF PRINCIPAL COMPONENTS

The technique of principal components analysis in multivariate statistics is nothing more or less than the statistical application of a well known method of mathematical physics, the so-called eigen-problems. The historical development of the subject has been reviewed earlier on in the book. In the present connection, we confine ourselves to a description of the mathematical principles involved.

Principal components analysis (PCA) is suitable for the analysis of the structure of multivariate observations, particularly from the standpoint of investigating the dependence structure occurring in a suite of observations, particularly when no *a priori* patterns of interrelationship can be suggested or are suspected. In PCA, the observable variates, such as morphological dimensions, are represented as functions of a smaller number of latent variates (i.e. the principal components). The principal components are expressed in terms of linear combinations of the observable variates and the analysis of the dependence structure amounts to the statistical estimation of these linear functions. The first principal component of the observations, X, is defined as the p-variate linear compound

$$Y_1 = a_{11}X_1 + a_{21}X_2 + \ldots + a_{p1}X_p. \tag{12.1}$$

In matrix notation this is written

$$Y_1 = \mathbf{a}_1' \mathbf{X}.$$

The sample variance is

$$s_{Y_1}^2 = \mathbf{a}_1' \mathbf{S} \mathbf{a}_1,$$

where S is the sample dispersion matrix. The coefficients a_1, of the first linear combination of PCA must satisfy the p simultaneous linear equations

$$(S - \lambda_1 I) = 0 \tag{12.2}$$

where I is the identity matrix and d_1 is obtained by solving the determinantal equation

$$|S - \lambda_1 I| = 0 \tag{12.3}$$

The vector a_1 is the first latent vector of the dispersion matrix, S, and λ_1 its first latent root. The foregoing presentation can be easily reworded for any of the subsequent latent roots and vectors of S.

The sign and magnitude of a vectorial element indicate the direction and importance of the contribution of a particular variable to a particular component. More precisely, the ratio,

$$\frac{a_{ij}\sqrt{d_j}}{s_i}, \tag{12.4}$$

yields the correlation of the i-th variable and the j-th component.

In morphometric applications of PCA, it is common practice to take the logarithms of the observations and to perform the calculations on these transformed vectors. Gould (1967) has pointed out that providing growth effects are by far the major source of variation in the material, the first latent vector of the dispersion matrix of logarithmically transformed observations of morphometric variables is intuitively acceptable as a "growth vector". In this case, the ellipsoidal cloud of points (ellipsoidal if the observations are multivariate normally distributed) is spread around a strongly elongated major axis (*viz.* the direction of the first principal component).

CASE-HISTORIES OF THE USE OF PRINCIPAL COMPONENTS

The First Principal Component as a Discriminator of Size

Two forms of thrips, so alike that serious doubts about the independent identity of the forms had been expressed, were studied by Ward (1968), who measured three characters: the lengths of the seta at the pronotal posterior angle of the tenth tergite; and of seta No. 3 on the ninth tergite. A principal component analysis of the dispersion matrix gave a first principal component which succeeded in ordering the thrips, apparently by size rather than shape, despite the unusual choice of characters. The distribution of scores for individual thrips along this first principal component was clearly bimodal. Misclassification by the vector (1,1,2) was less than by any other method tried.

That the first principal component often has the character of a size vector has also been demonstrated by Jolicoeur and Mosimann (1960), who studied the shape of the painted turtle. These authors emphasize that all coefficients must be of the same sign, whereas those of the other components must generally be of mixed signs for this interpretation to be valid. Rao (1964b) gives a mathematical justification for this reasoning.

Reyment and Sandberg (1963) measured four characters of fossil ammonites on a logarithmic scale (shell diameter; umbilical diameter; height of last whorl, and breadth of last whorl). One Cretaceous and two Triassic species were measured. In each instance, the first principal component of the dispersion matrix took up over 90% of the total variance and had all four coefficients positive. From the taxonomist's point of view, the interest in this case centres on the "shape" components. The second component consistently was of the form (0.2, −0.7, 0.3, 0.4), though with minor variations from this representation, implying that the umbilical diameter decreases relative to the height of the last whorl in ammonites of standardized size.

Reyment (1966b) also measured six characters on the tests of Paleocene Foraminifera and seven characters on a Recent species (1969c) and again found the first principal component of the dispersion matrix to contain nothing but positive coefficients. The subsequent components are shape vectors which are influenced by the proportions of gamonts and schizonts (sexual and asexual forms) in the population (Fig. 18).

Principal components analysis is also useful in the analysis of Foraminifera for the morphometric analysis of problems other than polymorphism. For example, Reyment (1969b) was able to demonstrate for the agglutinating living species *Textilina mexicana,* that variations in the dimensions of the test are not correlated with the diameter of the proloculus.

Fries and Matérn (1966) have also used the first principal component as a measure of size in forest trees; Amtmann (1966) has extracted a size (first) principal component and a shape (second) one in an investigation of the skulls of squirrels. Many other illustrations of this general representation could be found in the literature. In

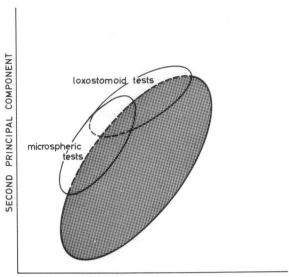

Fig. 18. Principal components chart of the first two transformed variables for *Afrobolivina africana*. Redrawn from Reyment (1966b). The shaded field denotes megalospheric tests.

some cases the first principal component takes up almost all the variation; Murdie (1969) measured eight characters on 80 aphids and found that the first principal component accounted for no less than 94.24% of all the variation in form, a reflection of the high correlations between the characters. We may also note that Jeffers (1967b) has published a series of case histories using principal component analysis in which this point is taken up. There is no *a priori* reason for the hypothesis that the first principal component has a "size" property and our present attitude is largely based on "the proof of the pudding is in the eating".

Although some workers tend to look upon the principal components analysis of size and shape variation with a certain degree of scepticism, it is, nevertheless, possible to produce some sort of empirical confirmation for this interpretation. Reyment (1966b), in connection with the morphometric study of the African Paleocene foraminifer *Afrobolivina africana* (Graham, de Klasz, Rérat), obtained an empirical confirmation of growth in accordance with a negative allometry coefficient. A set of trivariate data was constructed in which two variables grew in a relationship with a positive coefficient of allometry and one variable was made to grow in agreement with a negative coefficient of allometry (not, of course, to be confused with negative allometry). This led to a first latent vector with the elements (0.8, 0.5, −0.3). This latent vector was found to be associated with a latent root that accounted for 99% of the total variation; inasmuch as the usual type of growth relationship leads to positive coefficients in the first latent vector [exceptions do occur where the two latent roots are near each other in magnitude, as found by Reyment (1963b)], the appearance of a negative third element in the first latent vector is confirmation, in part at least, that a negative coefficient of allometry in a multivariate relationship can be picked up successfully by a principal components analysis.

Sometimes, especially in examples of what we have called "abstract morphometrics" such as marketing research (Gatty, 1966) it would be physically impossible for the first principal component to take on the function of a "size" discriminator. For instance, Baker (1954) collected the results of a number of taste and smell (organoleptic) experiments with wines which had been analysed by physical means. This body of information was partitioned into orthogonal "factors" (principal components) and in this case the first factor could not possibly relate to the size of the sample. Moreover, we shall see later (Chapter 16) that in certain phytosociological experiments, the concept of size is replaced by one of the general abundance of the plants, just as in rheology it is replaced by the "stiffness" of the material under test (Harper and Baron, 1951).

Variation in Scale Insects

The scale insects have proved difficult to identify because shape differences form an important part of the distinctions between species, yet the use of ratios and similar devices ("antennal formulae") has become discredited because the ratios changed in magnitude when larger individuals of a given species were compared with smaller ones.

Eighteen characters were measured by Blair et al. (1964) on 27 female individuals, all of which were taken from a single plant (because the shape of a scale insect can vary according to the plant from which it is taken). Of the 152 correlation coefficients between these 18 characters, no fewer than 35 were negative and only 52 were of such a magnitude that, when tested for significance as individual correlations, i.e. neglecting the fact that the correlations are not independent of one another, they were "significantly greater than zero". Few of the appendages grow isometrically with the rest of the body, none has a correlation of more than 0.47 with the length, nor more than 0.62 with the breadth. In general, the growth of the parts of the legs were most closely interlinked, that of the segments of the antennae the least so, to the extent that the first, fourth and fifth antennal segments each had eight negative correlations with other characters, which accounts for more than two-thirds of the negative correlations.

The first latent vector was easily identifiable as representing size variation, since it included only three negative coefficients and 15 positive ones. A vector with substantial positive coefficients for most of the characters usually takes up the greater part of the variation in a morphometric analysis because the size variation is often the most marked component. The second vector showed that the legs, in this pattern of growth, elongate disproportionately as compared with the rest of the body, since the parts of the legs have positive coefficients whereas the remainder mainly have negative ones. The third vector similarly shows that there exists a pattern of growth in which the distal antennal segments elongate disproportionately with the rest of the body. The fourth and fifth vectors seem to involve patterns of growth in which the development of the first antennal segment plays an important part, and the sixth vector represents "attenuation" that is, the elongation and (relative) narrowing of the body form, a type of variation which is known to occur when scale insects are transferred from one host-plant species to another. This last vector is thus a "signpost" to a source of variation which can be exploited by the insect in its adaptation to a new environment.

Boratynski and Davies (1971) have investigated the taxonomic value of male coccids by 12 different numerical methods using four Q-mode, five R-mode (see Chapter 13 for the distinction between Q and R-mode analyses) and three principal coordinate analyses, with 101 characters of various kinds. This is a paper whose importance goes far beyond the taxonomy of the coccids, but in that particular context they found generally good agreement between the numerical methods, not only with one another, but with conventional taxonomic ideas built up through investigations of the females and a more recent comprehensive investigation of the males. These authors stress, however, the great difficulty of deciding what constitutes the "best" taxonomy. They arrive at the tentative view that principal coordinate methods using non-parametric measures of association are perhaps best suited to the analysis of coded, multistate, data, though they add that ordinary principal component analyses are unlikely to yield seriously misleading results.

A striking finding of Boratynski and Davies was that, whereas their first two principal components separated the family Diaspididae into three major taxa (subfamilies) the third, fourth and fifth axes group the insects successively into associations of genera and species; thus the vectors describing successively smaller components of variation are also gradually descending the taxonomic hierarchy. A somewhat similar situation was found by Ivimey-Cook (1969) in his studies of the rest-harrows of the genus *Ononis*. These studies, therefore, usefully complete the comments made earlier in this chapter concerning the special attributes of the first principal component.

Cheetham (1968) published a thoughtful study of the taxonomic relationships within the Neogene cheilostomate bryozoan genus *Metrarabdotos* by a battery of quantitative methods, including principal components and clustering techniques. The resulting five phenetic groups (based on 23 characters) were related to a time-stratigraphic framework. Taxonomic interpretations were based on inferred phylogenetic relationships within and among groups. He concluded that the morphological overlap among groups has derived from convergent and parallel trends in size, positions, orientation and differentiation of avicularia and in denticulation of the secondary orifice in the American and Eurafrican stocks, considered to have been isolated through most of their history. The principal components analysis of 888 zooecia was enlightening and the first three components could be related to zooecial size, zooecial shape, and avicularian length relative to oral dimensions, respectively. The phenetic comparisons were made using a numerical code for both qualitative and quantitative characters, and analysing these by the methods of Sokal and Sneath (1963). Slightly more than half of the zooecial characters could be adequately expressed in two-state code, while the others needed codes running to as many as five states.

The indications yielded by the principal components study were used to weed out redundant variables from the numerical taxonomic treatment; for example, the means of the first three principal components were used in the stead of the original variables.

The Forms of the Common Salamander; Effects of Coding for Sex

Eighteen characters were measured on 527 salamanders of either sex. The correlation matrix of these characters was subjected to a principal components analysis on two occasions; on the first occasion the sex of the salamanders was coded +1 for males and −1 for females, making in effect a 19th character, whereas on the second occasion the coding for sex was dropped.

Inspection of the correlation matrix showed that most of the correlations between the characters were within the range 0.7 to 1.0 except for character 18 (the breadth of the tail) whose correlations with the remaining characters lay in the range 0.39-0.49, and character 17 (height of tail) in the range 0.60-0.78. We should, therefore, expect the first principal component to be essentially a "size" vector; in fact, the weights of the various characters in the first principal component are remarkably uniform,

$$(0.26; 0.25; 0.24; 0.24; 0.26; 0.25; 0.23; 0.24; 0.23; 0.24; 0.21;$$
$$0.25; 0.25; 0.26; 0.23; 0.24; 0.20; 0.13).$$

These weights are not sensibly affected by the inclusion of a coding for sex.

However, the sex coding, when included in the analysis, dominates the second principal component, to such an extent that it becomes virtually a unit vector; the range of weights for the other characters lies between -0.18 and $+0.15$. When the sex coding is removed from the analysis, the second principal component substantially maintains its quality of a unit vector, but it is the 18th character (breadth of tail) which predominates with a weight of $+0.91$. The rest of the characters have small weights. Since individuals with a code of -1 generally have broader tails than have individuals with a code of $+1$, it seems likely that tail breadth is taking over the function of a sexual discriminator along the second component. The sex-code itself seems to serve little useful purpose, and is virtually redundant. We may note that the first character, total length of the animals, is also redundant, since it is the sum of the second and third measurements.

We thus have a general size vector and a vector representing sexual dimorphism, taking up 79% and 4.4% respectively of the total sum of squares accounted for by the dispersion matrix. The only other root likely to generate a vector of consequence is the third, accounting for 3.3% of the total variation. The interpretation of this vector is rendered difficult because the relative weights of the characters change according to whether the sex-coding is included or not. When it is included the third vector is almost a unit vector reflecting the importance of the tail breadth in disclosing contrasts of form other than those associated with sexual dimorphism. When the sex-coding is omitted, so that the second vector expresses the capacity of the tail-breadth to reflect sexual dimorphism and other factors simultaneously, the third vector contrasts the 6th, 11th, and 17th characters with the 15th and 18th, essentially. This is a pattern of growth with no very straightforward interpretation.

It is true that a plot of the second and third principal components shows an almost complete separation between the males and females, but this separation is evidently due to the inclusion of the coding for sex, and thus simply reflects the information fed in to the analysis. If the coding is omitted, there is little sign of sexual dimorphism in the various plots of pairs of the first three principal components.

Another interesting feature of the salamander data worth a quick mention is the fact that the plots of the transformed observations for both sexes give neatly elliptical data, a reflection of the fact that the observations are multivariate normally distributed, a necessary prerequisite if the PCA is to include statistical tests.

We wish to express our thanks to Dr. Josef Eiselt of Naturhistorisches Museum, Vienna, for making this magnificent suite of observations available to us for analysis.

The Use of Principal Component Analysis for Clustering Data

Apart from the explicit use of clustering techniques, such as those discussed in Chapter 22, the "discriminatory power" of the principal components can serve as a clustering technique of great generality. Temple (1968) discovered the presence of two forms in the Silurian brachiopod *Toxorthis* by this means. It is a moot point whether these forms should be interpreted as morphs within a species or species of the same genus living in the same habitat, or even, conceivably, as an example of sexual dimorphism.

Temple took the view that the first principal component essentially reflected general growth, and plotted his material on charts constructed from the second and third principal components derived from a dispersion matrix (in terms of the correlations between five measurements of the pedicle valve and seven of the brachial valve). The pedicle and brachial valve were analysed separately. In the case of the pedicle valve the analysis split the data into two distinct groups, corresponding to the two forms as judged visually, but there was a very slight overlap. As Fig. 19 shows, there is complete separation between the two groups into two ellipsoids, based almost entirely on the discrimination afforded by the second principal component, which is probably close to the orientation of the optimum discriminant function. A further example of the use of the second principal component to distinguish two putative morphs within the species *Pionodema retusa* is also reported by Temple.

Because the initial visual discrimination of the two morphs was difficult and subjective, it is not surprising that the occasional

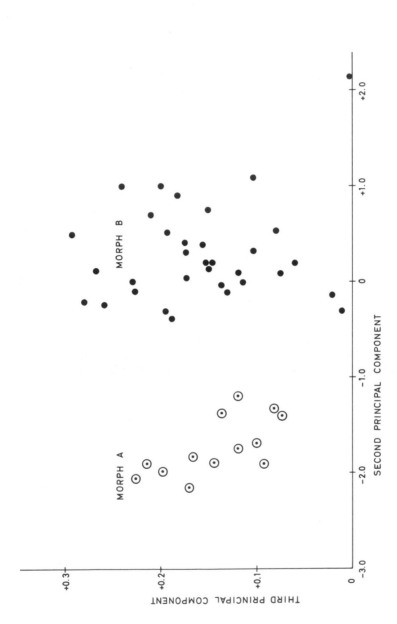

Fig. 19. The two morphs of *Toxorthis proteus* clustered by means of the second and third principal components of the matrix of the seven morphometric characters of the brachial valve. Redrawn from Temple (1968).

individual appears to have been wrongly placed in the initial assessment, upon which the principal component analysis in no way depends.

Sociological Components Differentiating British Towns

In an exceptionally well interpreted study of the sociological differences between 157 towns in Britain, which we offer here as an example of generalized or "abstract" morphometrics, Moser and Scott (1961) assessed 57 different characters of the towns. They were limited, naturally, to characters for which adequately comparable statistics had been published, and chose seven characters from the field of population size and structure, eight from that of population change, 15 reflecting the properties of the households and housing, ten economic characters, four social class indicators, five characters concerned with voting behaviour, six concerned with health statistics, and two with educational statistics.

These 57 characters were then submitted to a principal components analysis based on the correlation matrix. There are several features of this analysis which render it worthy of special attention. Firstly, there is no component representing the size of the towns, and indeed only one-fifth of the variance of "size" as a character is taken up by the first four components extracted. This finding should serve as a warning not to treat the first principal components of analyses as "size" components, unless the structure of the experiment clearly indicates such treatment. Secondly, the first four components only accounted for some 60% of the total variation, an unusually low figure, the first component only taking up 30% of the total amount. Thirdly, an interesting attempt at reification was made by splitting the towns into northern and southern locations, and repeating the analysis for each section. Similar results were obtained, which encourages confidence in the general veracity of the analysis. Fourthly, the analysis was repeated using the logarithms of the original data, and also with the rank order of the data. Again, closely comparable results emerged from these three analyses, giving one a feeling of confidence that the interpretation is "robust" and unlikely to be seriously perturbed by changes of scale or minor differences of method.

Moser and Scott's interpretation of the sociological nature of the four components that they studied in depth is as follows:

I. a vector differentiating social classes.

II. a vector representing population growth in the period between 1931 and 1951, when many suburbs were developing but the wool trade towns were declining.

III. a vector associated with the structure of the working population in 1951 together with population growth after that date. This vector reflects the "age" of the town, not in the historical sense but in the sense that an "old" town has a low proportion of new housing, a high proportion of women at work, a large proportion of households sharing dwellings, and so on.

IV. the last of the interpreted vectors reflects the existence of a strongly underprivileged fraction of the population concentrated in certain towns. Such towns have much overcrowding, their inhabitants tend to take little interest in elections, live in dwellings lacking piped water, and marry later. Overcrowding and its attendant features play a part in differentiating towns quite distinct from the social class structure of the towns, or of the "age" of the areas.

When these four components were used to classify the 157 towns they fell into clusters with recognizable qualities, such as seaside resorts, spas, commercial centres, railway centres, ports, textile centres, residential suburbs of various kinds, and the depressed towns of the north-eastern seaboard and the Welsh mining towns.

A Sociological Classification of the Regions of Italy: the Use of Principal Components as Discriminant Functions

Casetti (1964) has used principal component analysis to score various regions of Italy on a sociological basis, after eliminating size differences by dividing through by the working populations in each area. Out of a total of 25 components, derived from the matrix of 25 attributes of the regions, four were selected for interpretation. These, the first four components, took up 79% of the total variation in question. These were interpreted as reflecting (i) standards of living; (ii) a component of which one extreme represented districts with a high tourist attraction and mountains, the other pole representing areas of industrial agriculture; (iii) a vector whose high positive scores implied a predominance of intensive peasant agriculture, the low scores implying that of industrial agriculture; (iv) a component differentiating between regions of intense economic activity and those in a depressed condition. Casetti also used this technique to split up the North American region into climatic areas.

It should perhaps be pointed out, for the critically inclined, that we are fully aware of the fact that this is a statistically illicit application of PCA; significance testing is, however, not relevant to the interpretation given above, for even if the fourth root proved non-significant by the usual tests, its success in differentiating areas of high and low economic activity could readily be examined using non-parametric tests.

Components of Literary Styles in Latin Elegiac Verse

A great deal of mediaeval and more recent literary work consisted of attempts to write Latin elegiac verse in a style approximating as closely as possible to the verse of the Augustan age in Rome. This style, however, consisted of at least three components which were revealed by a principal component analysis (Blackith, 1963) of four attributes of the verse; the elision frequency (syllables written but not pronounced for reasons of euphony or the demands of the metres); the mean syllable number per word; the variance of the distribution of the syllable number; and the entropy of mixing of words of different syllable number. In this investigation the entity whose "structure" is being examined is style: we may look on it as "abstract morphometrics".

Since poetic composition is at least partly under conscious control, some attributes of the elegiac verse such as elision frequency were closely controlled by the authors, following the dictates of fashion since antiquity, and were almost uncorrelated with the other attributes. However, the third latent vector, in which the mean syllable number plays a predominant part, enables the analyst to determine the date at which a poem was written, in a way which is apparently independent of any attempts by the poet to write pastiche or even to forge texts. Anxiety on the part of the poet to write in an "acquired" style will influence the first two components of variation, but not the third, which appears to be hardly at all under conscious control.

Just how far this third component is influenced by subconscious stress is shown by comparing Ovid's elegiac verse written before and after his banishment by the Emperor Augustus to the Black Sea coast. There is a highly significant shift along this third latent vector, a shift which occurs suddenly at the moment of banishment, and not slowly, as one might expect if isolation and ageing were to be held responsible for the change.

The distinguished 20th century Latin elegiast, Henrici Paoli, has written verse which has scores along the first two axes which compare well with those of classical authors. Paoli's scores along the third axis are, however, dramatically different from any classical author, although they differ in a direction presaged by the shift along this axis of "Silver Age" poets as compared with those of the "Golden Age".

Cox and Brandwood (1959) studied the dating of Plato's works by multivariate discriminatory methods to good effect.

Quality Control by Analysis of Gas Oil Fractions During Refining

In a plant producing gas oils, within the sequence fuel oil-gas oil-kerosene, it is customary to measure 12 attributes of the oil as it is made; these attributes are the specific gravity; the distillation 10%, 50%, and 90% points; the flash point; the sulphur content; the kinematic viscosity; the cloud point; the pour point; the carbon residue; the aniline point; and the Diesel index. Although the space in which the principal components analysis of the correlation matrix was performed was of course twelve-dimensional, only five dimensions were needed for the adequate representation of the data, as shown by the fact that only five of the latent roots of the determinantal equation took up appreciable fractions of the total variation (Thomas, 1961).

The first principal component was

$$(0.88; 0.59; 0.95; 0.57; 0.54; 0.22; 1.00; 0.63; 0.88; 0.44; 0.14; -0.56)$$

which has positive weights for the first 11 attributes. In this case, it is not open to us to interpret this broadly positive assemblage as a "size" vector, Thomas, in fact, considers it as a measure of the general position of the sample of gas oil within the sequence from fuel oil to kerosene.

The second and third principal components, each of which has a mixture of positive and negative elements, seem to reflect different aspects of the narrowness of the fraction which is being sampled, that is to say its freedom from substances belonging to other heavier or lighter fractions.

The fourth and fifth principal components are virtually unit vectors, with small positive or negative coefficients everywhere except for the sulphur content (in component IV) and for the carbon residue (in component V).

Thomas stresses that, since the twelve-dimensional space reduces to a five-dimensional one, a considerable saving of time in the analytical laboratory could be achieved if only the five essential attributes were measured (including, of course, sulphur and carbon), the others could then be derived by regression techniques if needed. Spurnell (1963) and Jenkins (1967) make the same point in applying principal component analysis to a metallurgical process and to petroleum purification respectively.

In industrial applications of quality control, sets of measurements may have to be made repeatedly, and at great cost, in contrast to the overwhelming majority of biological applications of multivariate techniques where there is much less often a need to keep on measuring a set of variates. In the industrial context, therefore, a possible reduction of the numbers of variates is of interest. Two points seem to be worth emphasis in this connection.

(1) The omission of variables (characters) because they do not contribute significantly, on a superficial analysis, to the discriminating power of the functions computed, may be risky. If there are several fairly highly correlated variables, they may share an important fraction of the discriminating power, yet, because there are so many of them, this capacity may have been, as it were, so diluted that the contribution of only one variable may not reach "significance" in a particular context. The omission of all such variables could well lead to serious losses of information. The solution lies in a more gradual approach in which the analysis is rerun after dropping only one or two doubtfully useful variates. As in the more familiar multiple regression problem, tests of the significance of the coefficients of individual variables cannot be considered in isolation from the whole analysis. The various rules suggested for the elimination of variables have been critically considered by Weiner and Dunn (1966).

(2) Although the performance of a multivariate analysis ought to be reasonably objective, in that any operator should get the same numerical results with the same data and analytical programme, the interpretation of the results calls for a considerable knowledge of the potentialities of the analytical methods chosen and also of the structure of the problem being investigated. Here, the advice of an experienced multivariate analyst can be valuable and in the industrial context would probably be as worthwhile an investment as any other form of technical advice.

As an example, we may quote Draper's (1964) analysis of 13 tests of kraft paper sacking, in which he concluded that principal component analysis does not help to decide what properties to test. However, an experienced analyst (Jeffers, 1965) was able to relate this specific problem to a more general context and to make valuable suggestions.

The Growth of Fruit-trees

Two orchards of fruit trees were measured by Pearce and Holland (1960) in respect of four characters; the weight of the mature tree above ground; the basal trunk girth of the mature tree; the total shoot growth in the first four years of age.

The logarithms of these quantities were then converted first into a dispersion matrix and then into a correlation matrix, on the grounds that the dispersion matrix gives too much weight to those characters which happen to have high variances; the correlation matrix is in effect a dispersion matrix for standardized variates. The two matrices were:

1.000	0.951	0.596	0.517	and	1.000	0.939	0.266	0.178
	1.000	0.694	0.619			1.000	0.424	0.358
		1.000	0.898				1.000	0.835
			1.000					1.000

In each case, the principal components corresponding to the two largest roots of the characteristic equation were:

$$\lambda_1 = (0.489; 0.521; 0.507; 0.482)$$
and
$$\lambda_1 = (0.485; 0.551; 0.497; 0.463)$$

$$\lambda_2 = (0.564; 0.416; -0.439; -0.562)$$
and
$$\lambda_2 = (0.550; 0.409; -0.481; -0.547).$$

Pearce and Holland make the wise comment that one way in which one can try to make sense out of principal components is to see if they are at least consistent from one analysis to another. This consistency test is satisfied by the two first components, which essentially reflect "vigour" (size variation) and make an angle of only 2°9' with one another, and also by the two second principal components which make an angle of only 2°42' with one another. The second principal components are reflections of the "establishment" factor in tree growth, where a tree fails to live up to its early promise, or else exceeds it.

Pearce and Holland (1961) and Holland (1968a) followed this analysis of tree growth with a further principal component analysis of the five chemical measures made on the leaves of the trees; three of the components seemed to be of some consequence, and were reasonably consistent from year to year. Holland (1968b) has also analysed chemical data derived from groundnut and sugarcane plants by means of principal components.

An Essay in Geochemistry

Saxena (1969) analysed 93 pairs of samples of coexisting biotite and garnet in metamorphic rock formed under a variety of conditions. A matrix of the correlations between mole fractions of, say, iron, was made up of entries such as $X_{Fe} = Fe/(Fe + Mg)$ with similar entries for other elements, manganese, calcium, aluminium, and titanium. Although a regression analysis of the data added little to what was previously known, a principal component analysis of the matrix gave valuable results. The first principal component was the transformed general distribution coefficient and the second principal

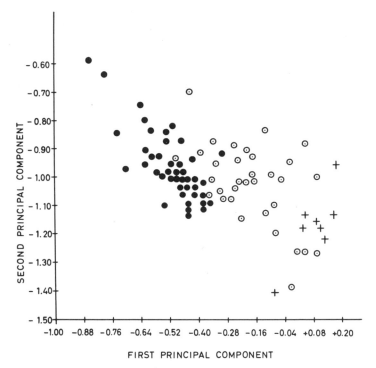

Fig. 20. The distribution coefficients of iron, manganese, calcium, aluminium and titanium in coexisting biotite and garnet, plotted as the first principal component against the second principal component of the matrix of correlations of these elements. Open circles, high amphibolite facies. Closed circles, staurolite zone and epidote-amphibolite facies. Crosses, charnockites. Redrawn from Saxena (1969).

component, though lacking any very obvious theoretical interpretation, had the useful practical function of helping to distinguish, by clustering, the parent rocks according to the pressures and temperatures at which they had been formed. Although it is reassuring to have a satisfactory theoretical interpretation for all the axes in a multivariate analysis, the practical utility of the analysis may sometimes outgrow the current theory.

Figure 20 shows that rocks falling into the staurolite zone and the epidote-amphibolite facies cluster into the region of low positive scores along the first component and small negative values along the second. The charnockites, on the other hand, form a cluster of rocks falling into the region demarcated by positive or near-positive scores along the first component but larger negative scores along the second. The high-amphibolite facies occupied an intermediate position along both axes.

Saxena went on to plot the transformed general distribution coefficient against the estimated temperature at which the various rocks were formed, yielding a roughly inverse linear relationship. A rough, but perhaps useful, "thermometer" is thus available to estimate the temperature at which any new sample of rock containing garnet and biotite was formed.

A Study of Limestones in the American Upper Ordovician

In what amounts to a principal component analysis Osborne (1969) analysed 12 sediment-petrographic variables for each of 65 calcarenites in order to bring about some form of multivariate classification. The variables studied were various constituents of limestones observable in thin sections, such as the relationship between the number of broken shells to whole shells of various taxonomic categories, as well as a few other variables. The study also included some other well-known multivariate techniques which were used to back up the results of the "factor analysis". The PCA produced clusters which make good sedimentological sense and Osborne was able to construct his further treatment of the data around the groupings obtained.

THE METHOD OF PRINCIPAL COORDINATES ANALYSIS

A useful way of considering classification procedures is that embodied in the concept of numerical taxonomy in the form presented by, for example, Sokal and Sneath (1963). An objection to this approach has been that the methods and procedures of numerical taxonomy lack, to a considerable extent, a firm mathematical basis. Gower (1966) has made an encouraging move in a direction towards strengthening the foundations of the subject and it is to be expected that more mathematical statisticians will in the future be drawn towards this area of research. As a complement to the method of principal components (an R technique; see Chapter 13 for the distinction between Q and R techniques) we shall consider the procedure termed "principal coordinates" by Gower (1966, p. 137) in the following terms.

In a numerical taxonomic study, there will be measurements on p variates for each of n individuals. The interrelationships between these variates will be estimated by means of some form of coefficient of association, a_{ij}, between all pairs of individuals. These form the "association matrix" $\mathbf{A}$. In many situations p will be greater than n.

We consider the symmetric matrix $\mathbf{A}$ of order n. The latent roots of $\mathbf{A}$ are $\lambda_1, \ldots, \lambda_n$ and its latent vectors are: $\mathbf{b}_1, \mathbf{b}_2, \ldots, \mathbf{b}_n$. These form the matrix $\mathbf{B}$ of order n. In applying this in numerical taxonomy, one takes the elements of the ith row as the coordinates of a point, T_i, in n-space. The distance, d_{ij}, between P_i and P_j is then given by

$$d_{ij}^2 = \sum_{r=1}^{n} b_{ir}^2 + \sum_{r=1}^{n} b_{jr}^2 - 2 \sum_{r=1}^{n} b_{ir} b_{jr}, \tag{12.5}$$

Thus, for rows 2 and 3

$$d_{23}^2 = \sum_{r=1}^{n} b_{2r}^2 + \sum_{r=1}^{n} b_{3r}^2 - 2 \sum_{r=1}^{n} b_{2r} b_{3r}. \tag{12.6}$$

The latent vectors of matrix $\mathbf{A}$ are normalized so that the sums of squares of their elements are equal to the corresponding latent roots

$$b_{ir}^2 = \lambda_r. \tag{12.7}$$

The distance relationship is

$$d_{ij}^2 = a_{ii} + a_{jj} - 2a_{ij}. \tag{12.8}$$

By this means, one may represent a multivariate sample of size n as points $T_1, \ldots, T_n$ in a Euclidean space. The relationship between the latent roots and vectors of the association matrix $\mathbf{A}$ are indicated below. Gower (1966) shows that these coordinates are referred to principal axes.

	Latent roots $\lambda_1 \quad \lambda_2 \ldots \lambda_n$		
Point	Latent vectors		
T_1	b_{11}	b_{12}	b_{1n}
T_2	b_{21}	$b_{22} \ldots b_{2n}$	
.	.	.	.
.	.	.	.
.	.	.	.
T_n	b_{n1}	b_{n2}	b_{nn}
Centroid	$\bar{T}$	$\bar{b}_1$	$\bar{b}_2$	$\bar{b}_n$

$$(12.9)$$

Gower (1966) has demonstrated that one may use the methods of principal components as a Q technique on the coordinates of the T_i to find the best fit in fewer dimensions. As is usually observed in principal components, a good representation of the set of points may be obtained in a reduced number of dimensions when some of the latent roots are small. That is if, say, λ_r is small, the contribution of $(b_{ir} - b_{jr})^2$ to the distance between T_i and T_j will be small. If λ_r is large, but the b_{ir} corresponding to it are not greatly different, then $(b_{ir} - b_{jr})^2$ will be small. Hence the only coordinates supplying much to the distances are those displaying a wide range of variation in the elements of the latent vectors and which are associated with a large latent root. In common with what is found in principal components analysis, the distances may often be adequately expressed by two or three vectors.

A further analogue with R-technique principal components is that the elements of the first latent vector may be found to have similar elements, relating to the mean value of all the elements of **A**. This mean value is not important to the problem at hand, as the distances are invariant for any constant added to **A**.

The addition of such a constant will, however, result in different coordinate values T_i and different latent roots and these new points are an orthogonal transformation of the original set after a change of origin. Interest attaches to determining which transformation gives the best fit with a reduced number of coordinates.

This mean value is unimportant in this connection as the addition of any constant to the association matrix **A** leaves the distance between T_i and T_j invariant. Gower (1966, p. 330) adjusts for the means in the following way. It is always possible to adjust matrix **A** so that it has a zero latent root without altering the distance between

T_i and T_j, for if $\bar{a}_i$ is the mean value of the ith row or column of **A**, and $\bar{a}$ is the overall mean, a matrix α_{ij} may be defined in terms of the elements

$$\alpha_{ij} = a_{ij} - \bar{a}_i - \bar{a}_j + \bar{a}. \qquad (12.10)$$

Inasmuch as every row and column of matrix α_{ij} sums to zero, α_{ij} has a zero latent root.

Gower (1968) has noted that when users of principal components analysis (including unrotated "principal components factor analysis") plot the values of the transformed variables, they seldom follow this up with a display of interest in the distances between sample points, which in PCA is beset with an arbitrary element.

When k multivariate universes are to be considered, the generalized distance can be used to bring about a generalization of the concept of principal coordinates. This distance can be found for every pair of universes to give a $k \times k$ symmetric matrix. A principal coordinates analysis of the matrix of D^2-values gives the coordinates of the universes referred to principal axes. This is consequently a sort of Q-mode canonical variates analysis.

CASE HISTORIES OF PRINCIPAL COORDINATES ANALYSIS

Study of Variation in a Species of Recent Marine Ostracod

The marine ostracod, *Buntonia olokundudui* Reyment and Van Valen, displays a type of meristic variation found not uncommonly among the cytherid ostracods. A principal coordinates analysis was made on an association matrix composed of the variables: length of the carapace and various counts on ornamental spines. The principal coordinates analysis of the 503 individuals of *Buntonia* disclosed the existence of several well-defined groups which could be related to the variation genetics of this species (Reyment, 1970b; Reyment and Van Valen, 1969).

There is a considerable concentration of the "variation" in the association matrix to the first latent root. This is a healthy sign and is a good indicator that the analytical method is going to produce some kind of intelligible result. Usually, one would expect such a division of the association variance in a situation in which the elements of the association matrix show a high level of correlation. This is not unusual in biological work.

There is also a considerable area of application of numerical taxonomic methods in such subjects as pedology and sedimentology (cf. an example for limestones by Osborne (1969)). Here, however, poor association is common. One can expect to find much near-random variation in such material which is reflected in the fact that the latent roots taper off very slowly. A survey of Recent bottom sediments in the Niger Delta by Reyment (1969e) did not disclose the presence of significant groupings in the chemical and granulogical data and the successive latent roots were found to decrease only slightly down the diagonal of the matrix of roots.

Analysis of a Cyclic Sedimentary Sequence

Baer (1969) made a study of the palaeoecology of a thin, well-exposed, section in the well-known Eocene Green River Formation of Utah. He found the 40m section to be divisible into five shale-carbonate sedimentary cycles, which could be further subdivided into categories related to the sedimentary environments deltaic, transitional, and lacustrine. The deltas proved to comprise organic limestones, siltstones and shales and low-grade oil shale units. The so-called transitional beds are finely laminated and there are calcite and aragonite paired laminae with occasional laminae which are rich in organic substance. The lagoonal sediments comprise lacustrine carbonates, shales and siltstones.

Baer made a great number of analyses of many kinds. Among his results there is an interesting table of determinations of the elements Si, Al, Fe, Mg, Ca, Na, K, Mn, Rb, Sr, U, and Th. The information is given in the form of mean values and, it was thought, might provide interesting material for a principal coordinates analysis. Programme PCOORD (cf. p. 176) was used on the data. Clearly, the various phases of the cyclicity are not exposed by the plot of the first two coordinate axes (Fig. 21), which seems to suggest that the chemical properties of the subdivisions at comparable levels do not differ strongly from each other. However, all the limestones cluster neatly to the right of the diagram and the argillaceous sediments rich in organic substance tend to group together, although organic content was not one of the things analysed by Baer in this connection. The lacustrine silts are also seen to group closely; the deltaic silt and shale form an isolated cluster well away from the other groups. It may also be of sedimentological significance that the rocks of the transitional beds lie spread over the centre of the diagram.

In connection with the foregoing example we have mentioned the importance of studying the "residual variance" of principal

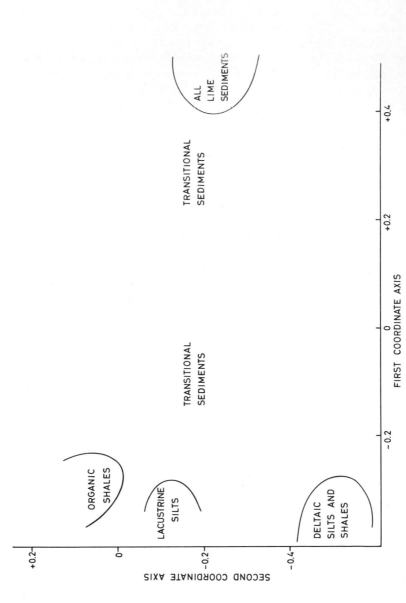

Fig. 21. A principal coordinate chart (first two axes) of chemical analyses for a cyclic-sedimentational sequence in the Lower Green River formation of Utah.

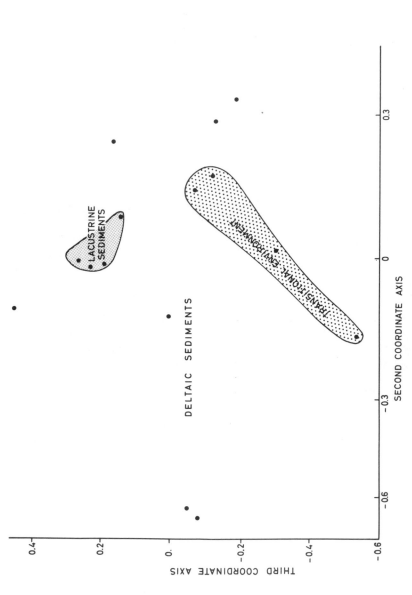

Fig. 22. A principal coordinates chart, of the second and third axes, of chemical analyses for a cyclic-sedimentational sequence in the Lower Green River formation of Utah.

coordinates analysis, as this gives a good indication of how important random variation is in the suite of observations; also, a case of pronounced random variation in sedimentological observations was discussed. The geochemical data here analysed show rather slight variation and the percentage residual, after the first two latent roots have been subtracted, is a low 38%. We dislike categoric statements about the reliability of analytical results in morphometrics but one would be tempted to place considerable faith in the groupings yielded by this principal coordinates study on what really is difficult material.

Let us now look at the plot of the second and third coordinates, given in Fig. 22. The results displayed here are of interest for they group the analyses in accordance with the environmental categories. This is quite a different picture from that shown in Fig. 21, in which the grouping was more in agreement with the nature of the sedimentary category than with the environment in which the sediment was formed. The second and third coordinates therefore reflect chemical differences arising from the particular conditions of the three sedimentary environments.

PRINCIPAL COMPONENTS, SIZE AND SHAPE

Let us just recapitulate some of the thoughts on principal components and a relevant alternative to the size and shape interpretation. If we regard the elements of the first latent vector, say, $b^{(i)}$, considered as an equation:

$$y_1 = b_1^{(1)} X_1 + b_2^{(1)} X_2 + \ldots + b_p^{(1)} X_{p'}$$

where $b^{(i)}$ is the latent vector and p the number of variables. If all the elements of this vector are positive, a unit increase in y_1 increases the value of each of the X_i and one could therefore think of the expression in terms of *size variation*. If, as is often the case for morphometric variables, some of the elements of the second, and subsequent, vectors are positive and some are negative, a unit increase in, say, y_2 increases the value of some of the X's and decreases the value of those with negative signs. As we already have indicated, such vectors might be thought of as shape factors. Rao (1964b) has approached the biometrical analysis of *shape* and *size* with respect to a given set of measurements $X_1, \ldots, X_p$ in terms of "instrumental variables", $Z_1, \ldots, Z_m$, where the latter may include some or all of the former. For example, it may be desirable to find a *size factor* of the head as characterized by Head Length (X_1) and Head Breadth (X_2) using $Z_1 = X_1$, $Z_2 = X_2$, and in addition thereto, Stature (Z_3) and Chest Girth (Z_4).

Let Σ be the dispersion matrix of variables X_1 and X_2, Ω the dispersion matrix between variables X_1, X_2, Z_1, Z_2, Z_3, Z_4, and Γ the dispersion matrix for variables Z_1, Z_2, Z_3, Z_4. As input information, one requires a vector $\mathbf{r} = (r_1, r_2)$, which represents the ratios in which X_1 and X_2 change for a unit increase in the shape or size factor. This introduces a certain element of arbitrariness, but this element may not be unpleasing to the biologist, who is thereby able to exercise some control over the investigation and may be able to supply useful selections with respect to the vector elements, based on particular biological knowledge or experience. If there are no special assumptions on which to base the choice, useful selections for the size factor are: $r_1 = 1$; $r_2 = 1$, or, $r_1 = \sigma_1$; $r_2 = \sigma_2$ where σ_1 and σ_2 are the standard deviations of X_1 and X_2. Similarly for the shape factor $r_1 = 1$ and $r_2 = -1$, or $r_1 = \sigma_1$, $r_2 = -\sigma_2$. The idea behind the extra variables, chosen on the grounds of expert biological knowledge, is that they should give a better estimate of the shape of the head. It may therefore be suggested that the linear function:

$$b_1 Z_1 + b_2 Z_2 + b_3 Z_3 + b_4 Z_4,$$

to be determined, is a better size-shape estimator than just $b_1 Z_1 + b_2 Z_2$. For the estimation procedures we refer to Rao (1964b, p. 345).

A PROGRAMME FOR CALCULATING PRINCIPAL COORDINATES

This programme was written by R.A.R. with the collaboration of Hans-Ake Ramdén, Uppsala University Computing Centre. It is in FORTRAN IV and should function on all computers with a suitable compiler. The programme computes an association matrix in accordance with Gower (1967a) and extracts a specified number of latent roots and vectors from it. An alternative to Gower's association matrix may be used if desired. The first three sets of principal coordinates produced from these latent vectors are plotted by the programme.

Remarks on PCOORD

The measure of association

The general applicability of the D^2-method is, to a degree, limited by the difficulty of satisfactorily using it in conjunction with discrete characters (cf. Kurczynski, 1970) and there are some other problems, such as the *a priori* establishing of the basic groups. Suggestions have been made that a possible approach is by means of an information-theoretic quantity. This requires the estimation of prior probabilities for character states, these determining the weights of the states. Other opinions with respect to the weighting dilemma are represented in the literature. We mention this in order to bring out the fact that several points of view are developing in the non-Adansonian sphere.

Observations on the concept of cluster analysis in numerical taxonomy have been most recently given by Gower (1967b). It is not proposed here to enter into a review of similarity coefficients but it does seem desirable to mention that Gower (1967c) has proposed a useful measure, which leads to a positive definite similarity matrix. For quantitative characters with values $x_1, x_2, \ldots, x_N$ of character k for the total sample of N individuals, Gower's coefficient is defined as

$$S_{ijk} = 1 - |x_i - x_j|/R_k. \qquad (12.11)$$

Here R_k is the range of character k. The matrix S_{ij} with elements s_{ijk} ranges in value between 0 and 1. This coefficient of similarity has been used in the present study and has therefore been mentioned in detail, although there are other coefficients equally worthy of consideration (cf. Sheals, 1964).

The computer program used for the calculations first forms the association matrix A, which is then transformed to matrix α_{ij}.

The latent roots and latent vector of matrix α_{ij} are computed and each latent vector is scaled so that its sum of squares is equal to the corresponding latent root.

The ith row (12.9) represents the coordinates of a set of points T_i whose distances apart are given by the best approximations to $(a_{ii} + a_{jj} - 2a_{ij})^{1/2}$ in the chosen number of dimensions.

The computed coordinates are plotted by a simple plotting subroutine included in the programme. Storage space is always a troublesome matter in a programme of the kind considered here. Our programme will take 120 individuals for any number of variables.

In summary, the method of principal coordinate analysis finds the coordinates of each individual of a sample, referred to principal axes, which preserve the distances, suitably defined, between the individuals.

PROGRAMME DESCRIPTION

Operating instructions

Card 1 Columns 2-3: Number of jobs
Card 2 Columns 1-72 contain the title of the job
Card 3 Columns 2-4: Number of individuals in the sample
        Columns 5-6: Number of variables
        Column 7: Number of latent roots required (maximum of 9)
        Columns 8-9: Number of quantitative variables
        Columns 10-11: Number of qualitative variables
        Columns 12-13: Number of dichotomous variables
        Column 14: If the association matrix available in this programme is desired, punch 0, otherwise a 1 is to be punched.
Card 4 The variable format card. If there are dichotomous variables present, following quantitative and/or qualitative variables, a second format card is to be inserted after card 4.
Card 5 Column 2 Punch 1 for quantitative variables, otherwise 0
        Column 3 Punch 1 for qualitative variables, otherwise 0
        Column 4 Punch 1 for dichotomous variables, otherwise 0.

The subsequent cards are the data cards, punched in the order: quantitatives; qualitatives: dichotomies. The observations must be so arranged that all of them pertaining to a particular variable are grouped together on the same card, or sequence of cards. There will thus be a sequence of cards for each variable. The programme accommodates up to 120 individuals. A means of adding further points to the plots is given in Gower (1968).

List of subroutines

The mainline programme PCOORD reads and prints on the formalized units LIN (read) and LUT (write). It computes "residuals" on the information returned by the subroutine HOUSER. These act as a guide to the efficiency of the coordinate analysis in ordering the individuals of the sample. If there is much near-random variation in the material, the residuals will be high.

Subroutine ASSOC permits the use of an alternative to the

association matrix of Gower (1967a). Subroutine SIMMAT acts as a martialling station for the calculations on the three categories of data. The essential information here is:

(a) quantitative characters are denoted by numbers in the usual manner:

(b) qualitative characters are coded as 0, 1, 2, . . .

(c) dichotomous characters are coded as +1 (present) and −1 (absent).

In this connection, the following method of storing large matrices was developed by Ramdén.

It happens very often that a problem is too big for a particular computer due to the need for big matrices. If these matrices are symmetric (which is not unusual) the following method can be used to reduce storage requirements. This means that a normally stored $n \times n$ matrix occupies n^2 memory locations but if the packed form is used the matrix will occupy only $n(n + 1)/2$ memory locations.

The packed form places the elements in a one-dimensional array AS.

A_{11}	A_{21}	A_{22}	A_{31}	A_{32}	A_{33}	A_{41}		A_{nn}
AS_1	AS_2	AS_3	AS_4	AS_5	AS_6	AS_7		

If we name the packed form AS, each element in AS has a correspondence in A.

For each row in A the method has a "zero index" in the array IX of dimension n to find the corresponding "row" in AS. The zero index for the first row is 0, for the second row is 1, for the third row is 3, for the fourth row is 6 and so on. The necessary N values in the IX array are created by

$$IX(1) = 0$$
$$DO\ 10\ I = 2,N$$
$$10 \quad IX(I) = IX(I-1) + I - 1.$$

Subroutines KVTKAL, KVALT and DICHOT compute the contributions of the quantitatives, qualitatives and dichotomous characters, respectively, which are required for developing Gower's association matrix. These subroutines are not called if this association matrix is not to be used in the calculations. Subroutine RANGE finds the ranges for each of the characters covered by subroutine KVTKAL. Subroutine MATOUT prints out the $N \times N$ matrix STMAT. MATOUT first tests whether STMAT is symmetric and if this is so, it prints only the lower left-hand triangle. All columns and rows are numbered. Subroutine MATUTS prints out a symmetric $N \times N$ matrix stored in packed form. Only the lower left-hand triangle of the matrix is printed. All columns and rows are tagged.

Subroutine PLT21 plots the coordinates $x_1 y_1, x_2 y_2, \ldots, x_n y_n$ on the printer. PLT21 first determines the scale for x-values (to be scaled for 100 printer columns) and then uses the same scale for the y-values. Subroutine HOUSER, for finding latent roots and vectors, has already been listed under programme ORNTDST.

```
      PROGRAM PCOORD
      DIMENSION TITLE(12)
      COMMON AS(7260),BS(7260),CS(960),E(120),F(120),R(120),X(120),
     1 FMT(24),IFMT(12),M,N,KLEV,KADD,MA,MK,MD,KVANT,KVAL,KDIK
      DIMENSION ASBS(14520),A(120,120),B(120,9)
      EQUIVALENCE  (AS,ASBS,A),(ASBS(14401),B)
      COMMON /INDEX/ IX(150)
      COMMON/UNITS/LIN,LUT,KDM,MONE
C----------LIN IS THE STANDARD INPUT UNIT (CARD READER)
      LIN = 60
C----------LUT IS THE STANDARD OUTPUT UNIT (PRINTER)
      LUT = 61
C----------KDM AND MONE ARE SCRATCH TAPE UNITS
      KDM = 59
      MONE=55
C------- THIS VERSION OF PCOORD USES WHEN POSSIBLE A PACKED FORM FOR
C------- SYMMETRIC MATRICES.THE COMMON ARRAY IX(150) IS BUILT UP FOR
C------- CONTROL OF THE INDEX FOR SUCH A MATRIX.
      IX(1)=0
      DO 10 I=1,149
   10 IX(I+1)=IX(I)+I
C----------------------
C---------               INPUT INFORMATION      ----------
C----------CARD 1)  THE NUMBER OF JOBS (1X,I2)
C---------- CARD 2) TITLE OF THE JOB IN COLUMNS 1-72
C---------- CARD 3) N,M,KLEV,MA,MK,MD,KASS IN THE FOLLOWING FORMAT
C----------          (1X,I3,I2,I1,3I2,I1)
C----------N = NUMBER OF OBSERVATION VECTORS (INDIVIDUALS)
C----------M = NUMBER OF VARIABLES
C----------KLEV = NUMBER OF EIGENVALUES NEEDED
C----------MA = IS THE NUMBER OF QUANTITATIVE VARIABLES
C----------MK IS THE NUMBER OF QUALITATIVE VARIABLES
C----------MD IS THE NUMBER OF DICHOTOMOUS VARIABLES
C----------IF  KASS = 1    ANOTHER ASSOCIATION MATRIX IS TO BE USED.
C---------- KASS = 0 ENSURES THE ASSOCIATION MATRIX OF THIS PROGRAMME
C----------WILL BE USED. FOR KASS = 1,MA,MK,MD MUST = 0.
C---------- THE FORMAT CARD THEN READS (1X,I3,I2)
C----------CARDS 4 AND 5 CONTAIN THE FORMAT OR FORMATS CORRESPONDING
C----------TO FMT(24). IF ONLY ONE OF THESE CARDS IS NEEDED, THERE MUST
C----------BE A SECOND BLANK CARD.
C----------IF THERE IS A SECOND KIND OF FORMAT TO BE READ IN USE
C----------FORMAT STATEMENT IFMT(12). THIS IS FOR DICHOTOMIES.
C----------IF NO SECOND FORMAT LEAVE BLANK CARD AT THIS POINT
C----------CARD 6. IS READ IN FROM SUBROUTINE SIMMAT AND DECIDES
C----------WHICH OF THE SUBROUTINES FOR QUANTITATIVES, QUALITATIVES AND
C----------DICHOTOMIES ARE TO BE CALLED INTO USE.
C----------CARD 7 AND SUBSEQUENT CARDS ARE THE DATA FOR QUANTITATIVE
C---------- VARIABLES. IF THERE ARE NO QUANTITATIVES THESE CARDS ARE
C----------LEFT OUT WITHOUT FURTHER PROGRAM INSTRUCTION
C---------- CARD 7  AND SUBSEQUENT CARDS - CONTAIN THE VALUES FOR THE
C----------QUALITATIVE VARIABLES. LEAVE OUT IF THERE ARE NONE
C----------8) THESE CARDS CONTAIN THE VALUES FOR THE DICHOTOMOUS
C----------VARIABLES IF THERE ARE ANY.
C----------                        ********              ----------
C-------- --INFORMATION FOR GOWER#S SIMILARITY MATRIX GIVEN
C----------UNDER SUBROUTINE ,SIMMAT,.

C                        ************************
      WRITE(LUT,300)
      WRITE(LUT,305)
      READ(LIN,7025) JOBS
      DO 7606 KLOBS = 1,JOBS
```

```
       WRITE(LUT,1235)
       READ(LIN,5)(TITLE(I),I=1,12)
       WRITE(LUT,5)(TITLE(I),I=1,12)
       READ(LIN,1)N,M,KLEV,MA,MK,MD,KASS
       WRITE(LUT,2)M,N
       READ(LIN,5)(FMT(I),I=1,12)
       READ(LIN,5)(IFMT(I),I=1,12)
       IF(KASS) 16,16,17
   16  CALL SIMMAT
       GOTO 18
   17  CALL ASSOC
   18  DO 15 I=1,N
   15  E(I)=0.
       SUM = 0.
       DO 25 K=1,N
       DO 20 J=1,N
   20  E(K) = E(K) + A(K,J)
   25  SUM = SUM + E(K)
C      SUM IS THE OVERALL MEAN FOR ALL VALUES
C      VECTOR E IS THE MEAN OF ALL COLUMNS
       DIV = 1./FLOAT(N)
       DO 26 J=1,N
   26  E(J) = E(J)*DIV
       SUM = SUM/FLOAT(N**2)
       WRITE(LUT,7)
       DO 35 I=1,N
       DO 35 J=1,N
   35  A(I,J) = A(I,J) - E(I) - E(J) + SUM
       WRITE(LUT,1927)
       WRITE(LUT,1051)(A(I,I),I=1,N)
       SUM=0.0
       DO 72 I=1,N
   72  SUM = SUM + A(I,I)
       WRITE(LUT,173) SUM
       CALL HOUSER   (N,KLEV,A,E,1,B)
       WRITE(LUT,45)
       WRITE(LUT,50)
   55  WRITE(LUT,60)(E(J),J=1,KLEV)
       DO 80 I=1,KLEV
   80  R(I) = SQRT(E(I))
   78  WRITE(LUT,1958)
       WRITE(LUT,1051)(R(I),I=1,KLEV)
       DO 81 J=1,KLEV
       DO 81 I=1,N
   81  B(I,J) = B(I,J)*R(J)
 7000  CONTINUE
       WRITE(LUT,91)
       DO 94 I=1,N
   94  WRITE(LUT,1051)(B(I,J),J=1,KLEV)
       WRITE(LUT,1234)
       CALL PLT21 (B,B(1,2),N)
       WRITE(LUT,227)
       CALL PLT21 (B,B(1,3),N)
       WRITE(LUT,225)
       CALL PLT21 (B(1,2),B(1,3),N)
       WRITE(LUT,229)
```

```
      WRITE(LUT,1235)
      WRITE(LUT,7)
      WRITE(LUT,110)
      TRACE = SUM
      DO 90 I=1,2
   90 TRACE = TRACE - E(I)
      GRACE=TRACE-E(3)
      TRACE = (TRACE*100.)/SUM
      GRACE = (GRACE*100.)/SUM
      WRITE(LUT,235)
      WRITE(LUT,2235) TRACE
      WRITE(LUT,2236) GRACE
 7606 CONTINUE
C-------- *************** FORMATS ***********************-----------
    1 FORMAT(1X,I3,I2,I1,3I2,I1)
    2 FORMAT(15H0   VARIABLES = I2,16H   INDIVIDUALS = I5)
    5 FORMAT(12A6)
    7 FORMAT(1H0)
   45 FORMAT(34H1     PRINCIPAL COORDINATE ANALYSIS)
   50 FORMAT(50H0     EIGENVALUES OF TRANSFORMED ASSOCIATION MATRIX     )
   60 FORMAT(F14.7)
   91 FORMAT(14H1  COORDINATES)
  110 FORMAT(30H0            VALUES OF RESIDUALS         )
  173 FORMAT(11H0   TRACE = E14.7)
  225 FORMAT(49H0     ABOVE IS PLOT OF FIRST AND THIRD COORDINATES     )
  227 FORMAT(50H0     ABOVE IS PLOT OF FIRST AND SECOND COORDINATES     )
  229 FORMAT(50H0     ABOVE IS PLOT OF SECOND AND THIRD COORDINATES     )
  235 FORMAT(21H0      ROOTS EXCEEDING,32H            PERCENTAGE RESIDUAL
     1   )
  300 FORMAT(40H0         PRINCIPAL COORDINATE ANALYSIS     )
  305 FORMAT(32H0          NUMERICAL TAXONOMY     )
 1051 FORMAT(11F12.5)
 1234 FORMAT(23H1  PLOTS OF COORDINATES         )
 1235 FORMAT(1H1)
 1927 FORMAT(40H0  DIAGONAL ELEMENTS OF TRANSF. MATR.         )
 1958 FORMAT(9H0  ROOT R)
 2235 FORMAT(8H0       2,18X,F12.2)
 2236 FORMAT(8H0       3,18X,F12.2)
 7025 FORMAT(1X,I2)
      END


      SUBROUTINE ASSOC
      COMMON AS(7260),BS(7260),CS(960),E(120),F(120),R(120),X(120),
     1 FMT(24),IFMT(12),M,N,KLEV,KADD,MA,MK,MD,KVANT,KVAL,KDIK
      DIMENSION ASBS(14520),A(120,120),B(120,9)
      EQUIVALENCE (AS,ASBS,A),(ASBS(14401),B)
      COMMON /INDEX/ IX(150)
      COMMON/UNITS/LIN,LUT,KDM,MONE
C---------THIS SUBROUTINE ALLOWS THE INCLUSION OF AN ALTERNATIVE
C---------KIND OF ASSOCIATION MATRIX TO THAT USED IN THIS PROGRAMME.
      DO 1 I=1,N
    1 READ(LIN,FMT)(A(I,J),J=1,N)
      WRITE(LUT,2)
      CALL MATOUT(A,N)
      RETURN
    2 FORMAT(25H0      ASSOCIATION MATRIX A     )
      END
```

```
      SUBROUTINE SIMMAT
      COMMON AS(7260),BS(7260),CS(960),E(120),F(120),R(120),X(120),
     1 FMT(24),IFMT(12),M,N,KLEV,KADD,MA,MK,MD,KVANT,KVAL,KDIK
      DIMENSION ASBS(14520),A(120,120),B(120,9)
      EQUIVALENCE  (AS,ASBS,A),(ASBS(14401),B)
      COMMON /INDEX/ IX(150)
      COMMON/UNITS/LIN,LUT,KDM,MONE
C----------THIS SUBROUTINE ARRANGES THE TREATMENT OF THE THREE
C----------KINDS OF VARIABLES
C----------DICHOTOMOUS VARIABLES.. IF PRESENT CODED AS ,+1,
C----------                        IF ABSENT CODED AS ,-1,.
C----------QUALITATIVE CHARACTERS ARE CODED AS 0,1,2,,.,...
C----------QUANTITATIVE CHARACTERS--AREDENOTED BY NUMBERS
C----------       ANY OTHER SIMILARITY MATRIX METHOD MAY BE USED
C----------       INSTEAD OF SIMMAT AND ITS SATELLITE SUBROUTINES
C----------         CODE FOR THE DATA TYPES
C----------KVANT = 1 MEANS THERE ARE QUANTITATIVE DATA
C----------KVAL = 1 MEANS THERE ARE QUALITATIVE DATA
C----------KDIK = 1 MEANS THERE ARE DICHOTOMOUS DATA
C----------A PRELIMINARY GROUPING OF THE DATA MUST BE MADE SO THAT
C----------THINGS OF THE SAME KIND ARE LOCATED TOGETHER
      KX=IX(N)+N
      DO 35 I=1,KX
   35 BS(I)=0.
      READ(LIN,1)KVANT,KVAL,KDIK
      IF (KVANT .GT. 0) CALL KVTKAL
      IF (KVAL .GT. 0) CALL KVALT
      IF (KDIK .LE. 0) GO TO 20
      REWIND KDM
      WRITE (KDM) (BS(I),I=1,KX)
      CALL DICHOT
      REWIND KDM
      READ  (KDM) (BS(I),I=1,KX)
      DO 15 I=1,KX
   15 BS(I)=AS(I)+BS(I)
C------- UNPACK THE MATRIX BS AND STORE IT IN ORDINARY FORM IN A
   20 DO 40 I=1,N
      DO 40 J=I,N
      K=IX(J)+I
   40 A(J,I)=BS(K)
      DO 41 I=1,N
      DO 41 J=I,N
   41 A(I,J)=A(J,I)
      WRITE(LUT,25)
      WRITE(LUT,26)
      CALL MATOUT(A,N)
      RETURN
    1 FORMAT(1X,3I1)
   25 FORMAT(10H1     +++++   )
   26 FORMAT(24H     ASSOCIATION MATRIX A          )
      END
```

```
      SUBROUTINE KVTKAL
      COMMON AS(7260),BS(7260),CS(960),E(120),F(120),R(120),X(120),
     1 FMT(24),IFMT(12),M,N,KLEV,KADD,MA,MK,MD,KVANT,KVAL,KDIK
      DIMENSION ASBS(14520),A(120,120),B(120,9)
      EQUIVALENCE (AS,ASBS,A),(ASBS(14401),B)
      COMMON /INDEX/ IX(150)
      COMMON/UNITS/LIN,LUT,KDM,MONE
C---------THIS SUBROUTINE COMPUTES THE SIMILARITY MATRIX ELEMENTS
C---------FOR QUANTITATIVE VARIABLES
C---------READING IN THE RANGES
C---------IF THERE ARE MORE VARIABLES THAN OBSERVATIONS AN EXTRA
C---------FORMAT CARD WILL BE NECESSARY - USE IFMT-SPECIFICATION
      CALL RANGE
      REWIND MONE
      WRITE(LUT,4)
      WRITE(LUT,1051)(F(I),I=1,MA)
      WRITE(LUT,224)
      KX=IX(N)+N
      DO 5 I=1,KX
    5 AS(I)=0.
      DO 25 KOUNT=1,MA
   50 READ(MONE)(X(I),I=1,N)
      WRITE(LUT,300)(X(J),J=1,N)
      DO 80 I=1,N
   80 X(I) = X(I)/F(KOUNT)
      DO 10 I=1,N
      DO 10 J=I,N
      K=IX(J)+I
   10 AS(K)=AS(K)+ABS(X(I)-X(J))
   25 CONTINUE
   40 FMA=MA
      DO 15 I=1,KX
   15 AS(I)=1.-AS(I)/FMA
      WRITE(LUT,70)
      CALL MATUTS(AS,N)
      DO 75 I=1,KX
   75 BS(I)=BS(I)+AS(I)
C--------- THIS PART OF ASSOCIATION MATRIX STORED IN B(I,J)
      WRITE(LUT,226)
      RETURN
C-------------------        FORMATS        ----------------
    4 FORMAT(30H0    RANGES FOR VARIABLES              )
   70 FORMAT(45H1  ASSOCIATION MATRIX FOR QUANTITATIVES            )
  224 FORMAT(50H0        DATA LISTING FOR QUANTITATIVES
  226 FORMAT(1H1)
  300 FORMAT(1X,16F8.3)
  505 FORMAT(12A6)
 1051 FORMAT(11F12.5)
      END
```

```
      SUBROUTINE RANGE
C------- COMPUTES THE RANGE FOR QUANTITATIVE VARIABLES
      COMMON AS(7260),BS(7260),CS(960),E(120),F(120),R(120),X(120),
     1 FMT(24),IFMT(12),M,N,KLEV,KADD,MA,MK,MD,KVANT,KVAL,KDIK
      DIMENSION ASBS(14520),A(120,120),B(120,9)
      EQUIVALENCE  (AS,ASBS,A),(ASBS(14401),B)
      COMMON /INDEX/ IX(150)
      COMMON/UNITS/LIN,LUT,KDM,MONE
      DIMENSION Q(120)
      EQUIVALENCE (Q,R)
      REWIND MONE
      DO 1 I=1,MA
      Q(I) = 9999999.
    1 F(I)=0.
      DO 2 KOUNT=1,MA
   10 READ(LIN,FMT)(X(I),I=1,N)
      WRITE(MONE)(X(I),I=1,N)
      DO 2 I=1,N
      IF(X(I).GT.F(KOUNT))F(KOUNT)=X(I)
      IF(X(I).LT.Q(KOUNT)) Q(KOUNT) = X(I)
    2 CONTINUE
      DO 15 I=1,MA
   15 F(I)=F(I)-Q(I)
      RETURN
      END



      SUBROUTINE KVALT
      COMMON AS(7260),BS(7260),CS(960),E(120),F(120),R(120),X(120),
     1 FMT(24),IFMT(12),M,N,KLEV,KADD,MA,MK,MD,KVANT,KVAL,KDIK
      DIMENSION ASBS(14520),A(120,120),B(120,9)
      EQUIVALENCE  (AS,ASBS,A),(ASBS(14401),B)
      COMMON /INDEX/ IX(150)
      COMMON/UNITS/LIN,LUT,KDM,MONE
C----------THIS SUBROUTINE PRODUCES ELEMENTS FOR QUALITATIVE VARIABLES
      WRITE(LUT,224)
      KX=IX(N)+N
      DO 1 I=1,KX
    1 AS(I)=0.
      DO 40 KFULL=1,MK
   50 READ(LIN,FMT)(X(I),I=1,N)
      WRITE(LUT,FMT)(X(I),I=1,N)
      DO 10 I=1,N
      DO 10 J=I,N
      K=IX(J)+I
   10 AS(K)=AS(K)+1.-ABS(X(I)-X(J))
C-----------PRODUCING THE ELEMENTS OF THE SCORES MATRIX
   40 CONTINUE
      WRITE(LUT,70)
      FMK=MK
      DO 15 I=1,KX
   15 AS(I)=AS(I)/FMK
      CALL MATUTS (AS,N)
      DO 55 I=1,KX
   55 BS(I)=BS(I)+AS(I)
C---------ASSOCIATION MATRIX VALUES ADDED TO B(I,J)
      RETURN
   70 FORMAT(45H1  ASSOCIATION MATRIX FOR QUALITATIVES          )
  224 FORMAT(50H1             DATA LISTING FOR QUALITATIVES          )
      END
```

```
      SUBROUTINE DICHOT
      COMMON AS(7260),BS(7260),CS(960),E(120),F(120),R(120),X(120),
     1 FMT(24),IFMT(12),M,N,KLEV,KADD,MA,MK,MD,KVANT,KVAL,KDIK
      DIMENSION ASBS(14520),A(120,120),B(120,9)
      EQUIVALENCE  (AS,ASBS,A),(ASBS(14401),B)
      COMMON /INDEX/ IX(150)
      COMMON/UNITS/LIN,LUT,KDM,MONE
C--------THIS SUBROUTINE PRODUCES ENTRIES IN THE SIMILARITY MATRIX
C--------FOR DICHOTOMOUS VARIABLES
C--------  +-VALUES ARE CODED AS +1
C--------  -VALUES ARE CODED AS -1
C--------
      KX=IX(N)+N
      DO 80 I=1,KX
      AS(I)=0.
   80 BS(I)=0.
      WRITE(LUT,224)
      DO 40 KFULL=1,MD
   50 READ(LIN,IFMT)(X(I),I=1,N)
      WRITE(LUT,FMT)(X(I),I=1,N)
      DO 10 I=1,N
      DO 10 J=I,N
      K=IX(J)+I
      IF (X(I)+X(J)) 8,8,7
    7 AS(K)=AS(K)+1.0
    8 IF (X(I)+X(J)) 10,9,9
    9 BS(K)=BS(K)+1.0
   10 CONTINUE
C------- NOW MATRIX OF WEIGHTS IS IN AS AND MATRIX OF SCORES IN BS
   40 CONTINUE
      DO 70 I=1,KX
      IF (BS(I)) 750,750,760
  760 AS(I)=AS(I)/BS(I)
      GO TO 70
  750 AS(I)=0.
   70 CONTINUE
C--------THIS PART OF ASSOCIATION MATRIX STORED ON A(I,J) AND RETURNED
      WRITE(LUT,700)
      CALL MATUTS (AS,N)
      RETURN
  224 FORMAT(50H1            DATA LISTING FOR DICHOTOMIES              )
  700 FORMAT(45H1  ASSOCIATION MATRIX FOR DICHOTOMIES          )
      END


      SUBROUTINE MATOUT (STMAT,N)
      DIMENSION STMAT(120,120)
      COMMON/UNITS/LIN,LUT,KDM,MONE
C------- LOOK IF STMAT IS SYMMETRIC. IF SO THEN PRINT ONLY THE LOWER
C------- LEFT HALF OF THE MATRIX.
      ISYMM=0
      DO 10 I=1,N
      DO 10 J=I,N
   10 IF (STMAT(I,J) .NE. STMAT (J,I)) GO TO 11
      ISYMM=1
   11 MEND=0
    1 MBEG=MEND+1
      MEND=MBEG+9
C------- PRINT MAXIMUM 10 COLUMNS EACH TIME
      IF(MEND.GT.N) MEND = N
      WRITE(LUT,101)(J,J=MBEG,MEND)
      K=1
```

```
      IF (ISYMM .EQ. 1) K=MBEG
      DO 2 I=K,N
      L=MEND
      IF (I .LT. L .AND. ISYMM .EQ. 1) L=I
    2 WRITE (LUT,102) I,(STMAT(I,J),J=MBEG,L)
C------- IF ALL N COLUMNS HAVE BEEN PRINTED THEN RETURN
      IF (MEND .EQ. N) RETURN
C--------OR ELSE PRINT NEXT 10 COLUMNS
      GO TO 1
  101 FORMAT (1H0,10I12)
  102 FORMAT (1X,I3,2X,10F12.5)
      END



      SUBROUTINE MATUTS (STMAT,N)
C------- MATUTS PRINTS OUT A SYMMETRIC MATRIX STORED IN A PACKED FORM.
C------- ONLY THE LOWER LEFT PART OF THE MATRIX WILL BE PRINTED
      DIMENSION STMAT(1)
      COMMON/UNITS/LIN,LUT,KDM,MONE
      COMMON /INDEX/ IX(150)
      MEND=0
    1 MBEG=MEND+1
      MEND=MBEG+9
C------- PRINT MAX 10 COLUMNS EACH TIME
      IF (MEND .GT. N) MEND=N
C------- PRINT OUT COLUMN NUMBERS
      WRITE (LUT,101) (J,J=MBEG,MEND)
      DO 2 I=MBEG,N
      J=IX(I)+MBEG
      M=MEND
      IF (I .LT. M) M=I
      K=IX(I)+M
    2 WRITE (LUT,102) I,(STMAT(L),L=J,K)
C------- IF ALL N COLUMNS HAVE BEEN PRINTED THEN RETURN
      IF (MEND .EQ. N) RETURN
      GO TO 1
  101 FORMAT (1H0,10I12)
  102 FORMAT (X,I3,2X,10F12.5)
      END




      SUBROUTINE HOUSER  (N,NEV,C,EV,VEC,V)
```

This subroutine for finding latent roots and vectors has already been listed in connection with the programme entitled ORNTDIST.

```
      SUBROUTINE PLT21 (X1,Y1,N1)
C-----------------------------------------------------------------------
      DIMENSION X1(10),Y1(10),ALINE(101)
      DATA(BLANK=1H ),(DASH=1H-),(UP=1HI),(DOT=1H.),(X=1HX),(AST=1H*),
     1(PLUS=1H+)
      LIN = 60
      LUT = 61
      XMAX = X1(1)
      XMIN = XMAX
      YMAX = Y1(1)
      YMIN = YMAX
      DO 1 I=2,N1
      IF(X1(I).GT.XMAX) XMAX = X1(I)
      IF(X1(I).LT.XMIN) XMIN = X1(I)
      IF(Y1(I).GT.YMAX) YMAX = Y1(I)
      IF(Y1(I).LT.YMIN) YMIN = Y1(I)
    1 CONTINUE
      KEY = 1
      ZMAX = XMAX
      ZMIN = XMIN
    4 RANGE = ZMAX - ZMIN
      SCALE = 1.E-9
    5 SCALE = 10.*SCALE
      IF(SCALE.LT.RANGE) GO TO 5
   50 ZMIN = ZMIN/SCALE
      ZMAX = ZMAX/SCALE
      MIN = 20.0*(ZMIN - 0.025)
      MAX = 20.0*(ZMAX + 0.025)
      BOTTOM = 0.05*FLOAT(MIN)
      TOP = 0.05*FLOAT(MAX)
      RANGE = 0.1*(TOP-BOTTOM)
    6 IF(BOTTOM.LE.ZMIN) GO TO 7
      BOTTOM = BOTTOM - RANGE
      GO TO 6
    7 IF(TOP.GE.ZMAX) GO TO 8
      TOP = TOP + RANGE
      GO TO 7
    8 YINC = 0.01*(TOP -BOTTOM)
      IF(KEY.EQ.2) GO TO 11
      KEY = 2
      ZMAX = YMAX
      ZMIN = YMIN
      XSCALE = SCALE
      XTOP = TOP
      XINC = YINC*SCALE
      XBOT = BOTTOM*SCALE
      GO TO 50
C
   11 YINC=XINC*10./(6.*SCALE)
      YLOW = TOP + YINC
      KEY = 5
      WRITE(LUT,1000) XSCALE,SCALE
   15 KKEY = 10
      IF(KEY.NE.5) GO TO 19
      DO 18 I=1,101
      IF(KKEY.NE.10) GO TO 16
```

```
      ALINE(I) = DOT
      GO TO 17
   16 ALINE(I) = DASH
   17 KKEY = KKEY - 1
      IF(KKEY.EQ.0) KKEY = 10
   18 CONTINUE
      GO TO 23
   19 DO 22 I=1,101
      IF(KKEY.NE.10) GO TO 20
      ALINE(I) = UP
      GO TO 21
   20 ALINE(I) = BLANK
   21 KKEY = KKEY - 1
      IF(KKEY.EQ.0) KKEY = 10
   22 CONTINUE
   23 CONTINUE
C     PUT POINTS ON ALINE
      YHIGH = YLOW
      YLOW = YHIGH -YINC
      YHS = SCALE*YHIGH
      YLS = SCALE*YLOW
      DO 24 I=1,N1
      IF(Y1(I).GT.YHS.OR.Y1(I).LE.YLS) GO TO 24
      INDEX = (X1(I) - XBOT)/XINC
      INDEX = INDEX + 1
      IF(INDEX.GT.101) INDEX = 101
      ALINE(INDEX) = AST
   24 CONTINUE
      IF(KEY.NE.5) GO TO 28
      WRITE(LUT,1001) TOP,ALINE
      GO TO 29
   28 WRITE(LUT,1002) ALINE
   29 CONTINUE
C
      KEY = KEY - 1
      IF(KEY.EQ.0) KEY = 5
      TOP = TOP - YINC
      IF(TOP.GE.BOTTOM) GO TO 15
      IF(KEY.NE.4) GO TO 15
      XINC = 10.*XINC/XSCALE
      ALINE(1) = XBOT/XSCALE
      DO 30 I=2,11
   30 ALINE(I) = ALINE(I-1) +XINC
      WRITE(LUT,1003)(ALINE(I),I=1,11)
      RETURN
C
C-------------                    FORMATS          ----------------
 1000 FORMAT(22H1SCALE FACTOR ON X IS ,1PE9.2,4X,21HSCALE FACTOR ON Y IS
     1 ,1PE9.2/1H )
 1001 FORMAT(F10.3,1X,101A1)
 1002 FORMAT(11X,101A1)
 1003 FORMAT(5X,11F10.3)
      END
```

A PROGRAMME FOR CALCULATING PRINCIPAL COMPONENTS

The type of sample considered in this connection is multivariate normal. A question frequently posed by the biologist concerns how many and which variables should be selected for a particular study. This is a question that cannot be answered satisfactorily by a statistician alone, involving as it does a combination of biological and statistical principles. From a statistical point of view, it is not particularly useful to base a multivariate analysis on a set of variables which are highly correlated with each other. This point becomes clearer if we consider a morphometric study involving, say, p variables. If one now adds a $(p+1)$-th variable, which is highly correlated with the original variables, the amount of additional information introduced is slight. In spite of this, however, there may be some overriding biological reason demanding the inclusion of this variable, which would not be apparent to a statistician. Because of their relative availability, morphometric variables have been most used in both neontology and palaeontology; nevertheless, publications are becoming more frequent in which physiological biochemical and ethological variables are employed. By the very nature of fossils, palaeontologists have to use more ingenuity in finding variables other than morphological ones. This is, however, not an impossible task and such information may be extracted from the shells of many fossils. For example, electron microscope and biochemical studies of organic material preserved in fossil shell substance may provide a source of useful data for statistical analysis, and it is also possible, for some groups of fossils, to make use of palaeoethological variables.

Principal components are valuable for the statistical analysis of the structure of multivariate observations. The observable variates (the measurements) are represented as functions of a smaller number of latent (= hidden) variates (= the principal components).

The method of calculation is, on a computer, quite a straightforward matter, otherwise it is a matter of quite stupendous difficulty. For most biological applications one extracts the latent roots and vectors of the dispersion matrix of the variables. The vectors, expressed in equation form, are

$$y_1 = a_{11}^1 x_1 + a_{12}^1 x_2 + a_{13}^1 x_3 + \ldots a_{ip}^1 x_p \qquad (12.12)$$

where the x_i $(i = 1, \ldots, p)$ are the original variables and y_1 is the first transformed variable, the principal components of the matrix. In general terms,

$$y_j = a_{1j} x_1 + \ldots a_{pj} x_p \qquad (12.13)$$

The correlation matrix is the more logical starting point if the data are not homogeneous (the homogeneous set of variates of morphological analysis) but rather a disparate mixture of size, weight and chemical analyses, for example. The corresponding latent root, λ_1, is the variance of the transformed variable y_1 and λ_j is that for y_j.

Details of PNCOMP

The programme was written by R.A.R. in collaboration with Hans-Åke Ramdén of the Computation Centre of Uppsala University.

The coefficients a_1 of the first linear combination of the principal component analysis must satisfy the p simultaneous linear equations

$$(S - d_1 I)a_1 = 0, \tag{12.14}$$

where S is the sample dispersion matrix, I is the identity matrix and d_1 is obtained by solution of the determinantal equation

$$|S - d_1 I| = 0. \tag{12.15}$$

The vector a_1 is the first latent vector of S and d_1 its first latent root. In general terms, the jth principal component of S is the linear expression

$$y_j = a_{1j}x_1 + \ldots + a_{pj}x_p,$$

where the coefficients are the elements of the latent vector corresponding to the jth latent root.

The sign and magnitude of a vector element indicate the direction and importance of the contribution of a variable to a particular component. More precisely, the ratio $a_{ij}\sqrt{d_j/s_i}$ yields the correlation of the ith variable and the jth component; here, s_i is the pertinent standard deviation.

Amongst other things, this programme also computes a test for isotropicity of the last $(p-k)$ latent roots (subroutine ISOTRP).

Another useful feature of this programme is that it computes confidence intervals for the latent roots of the dispersion matrix (subroutine CONINT).

DESCRIPTION OF THE PROGRAMME

Operating instructions

Card 1 The title card for the job with the title punched in columns 1-72.

Card 2 Columns 2-3: Number of characters:
Columns 4-6: Number of observations:
Column 7: 0 for the principal components of a correlation matrix, 1 for those of a dispersion matrix:
Column 8: 1 if checking of the latent roots is required, otherwise 0:
Column 9: 1 if the original observations are to be transformed to Briggsian (base 10) logarithms, otherwise 0:
Column 10: 1 if subroutine SCORES needed, otherwise 0:
Column 11: 1 if subroutine CORTST is needed, if not punch 0:
Column 12: 1 is punched for data in matrix and vector form; for raw data, punch 0:
Column 13: 1 if subroutine VECCHI is required, otherwise 0. Not available if column 7 contains a 0.

Card 3 The variable format card.

Card 4 Either the raw data (observational vectors) or matrices and vectors. In the latter case, the dispersion matrix is to be placed first, thereafter the vector of means.

Final Card There will be a final card of vectorial values if column 13 contains a one.

The mainline programme PNCOMP reads and writes on the formalized input and output units LIN and LUT, respectively. It will accept any number of problems and a maximum of $p = 55$ characters. In addition to computing some basic statistics, the mainline programme carries out a check on the latent roots should this be desired.

Subroutine MNTR performs the substitution of the original means into the equations formed from the latent vectors. MATOUT prints and labels matrices. Subroutine CHINCH carries out the transformation of the original observational vectors into the vectorial equations and displays the plots of the first three vectors through the medium of PLT21. ISOTRP computes a chi-square for p-q degrees of freedom for the last q latent roots of a large-sample dispersion matrix (here taken as exceeding 75 observations). CONINT is another large-sample procedure (only brought into the flow of calculations for more than 90 observations) for computing confidence intervals for the latent roots of a dispersion matrix. The subroutine SCORES finds the first four transformed variables (not available when the input information is in matrix form). These scores are particularly useful if the first four components account for most of the total sample variance, as they may be used in place of the original observations. CORTST computes correlation coefficients between the principal components and the original observations (not available when the input information is in matrix form); this offers a useful complement to the inspection of the matrix of latent vectors of the dispersion matrix. Subroutine PRINAX carries out a test of equality of the last p-1 roots of a correlation matrix. PLT21 and HOUSER are the same subroutines as used in programme PCOORD. Subroutine VECCHI tests a given principal component (latent vector) to see whether it is statistically identical with the corresponding latent vector of some dispersion matrix (Anderson, 1963). If this subroutine is required, the comparison vector is to be read in accordance with the variable format specification. The subroutine only functions for positive definite dispersion matrices.

```
      PROGRAM PNCOMP
C-----------                                          ----------
C-----PRINCIPAL COMPONENT ANALYSIS FOR CORRELATION OR COVARIANCE MATRICES
C
C-----------           ----INPUT SPECIFICATIONS----        ----------
C----------CARD 1)-- TITLE CARD FOR JOB
C----------CARD 2)--NUMBER OF VARIABLES = M IN 2 AND 3
C-----------           NUMBER OF OBSERVATIONS,N, IN 4 TO 6
C-----------           IN 7 KLIP = 0 FOR CORRELATIONS AND = 1 FOR COVARIANC
C-----------           IN 8 ICHECK = 1 IF CHECKING OF EIGENVALUES NEEDED
C-----------              OTHERWISE PUT IT = 0
C-----------           IN 9 KLOG PUT = 1 OF BASE 10 LOGARITHMS OF DATA
C-----------              OTHERWISE PUT IT = 0
C-----------           IN 10 PUT 1 IF SCORES (ISCORES) OF PCA NEEDED
C-----------              OTHERWISE A NOUGHT
C-----------           IN 11 ICORR PUT = 1 FOR CORRELATION OF PCA SCORES
C-----------              WITH DIMENSIONS NEEDED OTHERWISE 0
C-----------           IN12 MATT = 1 FOR MATRIX AND MEAN VECTOR INPUT
C-----------              OTHERWISE PUT IT = 0
C-----------           IN 13 IVEC = 1 FOR VECTOR DIRECTION TEST
C-----------              OTHERWISE = 0
C-----------        N,B, NOT AVAILABLE FOR KLIP=0
C----------- IF VECCHI USED, VECTOR FOR COMPARISON TO BE READ IN
C----------- ACCORDING TO VARIABLE FORMAT CARD
C----------------                                  ----------------
C----------CARD 3)--THE VARIABLE FORMAT CARD
C-------- CARD 4)--EITHER DATA CARDS (OBSERVATIONS) OR MATRIX INFORMATIO
C----------IN LATTER CASE READ COVARIANCE MATRIX FIRST THEN MEANS,
C
C----------           MAIN OUTPUT DETAILS           -----------------
C
C----------  CORRELATION AND COVARIANCE MATRICES , MEAN VECTOR AND
C----------STANDARD DEVIATIONS FOR EACH OF THE VARIABLES
C----------EIGENVALUES AND EIGENVECTORS FOR ALL
C----------EIGENVALUES AND EIGENVECTORS - THE PRINCIPAL COMPONENTS
C----------PLOTS OF TRANSFORMED VARIABLES FOR THE FIRST THREE
C----------TRANSFORMED VARIABLES
C-----------
C----------SUBROUTINE SCORES IS ONLY USED WITH THE COVARIANCE MATRIX
C----------- OPTION; CORRTEST IS USED WITH BOTH,
C----------                   ----------------
      COMMON    SX(55),SS(55,55),SSD(55,55),D(55,55),R(55,55),TITLE(12),
     1X(55),SD(55),FMT(12),XM(55),EGVAL(55)
      COMMON/UNITS/LIN,LUT,MTA
      LIN=60
      LUT=61
      MTA=56
  500 READ(LIN,10)(TITLE(I),I=1,12)
      IF(EOF,LIN) 3030,3031
 3030 CALL EXIT
 3031 CONTINUE
      READ(LIN,5)M,N,KLIP, ICHECK, KLOG, ISCORE ,ICORR,MATT,IVEC
      EN=N
      WRITE(LUT,55)
      REWIND MTA
      READ(LIN,10)(FMT(I),I=1,12)
```

```
      WRITE(LUT,10)(TITLE(I),I=1,12)
      WRITE (LUT,60) M,N
      IF (MATT.EQ.1) GO TO 611
      IF(KLOG) 310,310,305
  305 WRITE(LUT,300)
      GO TO 306
  310 WRITE(LUT,311)
  306 CONTINUE
      WRITE(LUT,345)
      DO 15 I=1,M
      SX(I)=0.
      SD(I)=0.
      DO 15 J=1,M
      SSD(I,J)=0.
      D(I,J)=0.
      R(I,J)=0.
   15 SS(I,J)=0.
      DO 330 NN=1,N
   20 READ(LIN,FMT)(X(I),I=1,M)
      WRITE(LUT,55555)(X(I),I=1,M)
      IF(KLOG.NE.1) GO TO 320
      DO 315 I=1,M
  315 X(I) = ALOG10(X(I))
      WRITE(LUT,55000)(X(I),I=1,M)
  320 WRITE(MTA)(X(I),I=1,M)
      DO 325 I=1,M
      SX(I) = SX(I) + X(I)
      DO 325 J=1,M
  325 SS(I,J) = SS(I,J) + X(I)*X(J)
  330 CONTINUE
   40 DO 45 I=1,M
      DO 45 J=1,M
      SSD(I,J)=SS(I,J)-SX(I)*SX(J)/EN
      SS(J,I) = SS(I,J)
   45 SSD(J,I)=SSD(I,J)
      WRITE(LUT,41)
      CALL MATOUT (SSD,M)
      DO 50 I=1,M
      XM(I)=SX(I)/EN
      IF(SSD(I,I).LE.0.) CALL EXIT
   50 SD(I)=SQRT(SSD(I,I)/(EN-1.))
      WRITE(LUT,55)
      WRITE(LUT,70)
      WRITE(LUT,75)(XM(I),I=1,M)
      DO 80 I=1,M
      DO 80 J=1,M
   80 SS(I,J)=0.
C----------CONSTRUCTING THE ELEMENTS OF THE COVARIANCE MATRIX
      DO 84 I=1,M
      DO 84 J=1,M
      D(I,J)=SSD(I,J)/(EN-1.)
   84 D(J,I)=D(I,J)
      GO TO 615
  611 DO 614 I=1,M
  614 READ(LIN,FMT)(D(I,J),J=1,M)
      READ(LIN,FMT)(XM(I),I=1,M)
```

```
      WRITE(LUT,70)
      WRITE(LUT,75)(XM(I),I=1,M)
      DO 6144 I=1,M
 6144 SD(I) = SQRT(D(I,I))
  615 CONTINUE
      DO 85 I=1,M
      DO 85 J=1,M
   85 R(I,J)=D(I,J)/(SD(I)*SD(J))
C---------CONSTRUCTING THE ELEMENTS OF THE CORRELATION MATRIX
      DO 86 I=1,M
      DO 86 J=1,M
   86 R(J,I) = R(I,J)
C---------DUPLICATION OF CORRELATION MATRIX
      DO 87 I=1,M
      DO 87 J=1,M
   87 SS(I,J)=R(I,J)
      WRITE(LUT,90)
      WRITE(LUT,75)(SD(I),I=1,M)
      WRITE(LUT,95)
      CALL MATOUT(D,M)
      STRACE=0.
      DO 101 I=1,M
C---------  THE TRACE OF THE COVARIANCE MATRIX
  101 STRACE=STRACE+D(I,I)
      WRITE(LUT,102) STRACE
      WRITE(LUT,105)
      CALL MATOUT (R,M)
      MMI=M
C
      IF(KLIP)1240,1240,1260
 1240 CALL HOUSER (MMI,MMI,R,EGVAL,1,,SSD)
      WRITE(LUT,121)
      WRITE(LUT,120)
      WRITE(LUT,75)(EGVAL(I),I=1,M)
      WRITE(LUT,125)
      CALL MATOUT (SSD,M)
      IF(MATT.EQ.1) GO TO 626
      WRITE(LUT,135)
      CALL CHINCH(SSD,M,N,X)
  626 CONTINUE
      IF(N.GE.90) CALL CONINT(N,EGVAL,M)
      IF(ICORR.EQ.1) CALL CCRTST (EGVAL,SSD,M,SD,KLIP)
      CALL PRINAX    (SS,M,N)
      IF(MATT.EQ.1) CALL MNTR(SSD,XM,SD,M)
      IF(ICHECK)160,160,165
  160 GO TO12266
  165 TRACE=0.
  151 WRITE(LUT,153)
      DO 166 I=1,M
  166 TRACE = TRACE + EGVAL(I)
      WRITE(LUT,399)
      WRITE(LUT,170) TRACE
      MCHECK=M
      IF(IFIX(TRACE+0.1)-MCHECK)175,176,177
  175 WRITE(LUT,180)
      GO TO 500
```

```
    176 WRITE(LUT,185)
        GO TO 195
    177 WRITE(LUT,190)
        GO TO 500
    195 WRITE(LUT,200)
        PROD=EGVAL(1)
        DO 210 I=2,M
    210 PROD=PROD*EGVAL(I)
        WRITE(LUT,215) PROD
        GO TO 500
C
   1260 DO 1250 I=1,M
        DO 1250 J=1,M
   1250 SS(I,J)=D(I,J)
        CALL HOUSER (M,M,SS,EGVAL,1.,SSD)
        WRITE(LUT,221)
        WRITE(LUT,120)
        WRITE(LUT,30)(EGVAL(I),I=1,M)
        WRITE(LUT,125)
        CALL MATOUT(SSD,M)
        PROD = EGVAL(1)
        DO 240 I=2,M
    240 PROD = PROD*EGVAL(I)
        IF(MATT.EQ.1) GO TO 640
        WRITE(LUT,135)
        CALL CHINCH (SSD,M, N,X)
    640 CONTINUE
        TRACE = 0.
        DO 251 I=1,M
    251 TRACE = TRACE + EGVAL(I)
        DO 993 I=1,M
    993 SX(I) = (EGVAL(I)*100.)/TRACE
   2222 CONTINUE
        IF (N ,LT, 75) GO TO 994
        CALL CONINT (N,EGVAL,M)
        IF (M ,GE, 2) CALL ISCTRP (N,EGVAL,M)
    994 IF(ICORR.EQ.1) CALL CCRTST  (EGVAL ,SSD,M,SD,KLIP)
        IF(ISCORE ,EQ;1) CALL SCORES(SSD,M,N,XM,X)
        IF(MATT.EQ.1) CALL MNTR(SSD,XM,SD,M)
        CONTINUE
C
        IF(IVEC.EQ.1) CALL VECCHI(M,N)
        IF(ICHECK,EQ,0) GO TO 12266
        WRITE(LUT,153)
        WRITE(LUT,215) PROD
        WRITE(LUT,153)
        WRITE(LUT,399)
        WRITE(LUT,170) TRACE
        WRITE(LUT,992)
        WRITE(LUT,75)(SX(I),I=1,M)
        STRACE = STRACE - TRACE
        IF(STRACE,NE,0.) WRITE(LUT,2266) STRACE
  12266 CONTINUE
        GO TO 500
C -------------------------        FORMATS        -------------------------
      5 FORMAT(1X,I2,I3,7I1)
```

```
   10 FORMAT(12A6)
   30 FORMAT(7F18,6)
   41 FORMAT(24H0   CROSS PRODUCTS MATRIX)
   55 FORMAT(1H1)
   60 FORMAT(26H0   NUMBER OF DIMENSIONS = I2/35H0   NUMBER OF OBSERVATION
      1 VECTORS = I3)
   70 FORMAT(18H0   VECTOR OF MEANS)
   75 FORMAT(1X,12F11,5)
   90 FORMAT(24H0    STANDARD DEVIATIONS/)
   95 FORMAT(22H1    COVARIANCE MATRIX/)
  102 FORMAT(32H0   TRACE OF COVARIANCE MATRIX = F18.7)
  105 FORMAT(23H0    CORRELATION MATRIX/)
  120 FORMAT(39H0    VARIANCES FOR PRINCIPAL COMPONENTS/             )
  121 FORMAT(33H0   PCA RESULTS USING CORRELATIONS)
  125 FORMAT(35H0    EIGENVECTORS/COLUMNWISE/             )
  135 FORMAT(26H1   TRANSFORMED VARIABLES/)
  153 FORMAT(25H1    CHECKING INFORMATION/)
  170 FORMAT(30H0   TRACE OF DIAGONAL MATRIX = F14.7)
  180 FORMAT(18H0   TRACE TOO SMALL)
  185 FORMAT(16H0   TRACE CORRECT)
  190 FORMAT(16H0   TRACE TOO BIG)
  200 FORMAT(24H0   CHECK ON DETERMINANTS)
  215 FORMAT(24H0   EIGENVALUE PRODUCT = E14,7)
  221 FORMAT(32H1   PCA RESULTS USING COVARIANCES)
  300 FORMAT(21H0   LOGARITHMS OF DATA)
  311 FORMAT(16H0   RAW DATA USED)
  345 FORMAT(13H0   INPUT DATA)
  399 FORMAT(10H0   CHECKER)
  992 FORMAT(25H0   EIGENVALUE PERCENTAGES   )
 2266 FORMAT(12H0   STRACE = E18.7)
55000 FORMAT(1X,13F10,4)
55555 FORMAT(1X,20F6,1)
      END



      SUBROUTINE MNTR(A,X,SD,M)
      DIMENSION A(55,55),X(55),B(55),SD(55)
      COMMON/UNITS/LIN,LUT,MTA
      WRITE(LUT,1)
      DO 3 I=1,M
    3 SD(I)=0.0
      DO 2 J=1,M
      DO 2 I=1,M
    2 SD(J) = SD(J) +X(I)*A(I,J)
      WRITE(LUT,4)(SD(I),I=1,M)
      RETURN
C-------------------------------   FORMATS   --------------------------
    1 FORMAT(20H0   TRANSFORMED MEANS         )
    4 FORMAT(1X,12F11,4)
      END
```

```
      SUBROUTINE MATOUT (STMAT,M)
      DIMENSION STMAT(55,55)
      COMMON/UNITS/LIN,LUT,MTA
      MEND = 0
    1 MBEG = MEND + 1
      MEND = MBEG + 9
C------PRINTS MAXIMUM OF 10 COLUMNS EACH TIME
      IF(MEND.GT.M) MEND = M
      WRITE(LUT,101)(J,J=MBEG,MEND)
      DO 2 I=1,M
    2 WRITE(LUT,102)I,(STMAT(I,J),J=MBEG,MEND)
C------IF ALL M COLUMNS HAVE BEEN PRINTED THE PROGRAM RETURNS
      IF(MEND.EQ.M) RETURN
C------OTHERWISE, PRINT NEXT 10 CCLUMNS
      GO TO 1
  101 FORMAT(1H0,10I12)
  102 FORMAT(2X,I2,2X,10F12.4)
      END


      SUBROUTINE CHINCH(A,M, N,X)
      DIMENSION A(55,55),B(600,3),X(55),Y(55)
      COMMON/UNITS/LIN,LUT,MTA
C------THE EIGENVECTORS ARE STORED IN MATRIX A
C------THE DATA COMPRISES MAXIMALLY 600 OBSERVATIONS
      REWIND MTA
      DO 14 I=1,M
      Y(I) = 0.
   14 X(I)=0.
      DO 20 IX=1,N
   15 READ(MTA)(X(I),I=1,M)
      DO 1 J=1,M
      DO 1 I=1,M
    1 Y(J) = Y(J) + X(I)*A(I,J)
      DO 30 K=1,3
   30 B(IX,K)=Y(K)
      DO 135 I=1,M
  135 Y(I) = 0.
   20 CONTINUE
      WRITE(LUT,60)
      DO 5 I=1,N
    5 WRITE(LUT,10)(B(I,J),J=1,3)
      WRITE(LUT,70)
      CALL PLT21 (B,B(1,2),N)
      WRITE(LUT,222)
      CALL PLT21 (B,B(1,3),N)
      WRITE(LUT,223)
      CALL PLT21 (B(1,2),B(1,3),N)
      WRITE(LUT,224)
      RETURN
C ----------------------     FORMATS     --------------------
   10 FORMAT(3F15.6)
   60 FORMAT(24H0   TRANSFCRMATION MATRIX)
   70 FORMAT(45H0   HEREAFTER FOLLCW PLOTS OF THESE VECTORS              )
  222 FORMAT(25H0               1 AND 2                    )
  223 FORMAT(25H0               1 AND 3            )
  224 FORMAT(25H0               2 AND 3             )
      END
```

```
      SUBROUTINE ISOTRP (N, R, M)
C----------ISOTRP COMPUTES A CHISQUARE FOR THE PROBABILITY OF EQUALITY OF
C----------THE LAST (M-K) EIGENVALUES OF A LARGE-SAMPLE COVARIANCE
C----------MATRIX,
      DIMENSION R(55),TDATA(55)
      COMMON/UNITS/LIN,LUT,MTA
      DATA(TDATA= 0,0,9,21,11,34,13,28,15,09,16,81,18,48,20,09,21,67,23,
     121,24,72,26,22,27,69,29,14,30,58,32,00,33,41,34,81,36,19,37,57,38,
     293,40,29,41,64,42,98,44,31,45,64,46,96,48,28,49,59,50,89)
      WRITE(LUT,40)
      IF (M-6) 15,2,3
    2 K=2
      GO TO 10
    3 K=3
   10 CHISQ = 0,
      CHISQ1 = 0,
      CHISQ2 = 0,
      KO=M-K
      FMK=KO
      DO 30 I=K,M
      CHISQ1 = CHISQ1 + ALOG(R(I))
   30 CHISQ2 = CHISQ2 + R(I)/FMK
      FNM1=N-1
      CHISQ=-FNM1*CHISQ1+FNM1*FMK*ALOG(CHISQ2)
      WRITE(LUT,20) KO,CHISQ
      KDF = (KO*(KO+1)/2) - 1
      WRITE(LUT,25) KDF
      TEST = TDATA(KDF)
      K = K+1
      IF(TEST.GT.30,) GO TO 150
      IF(TEST-CHISQ) 50,50,60
   50 WRITE(LUT,55)
      GO TO 150
   60 WRITE(LUT,65)
  150 CONTINUE
      IF((M-3) - K)10,15,15
   15 RETURN
C ------------------------        FORMATS      -------------------------
   20 FORMAT(25H0  SIGNIFICANCE FOR LAST I2,5HROOTS,9H CHISQ = F14,7)
   25 FORMAT(8H0  DF = I3)
   40 FORMAT(1H1)
   55 FORMAT(21H0  SIGNIFICANT RESULT            )
   65 FORMAT(18H0  NOT SIGNIFICANT            )
      END
```

```
      SUBROUTINE CONINT (N, EIGVAL, M)
C------CONFIDENCE INTERVALS FOR EIGENVALUES OF COVARIANCE MATRIX
C-------BASED ON ASYMPTOTIC THEORY FOR LARGE SAMPLES.
      DIMENSION EIGVAL(55)
      COMMON/UNITS/LIN,LUT,MTA
      WRITE(LUT,20)
      CONST=1.96*SQRT(2./FLOAT(N))
      DO 5 I=1,M
      CON1=EIGVAL(I)/(1.+CONST)
      CON2=EIGVAL(I)/(1.-CONST)
      WRITE(LUT,10) I, CON1
      WRITE(LUT,15) I, CON2
    5 CONTINUE
      RETURN
C ----------------------        FORMATS     -----------------------
   10 FORMAT(24H0  LOWER BOUND FOR ROOT I2,4H  = F14.7)
   15 FORMAT(24H0  UPPER BOUND FOR ROOT I2,4H  = F14.7)
   20 FORMAT(33H1  BOUNDS OF CONFIDENCE INTERVALS        )
      END




      SUBROUTINE SCORES (A,M,N,EMN,X)
C----------A IS THE MATRIX OF EIGENVECTORS
C----------COMPUTES THE TRANSFORMATION SCORES
      DIMENSION X(55),A(55,55),Y(55),EMN(55)
      COMMON/UNITS/LIN,LUT,MTA
      NN=N
      REWIND MTA
      WRITE(LUT,40)
      DO 190 IX=1,NN
      DO 3 I=1,M
    3 Y(I)=0
  100 READ(MTA)(X(I),I=1,M)
      DO 5 I=1,M
    5 X(I)=X(I)-EMN(I)
      DO 15 J=1,M
      DO 15 I=1,M
   15 Y(J) = Y(J) + X(I)*A(I,J)
C---------- THE FIRST FOUR COMPONENT SCORES
  190 WRITE (LUT,500) (Y(I),I=1,4)
  200 RETURN
   40 FORMAT(19H1  COMPONENT SCORES)
  500 FORMAT(15X,4F18.7)
      END
```

```
      SUBROUTINE CORTST (EIGVAL,EIGVEC,M,STD,KLIP)
      DIMENSION EIGVAL(55),EIGVEC(55,55),STD(55),A(55),B(55)
      COMMON/UNITS/LIN,LUT,MTA
C---------COMPUTES THE CORRELATIONS BETWEEN THE PRINCIPAL COMPONENTS
C---------AND THE ORIGINAL OBSERVATIONS
      IF (KLIP) 100,100,30
  100 DO 110 I=1,M
  110 STD(I) = 1.
   30 DO 5 IX=1,M
    5 B(IX)=SQRT(EIGVAL(IX))
      WRITE (LUT,15)
      DO 10 I=1,M
      DO 9 J=1,M
    9 A(J)=EIGVEC(I,J)*B(J)/STD(I)
   10 WRITE (LUT,20) (A(IX),IX=1,M)
      RETURN
   15 FORMAT(28H1  CORRELATION COEFFICIENTS
   20 FORMAT(1X,13F10.4)
      END




      SUBROUTINE PLT21 (X1,Y1,N1)
C------------------------------------------------------
```

A LISTING OF THIS SUBROUTINE IS GIVEN IN THE PROGRAMME PCOORD.

```
      SUBROUTINE HOUSER  (N,NEV,C,EV,VEC,V)
      COMPUTES EIGENVALUES-EIGENVECTORS---SYMMETRIC MATRIX
```

A LISTING OF THIS SUBROUTINE IS GIVEN IN THE PROGRAMME ORNTDST.

```
      SUBROUTINE VECCHI(M,N)
C---------THIS SUBROUTINE FUNCTIONS ONLY FOR POSITIVE DEFINITE
C--------- MATRICES
      COMMON     SX(55),SS(55,55),SSD(55,55),D(55,55),R(55,55),TITLE(12),
     1X(55),SD(55),FMT(12),XM(55),EGVAL(55)
      COMMON/UNITS/LIN,LUT,MTA
      WRITE(LUT,102)
      EM=M
      EN=N
C----------- READING IN THE COMPARISON VECTOR
      READ (LIN,FMT)(XM(I),I=1,M)
      WRITE(LUT,100)
      WRITE(LUT,101)(XM(I),I=1,M)
```

```
      UPP=0.0
      ON=0.0
      DO 1 I=1,M
      SX(I)=0.0
      DO 1 J=1,M
    1 SX(I) = SX(I) + D(I,J)*XM(J)
      DO 2 I=1,M
    2 ON = ON + SX(I)*XM(I)
      ON=ON/EGVAL(1)
      DO 25 I=1,M
      R(I,I)=1./EGVAL(I)
      DO 25 J=1,M
      IF (I.NE. J) R(I,J)=0
   25 CONTINUE
      DO 30 I=1,M
      DO 30 J=1,M
      D(I,J)=0
      DO 30 K=1,M
   30 D(I,J)=D(I,J)+SSD(I,K)*R(K,J)
      DO 35 I=1,M
      DO 35 J=1,M
      R(I,J)=SSD(J,I)
   35 CONTINUE
      DO 40 I=1,M
      DO 40 J=1,M
      SS(I,J)=0
      DO 40 K=1,M
   40 SS(I,J)=SS(I,J)+D(I,K)*R(K,J)
      CALL MATOUT (SS,M)
      DO 6 I=1,M
      SX(I)=0.0
      DO 6 J=1,M
    6 SX(I) = SX(I) + SS(I,J) *XM(J)
      DO 7 I=1,M
    7 UPP=UPP+SX(I)*XM(I)
      UPP = UPP*EGVAL(1)
      UPP = ABS(UPP)
      EN=EN-1.0
      CHISQ = EN*(ON+UPP-2.0)

      KDF = M-1
      WRITE(LUT,103) CHISQ,KDF
    8 RETURN
C----------------                   FORMATS          -----------------
  100 FORMAT(22H0   COMPARISON VECTOR         )
  101 FORMAT(F10.5)
  102 FORMAT(25H0    ISOMETRIC HYPOTHESIS      )
  103 FORMAT(17H0    CHISQUARE = F8.3,11H FOR DF = I2)
      END
```

```
      SUBROUTINE PRINAX    (R,M,N)
C----------CARRIES OUT A TEST OF EQUALITY OF THE LAST M-1 EIGENVALUES
      DIMENSION R(55,55),V(55)
      COMMON/UNITS/LIN,LUT,MTA
      WRITE(LUT,20)
      DO 3 I=1,M
      V(I) = 0.
    3 R(I,I) = 0.
      N=N-1
      EM=M
      EN=N
      DO 15 J=1,M
      DO 15 I=1,M
   15 V(I) = V(I) + R(I,J)
      DO 16 I=1,M
   16 V(I) = V(I)/(EM-1.)
      SUM=0.
      DO 25 I=1,M
   25 SUM = SUM + V(I)
      SUM = SUM /EM
      BETW =0.
      ADD = 0.
      DO 35 I=1,M
      DO 35 J=1,M
   35 BETW = BETW + (R(I,J) - SUM)**2
      BETW = (BETW-SUM**2*FLOAT(M))/2.
      FLAMBDA = 1.-SUM
      TOT = 0.
      DO 45 I=1,M
   45 V(I)=V(I)-SUM
      DO 46 I=1,M
   46 TOT = TOT + V(I)**2
      EMU = (((EM-1.)**2)*(1.-FLAMBDA**2))/(EM*(EM-2.)*FLAMBDA**2)
      CHISQ = (EN*(BETW-EMU*TOT))/FLAMBDA**2
      WRITE(LUT,50) CHISQ
      KDF =(M+1)*(M-2)/2
      WRITE(LUT,55) KDF
      DO 80 I=1,M
   80 R(I,I)=1.
      RETURN
C --------------------------         FORMATS      --------------------------
   20 FORMAT(67H0        TEST OF EQUALITY OF LAST M-1 ROOTS OF A CORRELATI
     1ON MATRIX                    )
   50 FORMAT(16H0   CHISQUARE  = F14.7)
   55 FORMAT(8H0   DF = I2)
      END
```

13
Factor Analysis

It is very hard to discuss factor analyses (for there are many different kinds) without generating more heat than light. They are the most controversial of the multivariate methods. Part of the difficulty stems from the widely different terminology of factor analysts as opposed to practitioners of other forms of multivariate analyses. Partly, confusion arises because principal component analysis grew up in the same context as factor analysis and most factor analysts consider it as a special case of their techniques; this situation makes discussion exceptionally difficult. For instance, when Ehrenberg (1962) pointed out that, after 50 years of the practise of factor analysis, it was still quite uncertain whether anything of value had emerged, he was answered by other contributors to the same symposium in terms of benefits accruing from the use of multivariate analysis in general, and it remains unclear to this day whether any considerable service is performed by factor analysis (in any strict sense of the term) that is not better performed by one or other of the multivariate techniques mentioned earlier in this book. Individual psychologists would, no doubt, claim that factor analysis had made possible an orderly simplification (Burt, 1941; Thurstone, 1947; Cattell, 1965a, b) of the vagaries of the workings of the human mind: this view we can gladly accept in principle, but there is a disquieting disagreement between schools of psychologists as to the nature of the structure of the mind, which suggests that the interpretative methods are something less than objective. Could it not be that factor analysis has persisted precisely because, to a considerable extent, it allows the experimenter to impose his preconceived ideas on the raw data?

There seem to be two quite serious difficulties with factor analytical methods in the strict sense: one is that they fail to include criteria for assessing the agreement of two or more sets of results, the other that they are much less amenable than principal components to the calculation of scores for individual organisms along the relevant vectors; whereas one can easily, and advantageously, compute scores along components, one can only estimate, by methods of doubtful utility, the corresponding factor scores. The first problem has been diminished, but not eliminated, by the maximum likelihood methods associated with Lawley (1940, 1958, 1960); Lawley and Maxwell (1963) and Jöreskog (1963), and the second remains. Readers who wish to pursue the topic may peruse the symposium comprising the following references: Lindley (1962, 1964); Warburton (1962, 1964); Jeffers (1962, 1964); Ehrenberg (1962, 1963, 1964). To this rather discouraging body of opinion should be added the papers by Rasch (1962), who summarizes at least one of the systems of factor analysis, and illustrates the summary with an analysis of 13 body measurements of cattle into three factors, as well as those of Cattell (1965a, b) and some thoughtful, and probably apposite, comments on the concept of simple structure by Sokal *et al.* (1961). One of the main ways in which factor analysis differs from principal component analysis is that factors may be rotated to determinable positions in which they are not necessarily or even generally, orthogonal. One way of determining these positions is to adopt "simple structure", described by Thurstone (1947) as one of the turning points in the solution of the multiple factor problem. Other solutions have been described with unusual clarity by Gould (1967).

Where the factors are not to be rotated, there is some evidence that despite the apparently distinct mathematical models, factor analysis and principal component analysis give closely similar outcomes; indeed, Gower (1966), who specifies some of the conditions under which this statement is approximately correct, considers that the meaningful results obtained when factor analysis is used under apparently unsuitable circumstances may well stem from the extent to which this technique can simulate a principal components analysis. We may note that Dagnelie (1965b) has presented some interesting comments on the use of the technique.

Seal (1964) has distinguished factor analyses from other multivariate techniques in two respects: that the p original characters are analysable into m $(m < p)$ orthogonal factors with uncorrelated residuals; and that these m orthogonal factors may be subsequently rotated to conform to a new set of factors, imposed by the experimenter, as a consequence of his theories about the natural processes underlying the measurements he has taken. In the earliest papers of Spearman (1904), the unique underlying factor (the g-factor of intelligence) was the only one considered, but, as special factors of various kinds were introduced, the distinction between imposing a factor model and allowing the data to decide how many factors were at work became blurred. Hotelling (1957) has also considered the relations between factor analysis and multivariate methods.

Far from principal component analysis being a special case of factor analysis, as one might think from the historical sequence of their respective developments, factor analysis is in fact a special (and, in Rao's (1964b) view, perhaps an unrealistic) case of principal component analysis. Rao notes that if Λ is the actual correlation matrix of the variables, we wish to choose a matrix $\mathbf{B}_q$ such that the off-diagonal elements of $\Lambda - \mathbf{B}_q \mathbf{B}_q'$ are as small as possible. Here q represents the number of factors by which the interrelationships between the variables are to be explained (there being, of course, fewer factors than variables) and $\mathbf{B}_q$ is the matrix whose i-th row is $(b_{i1}, \ldots, b_{iq})$, where b_{ik} is the standardized factor loading of the ith factor on the kth variable. As Rao remarks, the choice of $\mathbf{B}_q$ is not easy, there being still only iterative solutions.

Whether or not factor analysis really is a helpful way of viewing biological data, and in particular, morphometric data, is considered doubtful by Gower (1967a), as the same analysis of the Q-space obtained by a principal components analysis should give similar results.

The main motivation for indulging in the factor model seems to us to lie with the convictions of the would-be user: how strong his belief is in the existence of the postulated specific factors and how favourably these are interpretable in terms of his psychological theory.

Even among psychometricians, opinion is not wholly favourable to factor methods; Storms (1958) considers that where factor analysis gives results which differ from those found by multivariate methods in the sense used in this book, the factor analytical results represent a distortion, and even a falsification, of the truth in so far as this can be discovered.

The interpretation of specific factors becomes all the more difficult and tenuous when the matrix to be factorized, $\mathbf{S}$, is a Q-matrix of order n, where the elements represent comparisons between individuals of the sample, than when it is an R-matrix $(p \times p)$, where the elements are variances and covariances between the variables.

Attempts to use factor analyses have convinced us that the results rarely differ from those given by a principal components or principal coordinates analysis in any way that seriously influences a biologist or a geologist seeking to interpret a mass of data. A somewhat similar conclusion was reached by Crovello (1968c). Insofar as some differences appear, their induction and interpretation involve such a high degree of subjective judgement, often disguised from the user because he is led to believe that procedures advocated by other workers are justified mathematically when they are in fact arbitrary manipulations of the vectors, that we prefer not to give details which might be interpreted as recommendations. For the same reasons we do not give a computer programme, but refer the interested reader to Jöreskog's (1963) paper. We now offer some published examples of factor analysis that we would prefer to treat by principal components or principal coordinates analyses.

Middleton (1964) has investigated the interrelationships of ten chemical constituents in scapolites, beginning with a principal components analysis and continuing with a factor analysis, by rotation of the principal components to positions determined by the quartimax procedure (Harman, 1960); which has analogies with the least square technique for fitting a line through the points. The elements fell into two bands roughly at right angles, considered to represent the consequences of substitution in one of two series of solid solutions. The rotation of the axis does not seem to have done more than formalize conclusions apparent from inspection of the principal components analysis.

Some genera of fusulinid foraminifera were studied by a factor analysis, which involved the calculation of nine factors and the rotation of seven of these according to the varimax criterion (Pitcher, 1966). Only four of these factors were considered to be significant, and all four contributed to the separation of the various genera in successive two-dimensional charts (Fig. 23). Since the study of the characters which contributed most to the separation of the genera and species was conducted by means of factor scores, one might think that it would have been easier to interpret the results had a simple principal component analysis been performed since principal component scores are more readily computed and interpreted. Pitcher makes the useful general point that when a specimen remains within a particular grouping as the data are plotted on different axes of variation, the credibility of the proposition that the individual belongs to that group is enhanced: conversely, if replotting the data along new axes of variation shuffles the individual into another group, its classification must remain dubious.

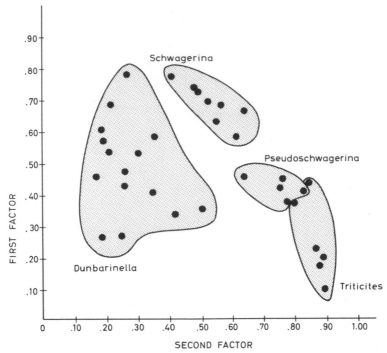

Fig. 23. Factor scores of representative species of four genera of fusulinid foraminifera showing clustering into generic groupings along the first two factors. Redrawn from Pitcher (1966).

The Case of the Bermudan Snails

Gould (1969) has uncovered a fascinating biological situation in Bermuda, where the snail *Poecilozonites bermudensis* underwent rapid evolutionary changes during the Pleistocene giving rise to four distinct paedomorphic lines, regarded by Gould as subspecies. A matrix of 33 variables and combinations of variables, measured on these pulmonate snails, was factorized, and then treated by varimax rotation. The first two axes divided the collection of shells into those of paedomorphs and those of non-paedomorphs, and separated one of the paedomorphic subspecies from the remainder. However, when the third and fourth axes of variation were made into the chart shown in Fig. 24, much finer resolution of the situation became possible. There is a general coarse division into Pleistocene paedomorphs, Recent paedomorphs, and the non-paedomorphic forms, and within the Pleistocene paedomorphs there is a distinct subdivision into the four subspecies as shown in the figure. Gould

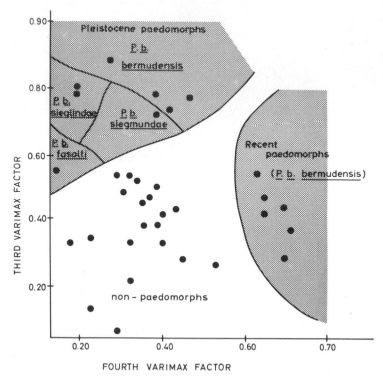

Fig. 24. Factor analysis of the paedomorphic and non-paedomorphic forms of the pulmonate snails of the genus *Poecilozonites* in Bermuda. Redrawn from Gould (1969).

took particular care to make his character combinations expressive of the shape of the shell and of its colouration.

We then tried a principal coordinates analysis on the same material as was used for Gould's factor analysis. Figure 25 shows the first two coordinate axes, taking up somewhat more than half of the "variation" in the latent roots of the association matrix. This does not show a clear separation into paedomorphic and non-paedomorphic fields [the plot of axes two and three does this (Fig. 26) well], but the various samples are clearly distinguished (Gould's axes III and IV did this); the subdivision of fossil and recent paedomorphs of *P.b. bermudaensis* stressed by Gould's third and fourth axes, is not apparent here.

We note *en passant* that Gould's specimen 05, which is on the edge of the charts for his axes I, II and III, IV, is centrally placed on all of the first four principal coordinate axes (this specimen is the Shore Hills paedomorph of Fig. 25): in view of its markedly isolated

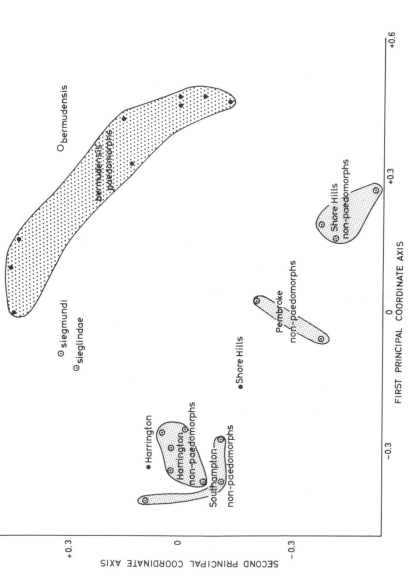

Fig. 25. A principal coordinates chart of the first and second axes for paedomorphs and non-paedomorphs of *Poecilozonites* in Bermuda.

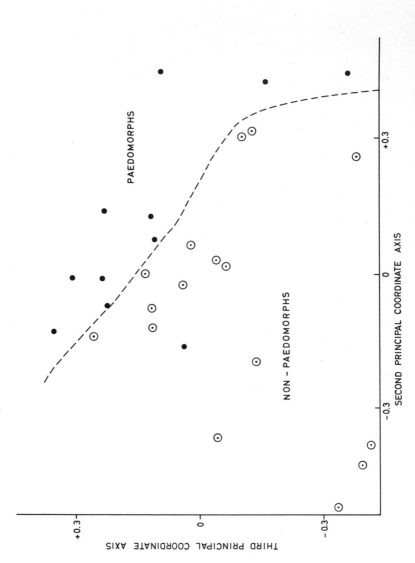

Fig. 26. A principal coordinates chart of the second and third axes for paedomorphs and non-paedomorphs of *Poecilozonites* in Bermuda.

position in Gould's plot of the first two factor axes, some mishap may have occurred in the processing by the factor programme. In Fig. 26, in which the first and third principal coordinates are plotted, it is, however, the only misplaced paedomorph.

Q- and R-Techniques of Analysis

Before we come to a discussion of the broader uses of multivariate taxonomy, a clear distinction has to be made between an analysis of the matrix of correlations between the characters (or character-states, attributes) of a set of organisms, and the analysis of the correlations between the organisms themselves. In ecological terms, to make the illustration more concrete, we always have the choice of regarding an organism as being the subject of the analysis, and analysing the quadrats in which it occurs to discover, for instance, how closely it resembles another organism ecologically, or alternatively, selecting quadrats of ground and measuring their affinities in terms of the numbers of organisms shared by two habitats, in which case the organisms are the characters of the habitat.

Q-techniques deal with the analyses in which the numbers of organisms in each habitat are studied, whereas R-techniques comprise those in which the habitats or other attributes of the organisms are correlated and analysed. To some extent these types of analysis are interchangeable. There is, however, a shift of emphasis in passing from one to the other, and most of the strictly morphometric work on shape and size described in Chapters 2-9 illustrates R-techniques, whereas a great deal of numerical taxonomic work involves examples of Q-techniques, as does the method of principal coordinates. It is often instructive to turn one's problem inside out in this way. Sokal and Sneath (1963) who discuss the differences between these two approaches at some length, point out the small amount of work, amounting at that time to a single paper by Stroud (1953), which has been performed by means of R-techniques at higher taxonomic levels, and have commented that information of interest to phylogenetically-minded taxonomists should be accessible by this route.

Orloci (1967b) discusses the degree of interchangeability between R and Q techniques in the context of data centering. He notes that when the two sets of techniques are used as alternative strategies, the various intermediate results should be directly transferable. Application of principal component analysis to a Q-matrix which has been generated by the R-expression of the correlation coefficient is inappropriate for the computation of component scores, although attempts to do this are frequently to be found in the literature.

The principal components of the R-matrix are not the same as the principal axes of the Q matrix when unstandardized data are used, but if the data are appropriately standardized, the two calculations become rearrangements of one another. There is, therefore, a real sense in which the argument about whether Q or R techniques are "better" is a non-question, a point which might be more widely appreciated than it appears to be.

Some information is lost in the course of each type of analysis, but the information is naturally qualitatively different, according to the nature of the analysis: in ecological studies, information about the relationships between the species is lost during Q-type analyses, and about relationships between the quadrats or stands during R-type analyses (Lambert and Williams, 1962).

Kendall and Stuart (1966) note that it would be desirable to distinguish between clustering of various kinds in the two cases. Clustering of individuals in respect of the characters that they possess would be called "classification analysis", whereas clustering of the characters in respect of the individuals in which they occur would be called "cluster analysis", on the system that these authors advocate. On this basis classification analysis corresponds to Q-type and cluster analysis to R-type treatments.

14
The Measurement of
Environmental Variation

In many animals and plants, the environment in which growth takes place determines to some extent the final form. It is remarkable how little this environmental variation has been studied, no doubt partly as a reaction against the "variety hunting" of the last century, when any variation which could be established was named and described. Environmental variation is, nevertheless, of interest in its own right, and may attain magnitudes comparable with specific or even generic variation. Examples are the effects of strong winds and grazing pressures on plants (Whitehead, 1959): the variation of locusts when reared in crowded habitats, the changes of shape being essentially the same in a wide range of species and genera (Blackith, 1962) and variation in Recent brachiopods (McCammon, 1970). Two kinds of variation were found in the brachiopods, one of which was due to physical restraints arising from overcrowding or growth between dead shells and rocks; this type of variation essentially resulted in asymmetry. There was another and perhaps more interesting form variation which was measured in terms of the principal components of the dispersion matrix of five characters: length; width; thickness; beak angle of pedicle valve; and the ratio of the areas of the two sections, termed the asymmetry ratio. The first principal component was found to represent size variation, having all its coefficients positive. The next two components represented shape variation of different kinds. Using these components as the axes of a chart on which the positions of the individual brachiopods of different species were plotted, it was evident that the costate species, even though not closely related to one another, exhibited one kind of shape change whereas the non-costate species exhibited another. As in research into the patterns of variation in locusts and grasshoppers, we have the curious result that species within the same genus may show quite distinct patterns of variation, yet much less

211

closely related species, sometimes allotted to distinct families, may show identical patterns. Here, apparently, is a widespread type of variation which has been neglected but for which the necessary techniques of study are now available in morphometric analyses. Yet the analysis may not be easy because variation of form may be strikingly distinct from variation in colour (Blackith and Roberts, 1958) so that similar colour forms are widely disparate morphometrically.

The anostracan brine-shrimp *Artemia salina* can live in a wide range of environments, and is tolerant of variations in salinity. For almost a hundred years now it has been known that the body proportions of this branchiopod are correlated with salinity. Gilchrist (1960) analysed brine-shrimps from water of two different salinities, and from three localities. She measured the total length; abdominal length; abdominal width; and the length of the caudal furca. Later Reyment (1966a) treated these data by means of principal components and canonical variates analyses.

The principal components, whether extracted from the dispersion matrix of the untransformed data or from the dispersion matrix of the logarithms of the data, did not alter appreciably with salinity or provenance. The general consistency of these components was of the same order as Reyment's earlier analyses of fossil and Recent ostracods had uncovered. In agreement with Gilchrist, Reyment concluded that the variability is partly of genetic and partly of environmental origin.

Rather more informative was a canonical variate analysis carried out on six variables, the original four together with the length of the prosoma and the number of setae per furca. Figure 27 shows the arrangement of the various samples along the first two canonical variates. Material of American origin tends to have low values of the first canonical variate, in which the length of the abdomen is important, but total length much less so. The width of the abdomen and the furcal length, though about half as important as the abdominal length, act in opposite senses. Material of Mediterranean origin tends to cluster along those parts of the first canonical variate where the score is highest. It is interesting, and rather strange, that differences of salinity affect the shapes of brine-shrimps much less than differences of geographical origin. There is, moreover, a pronounced sexual dimorphism which expresses itself in the fact that males have consistently lower scores along the first canonical variate than have females, just as brine-shrimps from 35% sea-water have consistently lower scores than have those from 140% sea-water.

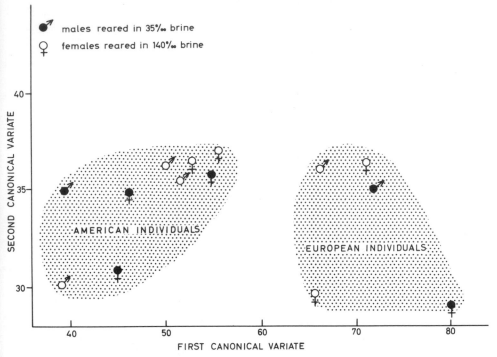

Fig. 27. The arrangement of samples of *Artemia salina* along the first two canonical variates. Redrawn from Reyment (1966a).

The living freshwater ostracod genus *Cypridopsis* has also been found to be influenced by the environment. The shape of the carapace differs slightly according to whether the water in which it lives is relatively rich in lime, stagnant, or fresh. Three characters were measured on the carapace, the height, length and breadth, and both generalized distances and canonical variates were computed from the means and dispersion matrices. Normal (i.e. freshwater) adults score most highly along the first canonical variate, which accounts for almost all the variation (99.8%) in conjunction with the second variate. It consists of the vector $(0.785, 0.428 - 0.449)$.

As Fig. 28 shows, for adults (which are stage IX of growth) and the last larval stage (VIII), there is a regular pattern by which ostracods grown in fresh water score more highly, not only along the first variate but also along the second, by comparison with the ostracods grown in lime-rich or stagnant water. Thus the effects of "contaminating" the water are almost identical for the two

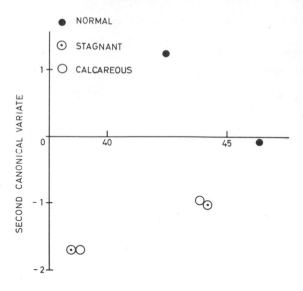

Fig. 28. Plot of the first two canonical variates for *Cypridopsis vidua* from "normal", "lime-rich", and "stagnant" laboratory environments. Redrawn from Reyment and Brännström (1962).

contaminants and form an axis of variation which lies obliquely across the first two canonical variates.

Changes of form in different geographical localities make up one of the most frequently encountered kinds of environmental variation. Sokal and Rinkel (1963) were forced to remark that "It may occasion some surprise that the statistical problems involved in the analysis of geographic variation have been almost untouched in spite of the amount of emphasis that has been given to this subject as a cornerstone of the New Systematics. With the exception of the valuable papers of Pimentel (1958, 1959) directions for statistical analysis of such data have been either erroneous or lacking altogether". This was something of an overstatement since Burla and Kälin (1957) had clearly shown how to combine seven morphometric characters of *Drosophila* into a discriminant function contrasting two geographically separated populations: nevertheless, the complaint was broadly justified. No doubt this state of affairs arose because of the need to handle a multiplicity of characters simultaneously. Sokal and Rinkel use factor analysis to reduce this multiplicity of geographically modified attributes to a smaller number based on the limited capacity of the organism to bring into play more than a few independent patterns of growth, much as Jolicoeur had earlier tackled geographic variation in the wolf (Jolicoeur, 1959, 1963c).

Other ways of tackling the study of geographical variation are by employing trend surface analysis (see Chapter 28) and by partitioning the area in question into definable categories; Gabriel and Sokal (1969) discuss these approaches for univariate and multivariate analyses.

Sometimes a species growing in two different localities shows a distinct pattern of growth in each. Malmgren (1970) measured the bryozoan species *Floridina brydonei* from the Danian of Denmark, and made a principal component analysis of the four measurements (length of zooecium, width of zooecium, and length and width of the aperture) taken on two samples, using programme PNCOMP. The sample from Ny Klostergaard (northwestern Jutland) contained 100 specimens and that from Klintholm (eastern Funen) contained 89.

As is often found, the first principal components for each sample are essentially size vectors with all the coefficients positive. For the Ny Klostergaard sample the first vector took up 51.7% of the total variation: this vector was (0.155, 0.576, 0.466, 0.653). For the Klintholm sample the first vector (0.150, 0.828, 0.191, 0.505) accounted for 60.7% of the total variability. A χ^2-test showed that these vectors, which are mutually inclined at 23°, differ significantly from one another in orientation. The two samples, however, do not differ significantly in the directions of the remaining principal components, which represent shape factors, being of unlike signs for the various coefficients.

The principal components measure the direction of evolutionary or environmental change under geographical isolation; canonical variates can be used to construct a diagram of the cross-section of this process actually attained by the organisms at any given moment. For instance, Delany and Healy (1964) measured ten characters on each of 156 specimens of the long-tailed field-mouse from eight islands, and two points on the mainland, of north-west Scotland. The first two canonical variates were used as the axes of a chart showing much overlap between the various populations. Indeed, the concept of subspeciation previously applied to these mice was shown to be dubious, quite apart from the generally low repute which the subspecies concept enjoys currently. It is interesting that the larger mice from the island of Rhum are discriminated along the second canonical variate and not along the first, although this situation would probably have been altered had more of the groups been involved in substantial size variation. It cannot be too often stressed that the order in which a particular biologically meaningful vector occurs in a multivariate analysis depends on the selection of the original material for study as well as on the intrinsic morphometric properties of the material.

This study has recently been extended by Delany and Whittaker (1969) who combined nine skull characters so as to obtain both canonical variates and generalized distances.

Environmental variation in the morphology of animals is generally thought of as adaptive, although the justification for this point of view is sometimes tenuous. Fujii (1969) has begun to explore the possibility of classifying animals into a hierarchy according to their ecological characteristics rather than their morphological ones. He employed 11 attributes such as developmental time, longevity, weight, and number of eggs per female, and obtained a hierarchy which distinguished the two species of bean weevil that he used, at low degrees of similarity, followed by clustering of the Japanese strains together, and the American strains together, before the remaining strains entered the clusters. Fujii also notes the possible advantages of representing the groups of weevils, as represented by their ecological characteristics, on a multidimensional ordination system rather than in a hierarchy.

Just as geographical variation among related organisms may involve only a limited number of common dimensions of variation, so may other types of variation such as sexual dimorphism (see Chapter 29). Geographical variation in grasshoppers has been studied by Blackith and Roberts (1958) and by Blackith and Kevan (1967), using multivariate methods of analysis.

Discrimination between closely related taxonomic forms may be rendered less accurate because of superimposed geographical variation. When Rempe and Bühler (1969) were distinguishing the mandibles of the water-vole *Neomys fodiens fodiens* from that of *Neomys anomalus milleri* by means of the three characters mandible length; height of lower jaw; and length of tooth row, geographical variation interfered with the comparisons along the discriminant function (−1.00; +2.58; +2.78). However, once a fourth character, the least height of the lower jaw, was included in a canonical variate analysis this interference was effectively suppressed. It seems from the data published by these authors that the geographical variation in mandible shape takes the form of an essentially north-south cline.

Palaeontologists are frequently faced with the problem of material which varies not only from place to place but from time to time in a geological sequence. There may be variation not only in the form of individual organisms but also in their relative abundances (Reyment, 1963b). The potential causes of such variation are many and various; among them are post-mortem sorting by currents of water; changes in the salinity of the aquatic environment, as noted above for ostracod carapaces; allometric growth leading to different shapes for individuals of the same species but of different sizes; changes in the climate leading to warmer or colder environments, and, in some cases, crowding effects when animals are bunched together because the available resting sites (rocks etc. on the sea floor) are in short supply.

Reyment (1971b) has partitioned the changes of form of the ostracod carapace into a size component based on the first principal component and a shape factor based on the second such component which he considers to be, in effect, a measure of the genetically determined variation, as opposed to the environmental variation, reflected in a broad sense by the first component. Of course, these partitions of the total variation (when there are only three variables measured) can only be interpreted in this way in a very general sense: the subject is taken up in Chapter 18.

Reyment uses measurements of ostracods taken at different horizons in the cores from bore-holes in a clay-shale sequence for his comparisons, and finds that there are indeed correlated responses of several species which, on an average, grow bigger or smaller when the environment becomes slightly more favourable or slightly less optimal. In this way some approach to the ecology of past environments may be made, and for this purpose those fossil forms which extend over a wide range of strata are most valuable, although such forms are often less well studied precisely because they are of less use as indicator fossils. The approximate nature of this approach is evident. There is an alternative interpretation, that all species react in the same way to selection pressure; this is less likely, as the type of slight change concerned is known to be normal for the reaction of the carapace to the environment (e.g. *Cypridopsis,* salt variability of the Jurassic ostracod *Fabanella* and living *Loxoconcha*).

Not only the organisms living in an environment, but the environment itself, can be examined for significant contributions to elements of the synoptic, multivariate, measures of climatic response. Hocker (1956) has used discriminant functions to distinguish the meteorological factors of consequence to the establishment of *Pinus taeda.*

15
Multivariate Studies in Ecology

From a measure of the amount of variation in an organism, according to its environment, to a measure of the features of the environment itself, is a modest step conceptually but one that requires some consideration at the biological level. If we leave aside the few remaining vestiges of the idea that morphological characters could be divided conceptually into adaptive and non-adaptive ones, and few ideas seem more dated and inappropriate to current ways of thinking about evolutionary problems, the ensemble of morphological characters can be considered as the direct result of environmental pressures on the genotype. We cannot have any such confidence when we come to consider the ensemble of environmental features that we are capable of appreciating, for in all but a few instances, we know next to nothing of the extent to which changes in any one of the features influence an organism living in the habitat. For this reason, the association of environmental features into vectors in terms of changes in the abundance, behaviour, or growth of the organism is even more tentative than in morphological studies, but this statement holds true, essentially, for all synecological investigations.

Buzas (1967) has used canonical variate analysis to uncover the features determining, or appearing to determine, the abundance of 45 species of foraminifera in eight areas of lagoon or open sea bed, off the Texas coastline. Bays on the lagoon were shown to have different abundances of foraminifers when the living populations are considered but not when the dead population is estimated. Three stations of the open ocean at which the samples were taken at depths of 0-30 m gave patterns of abundance which clustered together closely on a two dimensional plot of the first and second canonical variates, and were well separated from the deep-water station samples. Canonical variate 1 seems to reflect depth differences faithfully, whereas canonical variate 2 seems to reflect ecological differences other than depth and associated features of the environment. Along this vector, populations from the bays seemed more like the deep-water samples than did those from the shallow ocean, whereas on variate 1 there is a steady progression from bay samples to shallow water samples to deep water samples. The samples from the moderately deep oceanic water seem to be aberrant when assessed along the second and third variates. Evidently factors other than depth are of consequence to foraminifera, and Buzas suggests that canonical variate analysis should help ecologists to distinguish between such basically distinct combinations of factors as those associated with depth and those not primarily associated with depth.

Cassie and Michael (1968) have applied principal components and canonical correlation analyses to the study of the invertebrate fauna of a silty intertidal mud flat in New Zealand. They were able to discern four communities of species by means of the principal component analysis. The community into which any given species falls is determined by the principal component to which it contributes the highest "factor loading" (score along the principal component). The distribution of the various communities seemed to be determined by the coarseness of the underlying sediments, as expressed by a linear function of the eight fractions into which the particle size range of the sediment had been separated. This linear fraction proved to be about twice as effective as the simple arithmetic mean particle size, and its more widespread use may help to avoid some of the loss of information associated with the use of the mean particle size for sediments where the grain size has a skewed distribution.

Cassie (1962) used principal components to distinguish a group of four species of plankton whose behaviour was mainly determined by the mixing of two water masses in an estuary, and a smaller group of two species whose behaviour did not appear to be governed by this factor.

Pelto (1954) has used the fact that in an association which is virtually a pure stand, the entropy of mixing of the constituent elements is low relative to that of an association where many elements are roughly equally represented. Thus high entropy delimits the zones of intergradation between two associations or communities (Howarth and Murray, 1969). This concept has been used by McCammon (1966) to provide a criterion by which principal components could be rotated to positions such that the entropy of the system is minimized, when the components might be expected to link the centres of associations. More recently, McCammon (1968a) has compared this approach with several others, and shows that his multiple component analysis produces groupings of the elements of Bahamian sediments which have a lower intra-group variance than other candidate methods, including rotated principal components, although the improvement on the principal component solution was not large. In principal component analysis, McCammon first clusters the data using the unweighted pair-group method of Sokal and Sneath (1963). The number of distinct facies is then estimated from the resulting dendrogram, thus allotting the original data (variables of the facies) to definite groups. The correlation matrix is then partitioned into submatrices in accordance with the decision taken as to the number of distinct facies present. Principal components are then taken for each of the submatrices, and the scores of the samples are computed and converted into a facies map showing the distribution of each of the facies over the area sampled. The hypothesis is, essentially, that to each of these principal components corresponds a particular type of environment or facies. Each sample is then classified according to the component it most closely resembles, and then allocated to the appropriate environment.

As an example of the use of multivariate techniques in an ecological problem we take the case of rotifers living in a series of inland waters in Europe, which have been examined in detail by Dr. A. Nauwerck. Four characters of the rotifers, the length and breadth of the body and the lengths of the anterior and posterior spines, were measured on samples of the order of 50 individuals from ten different inland waters, of which three, the Wolfgangsee in Germany, the Vierwaldstaettersee in Switzerland and Lago Maggiore in N. Italy, were chosen for more detailed morphometric analysis. We are most grateful to Dr. Nauwerck for permitting us to make use of this extensive body of material.

As a first step, a canonical variate analysis was run on the measurements of *Kellicottia longispina* and *Keratella cochlearis* from the three localities to see whether the variation of form from lake to lake was consistent from one species to another. Figure 29 shows the analyses for each species, which in fact show a remarkable degree of parallel variation: the first canonical variate consistently distinguishes material from the Wolfgangsee from that taken in the other two lakes, whereas the second canonical variate distinguishes the Vierwaldstaettersee material from the remainder. The first variate takes up 90% of the variation in *Kellicottia* and 66% in *Keratella*, the second variate taking up all the remaining variation in each case.

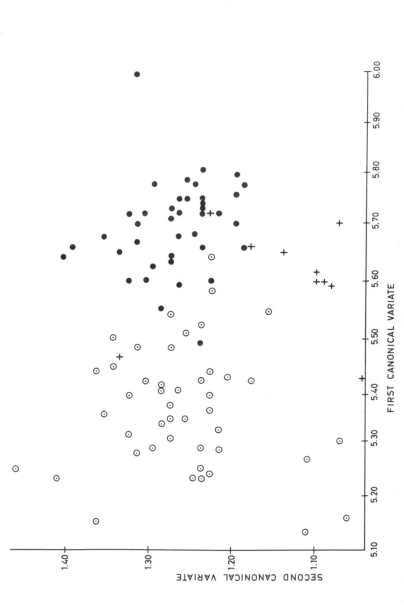

Fig. 29. Principal coordinate chart showing the relative morphometric distances between individual rotifers of the species *Kellicottia longispina* from the Wolfgangsee (open circles), the Vierwaldstättersee (crosses) and Lago di Maggiore (closed circles). Data of Dr. A. Nauwerck.

Both these vectors correspond to significant roots, a statement which translates itself into Fig. 29, which shows that there is a very substantial, but not complete, degree of separation between the three sets of points relating to *Kellicottia,* so that the reality of the ecological differences between the forms of the rotifers from the three lakes cannot be doubted.

The first canonical variates are:

Kellicottia	$(+24.5, -15.1, -1.4, +7.0)$
Keratella	$(+13.0, -3.4, -8.8, -3.8),$

which make an angle of $43°$, whereas the second canonical variates are:

Kellicottia	$(+10.8, -1.1, +6.9, -14.0)$
Keratella	$(+10.2, +0.9, -1.5, -2.2)$

making an angle of $51°$.

A second canonical variate analysis was run on the combined populations of *Kellicottia* and *Keratella* from the three localities, and the results are shown in Fig. 30. The extent to which the second canonical variate separates the material from the three lakes is unaltered, but there is a curious reversal of the direction of the separation. In the case of *Kellicottia,* the specimens from Lago Maggiore have the smallest negative scores, whereas in material of *Keratella* they have the largest. The clear separation between the two species, entirely on the basis of the first canonical variate, is apparent. A third root, taking up some 5% of the total variation, also separates the localities but the physical interpretation of this separation is not apparent.

· The seven remaining lakes for which samples were available were more or less intermediate between the three chosen for a detailed analysis, so that by studying the physical properties of the three exemplar lakes we may gain some clues as to the reasons for the form variation in the rotifers. The first canonical variate seems to correspond fairly closely to the pH of the water (Wolfgangsee, 7.78: Vierwaldstaettersee, 7.47: Lago Maggiore, 7.14) but it has not proved practicable to reify the nature of the second canonical variate, which seems not to depend at all closely on temperature, or on conductivity. The changes of form of living organisms in a lake may, of course, respond to changes in trace elements or in other factors which one is unlikely to measure in the absence of some definite reason for so doing.

Turning to the principal components as indicators of the nature of

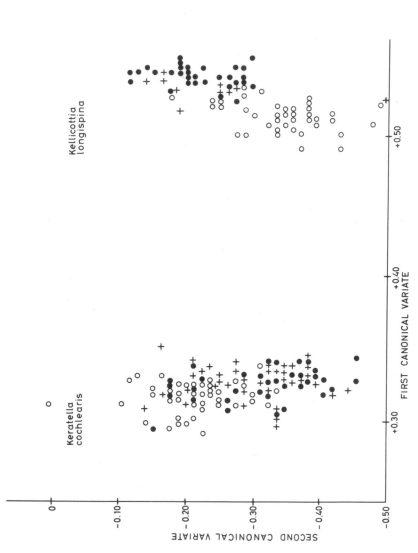

Fig. 30. Canonical variate chart for the populations of *Kellicottia longispina* and *Keratella cochlearis*, based on four morphometric characters. Symbols and localities as in Fig. 29. Data of Dr. A. Nauwerck.

the morphometric variation within samples, we find that the first principal components for *Kellicottia* in the three localities are:

Wolfgangsee	(+0.41, −0.07, +0.88, +0.23)
Vierwaldstaettersee	(+0.20, +0.43, −0.26, +0.84)
Laggo Maggiore	(+0.42, +0.56, +0.67, +0.24)

and for *Keratella* in the same three localities:

(+0.39, +0.26, +0.44, +0.76)
(+0.10, +0.07, +0.51, −0.85)
(+0.44, +0.31, +0.76, +0.34)

The first principal component is thus to some extent a measure of size differences (normal growth) for both species in Lago Maggiore and for *Keratella* in the Wolfgangsee. Moreover, the second principal component for *Keratella* in the Vierwaldstaettersee is a size vector. We may thus interpret the growth of *Keratella* as being more stable than that of *Kellicottia* in that it is only moderately perturbed in one locality, whereas *Kellicottia* seems to be disturbed in two of the three lakes to the point where it exhibits no general growth but only subsidiary patterns of growth. By the same sort of reasoning, which admittedly has a large subjective element in it, we may consider Lago Maggiore as the most "normal" habitat for the two species, i.e. where the growth is most consistently normal, and the Vierwaldstaettersee as the least normal of the three. Of the remaining lakes examined, the Ostersee and Mondsee give normal (size) first principal components for both species, whereas the Schlusssee gives "abnormal" first vectors for both species. There is, again, no very apparent physical meaning behind these results, which Dr. Nauwerck suggests may prove to be associated with disturbance of the water body.

Cassie (1967), in a principal component analysis of the plankton of Lago Maggiore, has shown that there is no "winter" component of the major groups of plankton ceding to a "summer" grouping: the larvae of *Kellicottia* were found to occur in Cassie's group 1 (a 12-species group including *Eudiaptomus vulgaris, Cyclops* larvae and other rotifers) throughout the year.

16
Phytosociological Studies

This and the following chapter, respectively botanical and zoological, deal with the comparison of floral and faunal elements and the associations into which they might be thought to fall. These examples of "abstract morphometrics" are broadly complementary.

Many of the problems with which phytosociologists have to deal are essentially morphometrical; but it is a moot point whether phytosociological studies as a whole are properly to be subsumed under the heading "morphometrics". In general, the presence or absence of plant species in a quadrat constitutes a "character" of that quadrat with the aid of which (together with the corresponding information about many other species) quadrats can be clustered or otherwise investigated. Nevertheless, there are special concepts in phytosociology, such as "fidelity" and "indicators", which can be incorporated with what we ordinarily treat as morphometric studies only at the risk of taking us far from topics familiar to the authors of this work. We content ourselves, then, with reminding readers that there are some excellent reviews of the topic which it would be foolish to try to summarize here. Outstanding are those by Dagnelie (1965a, b), who, after presenting the basic theory, goes on to give an example of factor analysis applied to a matrix of correlations between the species and to a matrix of correlations between the various ecological variables measured. A factor reflecting the influence of physical variates was extracted, as was one reflecting the influence of chemical variates. Yet there was much variation in the distribution of the plants which was not accounted for in these terms.

Detailed accounts of multivariate methods have been written by van Groenewoud (1965) and Gittins (1969) with phytosociologists particularly in mind.

Orloci (1968b) has commented that the application of information analysis in phytosociological work is rapidly gaining momentum, and it may be that for special purposes information analysis is more suitable than are multivariate methods. However, there are constant swings of the pendulum of fashion between the entropy-information approach to classification and the more strictly multivariate approach, and it is unwise to ignore the potentialities of either.

The very idea of ordination stems from the essentially multivariate approach to phytosociology of Goodall (1954a, b) whose work has been of immense influence in this field. Although there has been much valuable work done using analytical methods constructed particularly with phytosociological investigations in mind (Williams and Lambert, 1959) there is no real evidence that it is necessary to leave the general framework of multivariate techniques as we have outlined it in this book. Harberd (1962) has used the generalized distance and canonical variate techniques quite successfully to examine plant communities, even with qualitative data. Whilst no one could deny that special problems may require special analytical methods, it seems open to question whether the enormous number of special methods devised for particular fields of study is at all necessary; in fact, the construction of special methods often reflects a restricted knowledge of what is going on outside the special field.

Phytosociology offers an example of a field of study where it is quite hard to escape from traditional patterns of thinking. The characterization of plant communities on the basis of indicator plants, whose presence or absence in a community serves to define its nature, is so deeply ingrained that Williams and Lambert have reconstructed the analytical system for splitting up an area into plant communities so that the divisions take place primarily on the basis of the presence or absence of indicator plants and only secondarily on the finding of divisions which maximize the variation between groups relative to that within groups, a practice in line with the development of phytosociological theory. Whether it is desirable to make divisions on this basis is debatable, but Dagnelie has shown that useful results can be obtained without it.

Gittins (1965a, b) has shown that there are benefits in applying both Q-type and R-type analyses to phytosociological observations, and has made the valuable recommendation that ecological and environmental variates should be included in the analysis to help in the reification of the various axes of variation that can be extracted. He also points out the advantages to be gained from the very detailed analyses possible when the monothetic association analysis can be applied both in the direct and inverse forms (the nodal analysis of Lambert and Williams, 1962). Despite our general preferences for methods of analysis that fall into the framework of multivariate techniques outlined in this book, and a certain suspicion of special techniques invented to solve particular problems, there can be little doubt that the intensive study of definite aspects of a problem by workers deeply familiar with it and who have learnt to exploit the potentialities of the method of their choice, will often produce more meaningful results than the casual application of even the most powerful analytical techniques by workers whose grasp of the problem, and perhaps also of the technique for its solution, is imperfect. Jeffers (1967b) has pointed out that when multivariate methods have been discarded as failing to contribute to the solution of a problem, an interpretation by a more experienced worker will often show that the initial use of the technique was poorly judged.

In general, phytosociological research has exploited the battery of special techniques which it has invented with flair and persistence, and one might think that the study of animal communities would be further advanced if it had been prosecuted with equal devotion. As always happens, there is a tension between the advantages of maintaining special techniques in terms of which groups of workers have learnt to conceptualize, and the advantages of using general methods of analysis that most workers can discuss in the reasonable hope of being mutually intelligible.

That general methods can be applied to phytosociological problems to good effect has been shown by Orloci (1966), who made a principal component analysis of the frequencies with which certain plants occurred in nine distinct habitats within a dune-and-slack vegetation complex. The ordination that was produced by the first two of these components is shown in Fig. 31: a reasonably clearcut separation of the habitats is evident. The efficiencies of principal components, and of two other methods of ordination, were compared, and principal components proved to be materially more efficient in accounting for the interstand distances, whether raw frequencies, frequencies transformed by the arcsine square root transformation, or simple presence or absence data were involved.

To provide for views to the contrary, we must quote Ivimey-Cook and Proctor (1967) who found that a varimax rotation of principal components axes gave a more informative picture of floristic relationships within an east Devon heath than did the unrotated principal components analysis. A principal coordinates analysis might have been the more appropriate multivariate method to compare with the varimax factor analysis, but one can see the point that the authors were trying to make, that the components analysis revealed

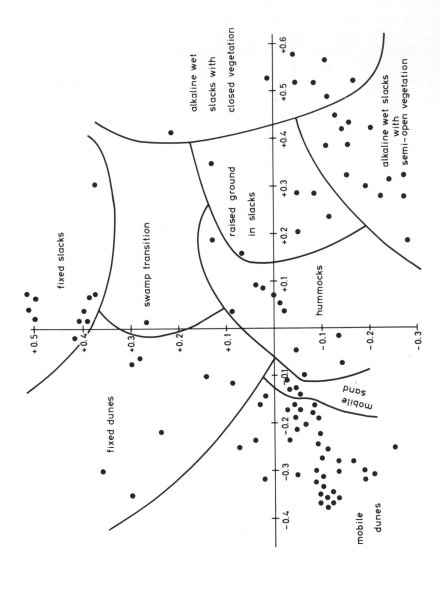

Fig. 31. Principal components ordination, based on frequency data, of plant habitats in "dune-and-slack" vegetation. Redrawn from Orloci (1966).

axes oriented along such polarities as abundance-scarcity: soil moisture-dryness: and base status, whereas the factor analysis reoriented these axes to polarities based on floristic relationships in which the experimenters were more immediately interested.

On the other hand Proctor (1967) found that ordinations of some British liverworts gave readily interpretable polarities. The first axis, as usual in phytosociological studies analysed with principal components, reflected abundance (analogous to the "size" component of zoologists) whereas the second axis contrasted the mountainous Northwestern regions with the Southeastern lowlands. A third axis was polarized between those areas with oceanic and montane plant distributions. Noy-Meir and Austin (1970) note that ordinations may well be non-linear; rotation of the principal components would then lead to serious difficulties.

Phytosociologists currently emphasize the contrast between the classification of a community (Orloci, 1967a) and its ordination, the former being achieved by various techniques such as association analysis and the latter by more obviously multivariate methods such as principal components analysis (Greig-Smith, 1964) and principal coordinates analysis. However, the objectives and resources of the experimenter are usually mixed, rather than confined to one of these approaches. It is entirely understandable that differences of emphasis should exist. For instance, in two papers on the analysis of data from tropical rain forests, Webb et al. (1967) consider that a classification should be applied first, and ordination invoked only if the classification proves unprofitable: they commend Gower's ordination procedure (Gower, 1967a) in such cases. Writing from a different standpoint, Greig-Smith et al. (1967) make the point that classification is more likely to prove satisfactory at high levels of vegetational variation and ordination at low levels: they found that ordination within the forest types revealed correlations between composition and environment not otherwise apparent.

Ordinations of grassland vegetation and associated soil attributes are discussed by Gittins (1969) and by Ferrari et al. (1957), whereas the soil characters are ordinated with beechwood vegetation by Barkham and Norris (1970).

A valuable comparison of the possible ways of examining floral assemblages has been published by Moore et al. (1970); this study shows that the classical techniques associated with the Montpellier and Uppsala schools constitute a compromise between a thoroughgoing ordination and a thoroughgoing classification. In this way much of the information which is lost when one or other extreme approach is adopted can be retained, and this preservation of information is much to be desired; it corresponds to a considerable extent with the practice of reification (identification of the physical, biological or geological meaning of the axes of variation) that we have recommended elsewhere in this book, since reification amounts to writing back on to an ordination the information that would have been preserved in a classification, or conversely. Biometricians from several disciplines are moving towards this less partisan and more broadly-based view of the classification *versus* ordination controversy, as Clunies-Ross (1969) has noted for some methods of handling market research surveys.

Nevertheless, the comments that Moore *et al.* make on the time taken to compute the various sets of results apply to work on a computer of an obsolete generation; so far as the principle components analysis is concerned, the time seems to be one or two orders of magnitude greater than would now be needed, computer time is rarely a limiting restraint on the choice of multivariate methods except for the special case of polythetic divisive classification (see p. 290). It is interesting that Moore *et al.* found a number of types of variation represented along each of their component axes, amounting to what we have termed symbatic components of variation.

17
The Quantitative Comparison of Faunal Elements

Before considering the assessment of faunal similarity, we need to consider the choice of a suitable measure of the extent to which two faunal assemblages resemble one another. There are in the literature numerous coefficients of association which measure similarity (similarity indices) (the Jaccard coefficient, Sorensen's QS, Mountford's I, etc.; see Chapter 22, on cluster analysis for the properties of several such coefficients). Many of these coefficients, which are basically a count of the number of species common to both assemblages, can be adjusted to allow for the total number of species present in each assemblage. Now, there are certain occasions when such an adjustment is desirable, and certain occasions when it is not; the choice is illustrated by the following example.

Consider the faunal assemblages in two habitats of which one (A) contains 40 species and the other (B) 10, all the ten in B being included in the forty in A, the remaining thirty of the species of A being absent from B. The question now arises as to why these 30 species are absent from B, to which there are two broad categories of answer. Either the 30 species do not occur in B because that assemblage inhabits a region inimical to the 30 species, i.e. the absence affords positive information about the nature of B relative to A, or the 30 species do occur in B at such low densities that they do not happen to have been found in the samples taken, i.e. the absence tells us little or nothing about the differences between A and B. We assume, for simplicity, that the sampling effort and sampling efficacy are comparable for the two habitats. If an adjustment is made to the measure of similarity, to allow for the different numbers of species in the two assemblages, the implication is that absences are telling us little or nothing; if we consider the absences to be informative, we shall not want to make adjustments of this kind. Presences and absences may, in fact, provide distinct, even contradictory, evidence about ecological relationships, as Field (1969) has shown, in a study in which joint absences were sometimes treated on the same footing as joint presences and sometimes distinguished from them.

In many practical situations our knowledge of the reasons for absences will be inadequate for a firm decision on the matter. Nevertheless, in zoogeographical investigations, and much ecological work, one might think that absences from faunal assemblages are too useful to jettison, since in such cases we would not usually attribute absences to the vagaries of sampling variation.

The simplest unadjusted measure of the difference between two assemblages consists of the count of the number of dissimilarities, that is to say the number of species occurring in either assemblage but not in the other. This measure of dissimilarity is, of course, the square of the generalized distance between the assemblages when the data are dichotomous and correlations are neglected. The advantages of its use, however, extend beyond that of bringing measurements of faunal (and perhaps floral) affinity into the same theoretical framework as the other topics mentioned in this book; use of the generalized distance has the powerful property of being invariant under scale changes, which in practical terms means that generalized distances are not altered when species are added to, or subtracted from, assemblages unless these modifications alter the number of dissimilarities between the assemblages. In this way we can use the generalized distance to measure affinities between assemblages irrespective of the total numbers of species in each assemblage, a considerable advantage in zoogeographical work, where the inclusion of an atypical assemblage, most of whose elements are absent from the remainder, can be a nuisance if similarity coefficients are employed, seriously disrupting the additivity of the affinities. In fact, it is only with generalized distances that additivity can be fully preserved by virtue of the theorem of Pythagoras. As we shall see later in this chapter, the generalized distance as defined here indicates the direction, as well as the magnitude, of faunal change by virtue of its vector properties.

It is, then, for reasons of the kind enumerated above that we prefer to work within the framework of the generalized distance for zoogeographical and phytosociological purposes rather than to adopt the special coefficients whose use, though justifiable in special circumstances, is likely to waste expensively acquired information. A history of some of the various coefficients invented (and reinvented) in zoogeographical work is given by Udvardy (1969).

One group of coefficients of similarity (or, more appropriately, dissimilarity) not mentioned in Udvardy's review is the distance measurement, to the advantages of which we have already called attention above; for the generalized distance which we recommend is not the only distance measure in zoogeographical work. Matsakis (1964) used a distance coefficient which contains the sums of the dissimilarities between assemblages each standardized by the number of species "at risk" averaged over the two habitats compared. The important advance made by Matsakis is to concentrate attention on the dissimilarities rather than on the similarities, since we learn how habitats differ by studying the faunal elements which do not occur in one of them but do occur in others. Species common to, or absent from, both of two habitats tell us nothing about the comparison of these habitats; this is a source of vagueness when attention is focused on similarity indices. Matsakis' coefficient (which will converge to a constant fraction of the generalized distance when the numbers of species in the two habitats are equal) has the disadvantage of not being additive, in the sense that a comparison between habitats A and B cannot be directly assessed against that between B and C, and to that extent is less suitable for zoogeographic work; however, it has the property that two habitats each of which is almost barren, and hence with very few species, but barren for different reasons, for instance the one a saline marsh, the other an arid desert, will appear to be quite dissimilar on Matsakis' computation but quite similar on that using the generalized distance. Again, it is a matter of deciding what kind of information one wishes to preserve in the calculation which is one reason why no finality in the matter of coefficients of similarity is to be expected. Matsakis' coefficient has been used to good effect by Cassagnau and Matsakis (1966) in studies of collembolan ecology.

All these coefficients are either scalar quantities or vectors used as scalars. Since faunal or floral change has a definite direction as well as magnitude, it seems worthwhile exploring the possibilities of using overtly vectorial techniques.

It seems to be generally agreed that the elements of a fauna are less closely associated into communities than is often thought to be the case with the floral elements. However, for many purposes the quantitative comparison of faunal elements is of interest; for instance, the effects of pollution, or of agricultural practices, on the animal associations may need to be assessed. Where these elements are scored as being present or absent, on a 0,1 basis, the problems involved are similar to those met with in numerical taxonomy, each habitat being regarded as an "organism" with the various faunal elements as its "characters".

Whereas in numerical taxonomy the characters are chosen so as to be invariate within the taxa studied (Operational Taxonomic Units), in ecological studies the organism may vary in presence or absence from one sample to another within a habitat: for it is inappropriate to take the replicate samples from any one place within the habitat when two distinct areas are being assessed, what is required is the general measure of variability within the habitats as a whole, as estimated by taking samples from all appropriate localities within the habitats. Moreover, whereas in numerical taxonomy and in zoogeographical work the characters are rarely assessed quantitatively, in ecological (biocoenotic) investigations the abundance of the animals may be of interest, as well as their presence or absence. A comparison of analyses using abundances with those using the 0,1 scoring is illustrated here, because such comparisons are very rare in thy literature, although suitable data must exist in most ecological laboratories.

We use a typical set of data derived with his kind permission from the work of Dr. R. D. Goodhue, of Trinity College, Dublin, who sampled a series of streams on a hillside in Co. Tipperary, south-western Ireland. Three of the sampling points were in streams suspected of being polluted as a result of local silver mining (in this context pollution is understood to include silt and other detritus from the operations as well as heavy metal toxicity). Eleven samples were taken for comparison from the hillside west of the area in question, and nine samples from the hillside to the east of the area. Twenty faunal elements (invertebrate) were counted in these 23 samples, which were taken so that a representative assessment of the fauna in the water, in the vegetation, and on and in the bottom sediments and rocks could be made.

Ideally, each faunal element should be identified to the species level for comparisons of this kind, since the ecological reactions of the different species may vary considerably. However, apart from the practical difficulties of such detailed identification, there may also be problems because some species occur infrequently, thus adding to the background "noise" in the experiment, although when pooled into, say, generic or family levels, they may provide useful information. In this example the less common species were accumulated into such general categories as "miscellaneous annelida" etc., although identification had often been made to lower taxonomic levels.

Two principal coordinate analyses were performed on the data, the first on the untransformed counts, and the second on the presence-or-absence data for each type of organism. In both cases, the roots of the determinantal equations fell off very slowly, suggesting that a large part of the variation between the localities was "random noise": some three-quarters of the total variance was not taken up by the two first roots. The prospects for any very detailed comparison of the techniques are, therefore, poor in this example.

Table 3 shows the rankings of the 23 localities along the first two axes of variation for both counts and dichotomies. The first conclusion is that there is no evidence for the suggestion that the

Table 3

Rankings of the 23 localities in Tipperary, Ireland, compared in respect of 20 groups of invertebrates with three samples from a place potentially subject to pollution (P_{12-14}). Localities W_{1-11} lie immediately to the west of the suspect area. Localities E_{15-23} lie immediately to the east of the suspect area

Locality code	Ranks along axis I Assessment by:		Ranks along axis II Assessment by:	
	Counts	Presence/ absence	Counts	Presence/ absence
W_1	3	8	5	23
W_2	8	6	18	18
W_3	1	18	4	19
W_4	9	1	12	11
W_5	23	10	10	6
W_6	10	4	6	13
W_7	14	9	21	5
W_8	13	19	19	9
W_9	17	2	16	14
W_{10}	19	23	22	15
W_{11}	12	5	9	17
P_{12}	11	20	14	10
P_{13}	4	15	23	3
P_{14}	7	21	8	21
E_{15}	22	13	1	7
E_{16}	21	3	3	2
E_{17}	16	14	20	1
E_{18}	20	16	11	8
E_{19}	2	12	2	22
E_{20}	15	11	7	12
E_{21}	18	22	17	4
E_{22}	5	17	15	16
E_{23}	6	7	13	20

three putatively polluted localities ($P_{12\text{-}14}$) differ in any significant way from those to the east or west of the area at risk. This conclusion differs from the result of a subjective first appraisal of the raw data. The second conclusion is that there is no rank correlation, as judged by Spearman's ρ, significant of any structure in the analysis using direct counts that can also be found in the analysis using presence-or-absence data. However, it would be rash to assume that this result indicated a serious loss of information due to the use of presence-or-absence data, rather is it likely that there was little or no structure initially in the experiment to be preserved. Such a high "noise" background seems to be characteristic of ecological experiments with animals, and no doubt contributes to the less well developed concept of animal, as opposed to plant, communities.

A high "noise" background is also a frequent component of studies involving sedimentary data. This was found to be so in a study on sediments in the Niger Delta (Reyment, 1969e) in which the latent roots of the association matrix drift off distressingly slowly.

One way of cutting down the noise would be to use the rank order of the localities in respect of the individual species abundances instead of their absolute abundances, as a compromise between including abundances with their high "noise" content and rejecting information about it altogether in favour of presence-absence data. The use of order statistics in multivariate studies has been developed by Mantel and Valand (1970).

A quite different possibility is that these experiments are, in an extreme sense, hypermultivariate, in that each faunal assemblage differs from the remainder along an axis of variation peculiar to it; such a situation would be hard to distinguish from that in which a high noise background was present. In zoogeographic work attempts to measure abundances are almost meaningless, giving place to attempts to assess frequencies where the data are sufficiently extensive to allow this luxury. For most purposes, however, we have to make do with presence or absence data for zoogeographical comparisons, and it is convenient that we can measure the square of the generalized distance between regions by adding up the number of species which occur in each region but not in the other, subject to the restrictions noted by Kurczynski (1970). Comparisons based on dissimilarities afford a healthy check on conclusions drawn from similarities, for we might remember, from time to time, that where there is a large number of mathematical tools available there is the temptation to select the one that throws one's pet ideas into brilliant relief; an equally apposite tool that makes use of different information is then much to be desired.

If one wished to illustrate the hypothesis that there was a strong "southern" element in the collembolan fauna of the blanket bogs of Co. Mayo, Ireland, one could point to the fact that about half the 70 species living there also occurred in Portugal, and about one-third were even to be found in North Africa. This impression could quite properly be enhanced by adding that only about one-sixth of the Irish blanket bog species had been found in fenland in south-east England, leaving the reader to infer that the Portuguese fauna was much more like that of the Irish blanket bog than is the English fen fauna. However, a study of the dissimilarities shows that there are only 43 species occurring in the English fen but not in the Irish blanket bog and conversely. In the case of the Portuguese fauna there are 188 dissimilarities. The impression given by the dissimilarities is thus quite different from that given by the similarities. Both are necessary for a proper understanding of the zoogeographical situation; there is an important "southern" component of the Irish collembolan fauna but it is also true that the Portuguese fauna is much richer, and in that sense dissimilar to that of the blanket bogs of Ireland.

Whereas these dissimilarities add up to give an ecological distance (more strictly, a squared distance) between the regions, the vital point to appreciate is that they also form a vector describing the nature of these differences, a vector whose coefficients are 1 whenever two regions do not share a species and 0 when they do. Only dissimilarities, then, tell us about the direction of faunal changes, similarities being quite uninformative on this topic. The angle between the two vectors tells us how the direction of faunal change is progressing. An advantage of working with ecological distances is that we are accustomed to associating distance and direction in everyday life; the distance to a place only takes on its full meaning if we know its direction. Thus we are led to understand that we cannot equate the ecological distance between Ireland and Portugal with that between Ireland and England; the vectors that describe the directions of faunal change make an angle, 74°, which is sufficiently large to release us from the temptation to make snap judgements based on scalar measures of affinity.

The cosine of the angle is a fraction whose numerator consists of the sum of the number of species occurring in the Irish but not in either the English or Portuguese faunal lists (20 species), together with the number not occurring in the Irish list but present in both the others (four species). The denominator of the fraction is the square root of the product of the number of species dissimilarities (188 for Ireland-Portugal; 43 for Ireland-England). The calculation and use of angles between vectors is discussed in Chapter 2.

If we then expand our study to include many different regions we find that we are, indeed, in a hypermultivariate situation; and only when we begin to make sense out of the direction of faunal change can we reasonably hope to place zoogeography on a biologically meaningful quantitative basis. The application of both Q and R analyses to presence or absence data of collembolan distributions, by Bonnet *et al.* (1970), shows how fruitful such an approach can be.

As an example, let us look at the change in the collembolan fauna with latitude, in the light of these multivariate techniques.

We can compute the angles between the vectors describing the contrasts between the Irish bog fauna on the one hand and the separate faunas of Iceland (Bödvarsson, 1967), Portugal (Da Gama, 1964), and the Maghreb of North Africa (Cassagnau, 1963; Lawrence, 1963) on the other. All these regions lie roughly on a north-south axis, but the faunal changes that occur as we move south cannot be represented on a single plane still less on a single line. The extreme vectors representing faunal contrasts between Ireland-Iceland and Ireland-Maghreb include an angle of about $64°$, whereas the coincidentally equal intermediate stages of the "swing" of the vector are Ireland-Iceland to Ireland-Portugal, $74°$; Iceland-Portugal to Ireland-Maghreb, $74°$. The fact that the vectors are not coplanar is shown by the fact that the angles formed by the intermediate stages add up to more than the full sweep of the moving vector which represents faunal change; the movement of this vector is thus analogous to that of a windscreen-wiper on a strongly curved windscreen. The main sweep of the vector across the windscreen represents the continuous faunal changes with latitude, whereas the curvature of the sweep, corresponding to the depth of the curved windscreen, represents the growth and decline in "richness" of the fauna as one moves from the relatively barren Icelandic habitats, which are cold and wet, to the Maghreb, which is also relatively barren through being hot and dry, by way of the cool temperate climate of Ireland and the warm temperate climate of Portugal. However, we must not forget that in addition to the changes of direction of faunal composition, we have also to deal with the magnitude of the changes; as we move southwards from Iceland, the amount of change increases progressively, imposing a third dimension on our overall picture, and corresponding to a lengthening of the arm of the windscreen-wiper as it pursues its stroke. The net result is a spiral track as illustrated in Fig. 32. The three basic dimensions of variation are:

(i)   The dimension of faunal diversity (numbers of species) already mentioned, contrasting Iceland and the Maghreb with Ireland and Portugal.

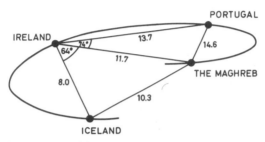

Fig. 32. The spiral pathway of faunal change in the collembola between Iceland and the Maghreb (N.W. Africa) based on the generalized distances between the faunal lists for the four regions.

(ii) A dimension of faunal balance, represented by a shift in the relative importance of the collembolan families Onychiuridae and Isotomidae, more diverse in the northern latitudes, in favour of the Entomobryidae, whose diversity tends to expand southwards. This dimension contrasts Iceland and Ireland with Portugal and the Maghreb.

(iii) A dimension which sets off the Maghreb countries against the rest, and may be interpreted variously as an indication of the beginnings of the African fauna, as distinct from the European, or as the specific effect of hot dry climates as against cold wet ones. The two interpretations are by no means exclusive. This dimension is parallel with the axis of the spiral.

The diversity of faunal change with descending latitude could not have been inferred from a study of the similarity coefficients; use of the Jaccard coefficient suggests that Iceland is faunistically nearer to Ireland than are either Portugal or the Maghreb, and that the last two are about equidistant from Ireland. One can see how these coefficients are consistent with the spiral relationship between the faunas of the four places, but not how that relationship could have been deduced from them. The actual dimensionality of the diagram is of interest; the fact that we can describe such extensive faunal changes in a space of only three dimensions testifies to the extremely wide range of many collembolan species. We thus have a useful measure of the general tendency of any group of organisms to respond to geographical dispersion by speciating, a virtually unstudied aspect of zoogeography.

More sophisticated readers may have noted that the total faunal change would require a tensor for its representation and also that there is hardly any background noise to disturb this analysis, because only presence-or-absence data have been used.

This example has, naturally, been much oversimplified in this morphometrical context; much more would need to be said and done on the biological side to make it an acceptable exercise in zoogeography. Nevertheless, it does illustrate the fact that zoogeographical changes can be analysed into vectors which have definite biological interpretations; in this way the whole of the fauna can be used in so far as this is known for any given region; there is no need for the highly subjective traditional forms of zoogeographical analysis, in which some species are discarded as being "ubiquitous", others because they are held to be "accidentally introduced" (by no means always on objective criteria); the resemblance of zoogeographical problems to some of the growing pains of taxonomic theory is striking, and parallels can also be found in phytogeographical studies. Perhaps some of the reasons for the vagueness of zoogeography lie in the fact that so much can so obviously go wrong with comparisons of faunal lists; unremarked misidentifications and name changes, disparities of sampling effort and effectiveness, difficulties associated with the non-operational nature of the species concept, exacerbated when animals cover a wide range, all these and more may seem to serve as reasons why one should not take quantitative zoogeographical comparisons too seriously. But later in this book (Chapter 25) we shall see that a new element is entering into considerations of this kind, based on the robustness of multivariate methods; at the moment this robustness has only been examined within the taxonomic context, but there is every reason to hope and expect that it will apply to zoogeographic analyses as well as to taxonomic ones.

18
Genetical Aspects of Morphometrics

There have been relatively few well-conducted multivariate morphometric investigations with a genetic basis, considering the importance for plant and animal breeders of the conformation of the organism. Cock (1966) and Rouvier and Ricard (1965) have summarized most of what has been done in this field, and Cousin (1961) has long insisted on the need for studies of the inheritance of shape and size to be placed on a sound footing. Lefèbvre (1966) has given a detailed review of the studies (multivariate and otherwise) devoted to the growth and conformation of cattle.

Occasionally a direct relationship has been suspected between the action of a single gene and the promotion or suppression of growth along one or other of the principal components of the dispersion matrix of a number of measured variables. For instance, Kraus and Choi (1958) extracted principal components which reduced 12 measured characters of the human foetus to four patterns of growth. Because they were able to obtain teratological specimens in which the failure of a gene, by mutation, entrained the suppression of the pattern of growth constituting its phenotypic expression, they were able to speculate that each of the observed patterns of growth of the long bones of the human foetus were controlled by a single gene. A consideration of matters of this kind is doubly useful; it illuminates genetics by enabling the experimenter to handle large numbers of "unit" characters in meaningful groups, and it illuminates morphometrics and in particular the subdivision known generally as numerical taxonomy, by throwing light on two fundamental hypotheses. These hypotheses are, firstly, the nexus hypothesis, which assumes that every character is likely to be affected by more than one gene (or, conversely, that most genes affect more than one character, by virtue of pleiotropism) and secondly, the non-specificity hypothesis, that no distinct classes of genes affect one

241

class of characters exclusively. Sokal and Sneath (1963) who are principally responsible for the enunciation of these hypotheses, note that one relies on the non-specificity hypothesis to justify taking characters for a morphometric study from, say, the external morphology of animals without needing to examine the internal anatomy. However, there is now adequate evidence that the non-specificity hypothesis is only partly valid (Michener and Sokal, 1966) so that the experimenter's choice of characters should be spread as widely as possible.

Actual genetically oriented multivariate experiments are few. Eickwort (1969) has commented that multivariate methods are particularly well suited to the task of following the results of genetical experiments on form, and has used ten characters to examine the differential variability of male and female wasps, males being haploid in these insects. The results agree with those of Blackith (1958) in showing that males are more variable than females. White and Andrew (1959) found that in an Australian eumastacid grasshopper which is, in some populations, polymorphic for pericentric inversions in two autosomes, the "standard" sequences serve to augment the weight of the adults, whereas the "Blundell" and "Tidbinbilla" sequences served to decrease the weight. Somewhat surprisingly, these effects were consistent in four separate populations. As White and Andrew comment, this finding implies an amazing evolutionary stability for the size-determining gene complexes over a period probably exceeding a million years.

One of the most striking features of morphometric analysis is its ability to uncover patterns of growth, involving all or at any rate some large part of the body, which must have persisted unchanged over such long periods of time that many other evolutionary processes have gone on in the animals concerned without disturbing these stable patterns. A good example is that of phase variation in locusts (see p. 41) where the change of adult form when the immature stages are reared in crowds is almost identical in not very closely related genera, even though each genus may contain species which do not show this phenomenon (Blackith, 1962). It is of course open to question that phase variation has arisen many times in the course of the evolution of the Acrididae, the perpetual choice between a monophyletic and a polyphyletic concept is never wholly to be forgotten, but at least the capacity to undergo such a change may have arisen but once; it is known to be associated with the sex chromosomes. Parallel, or homologous, variation, must often present this dilemma (Blackith and Roberts, 1958) and is of great concern to geologists (Reyment and Van Valen, 1969).

Very consistently, the number of independent axes of variation disclosed by multivariate studies is less, usually much less, than the number of characters employed in the analysis. It is interesting that an assumption closely related to this proposition has been made by Lawley and Maxwell (1963) in order to make progress with maximum-likelihood factor analysis. Perkal and Szczotka (1960) invented a method of analysis which picks out such correlated characters but, as Jeffers (1965) points out, fails to preserve the vectorial nature of the full analysis. There is a certain amount of evidence that natural selection does not operate directly on individual characters but on the degree of interlocking between one dimension of variation and another, the suite of characters undergoing change as a unit in a fashion somewhat analogous to the evolution of "supergenes". This view, of course, is widely held by conventional taxonomists, though it is usually expressed in other terms. Even characters which vary substantially independently within a homogeneous population (and thus give rise to what are virtually unit vectors in a principal component analysis) form part of a more complicated contrast of form as reflected in the discriminant (or canonical variate) linking two distinct groups of organisms. Characters giving rise to unit vectors within, and non-unit vectors between, homogeneous groups of organisms include the weight, number of antennal segments, and elytron length of acridid grasshoppers (Blackith, 1960) and the elytron length of a pyrgomorphid grasshopper (Blackith and Kevan, 1967). Sokal (1962), Grafius (1961, 1965), Olson and Miller (1958) and Schreider (1960) have all, in their different ways and in widely separated fields of study, reached the conclusion that natural selection operates on characters in stable groups which may be discerned by a multivariate analysis. These stable groups are not necessarily the correlated responses to selection that arise in many genetical experiments (see, for example, Scossiroli, 1959) but must often include them. Bailey (1956) has extracted principal components corresponding to the genetical and environmental components of morphogenesis in mice but this distinction may not always be easy to make: Hashiguchi and Morishima (1969) discuss the estimation of such principal components. The subject is of consequence not only because of its interest to animal breeders but also because of the distinctions we draw later in this book (Chapter 29) concerning the possibility of distinguishing "genetic" and "environmental" principal components in palaeoecology.

Even such "obviously" adaptive traits as robust limb-bones in heavier terrestrial animals may, as Jolicoeur (1963a) has found, be associated with a general, probably genetically determined, process throughout the body rather than by adaptive growth as modified by environmental influences. Jolicoeur found that once the principal component reflecting size differences had been removed from the analysis, the remaining "shape" components had markedly unequal variances, and he suggests that those with the lowest variances represent highly canalized patterns of development, which are well buffered against environmental influences.

There is very little evidence for some traditional ideas, such as that changes in body form are mediated through stresses felt in early life before growth has ceased. In the case of Jolicoeur's martens, there seemed little support for an hypothesis of Scott (1957) according to whom robustness depended on the degree of use of the musculature and on the action of sex hormones. Similarly, when Blackith and Kevan analysed growth in a pyrogomorphid grasshopper, they found that the inverse relationship between the width of the mesoternal interspace and the elytron length formed part of a genetically determined pattern of growth and was not directly influenced by differential muscular stresses in long- or short-winged forms.

That body conformation as a whole tends to be inherited is known to any animal breeder and can be followed quantitatively, as revealed by Lefèbvre's (1966) multivariate studies on Norman cattle. He made 14 measurements on the progeny of 37 bulls. The shapes of the progeny clustered strongly, on a chart whose axes were the first two canonical variates extracted, according to the sire, but a maternal effect of the same general magnitude as that of the sires was noted.

Lefèbvre suggests that the groups of progeny also tend to form clusters representing "typical" conformations equivalent to such concepts as "stocky", "lean" etc. in humans. It may be that different breeders work towards ideal forms and since it is notorious that breeder's concepts of ideal form have had little to do with the economic value of the product and a great deal to do with fashion, it seems plausible that different breeders will be working towards rather different ideal forms. The work of the Centre National de Recherches Zootechniques in France indicates an economically important field for morphometrics in helping to rationalize breeding patterns for stock (see for example, Rouvier and Ricard, 1965).

Multivariate discriminants for the selection of animals and plants were among the first developments of morphometric techniques and reached an advanced stage early in the history of the subject (Smith, 1936; Hazel, 1943; Rao, 1953). Still further progress has been made by Rouvier (1969).

Multivariate methods of analysis have also been able to clarify the situation in the polecats *Mustela putorius* and *M. eversmanni,* which are sympatric, and between which some hybridization has been alleged. Rempe (1965) showed that this suggestion could not be supported from a generalized discriminant analysis of the skulls.

Hybridization between the voles *Apodemus sylvaticus* and *A. tauricus* has been studied by Amtmann (1965) using 17 morphometric characters, of which five accounted for 88% of the discriminating power in the males and 50% of it in the females. Hybrid individuals were few in the zones of contact, which was quite narrow in relation to the mobility of the animals. In such work as this the careful morphometrical studies could perhaps have been enhanced by cytogenetic ones, since it is known that the genus *Apodemus* shows chromosomal rearrangements.

A full study of the nature of hybridization and the selective forces operating at tension zones where two species meet is of the greatest value in understanding the evolution of the species involved: the morphometric aspects of the work are then useful but where a cytogenetic analysis is possible the combination of the two approaches is greatly to be desired. This statement also holds good for experimental crosses, and Lecher's (1967) cytogenetic analysis has added to the value of the morphometric studies of Bocquet and co-workers (Bocquet and Solignac, 1969). For want of the necessary cytological skills, or because the material to hand does not show suitable chromosomal differences, the morphometric approach may be the only one available. Rising (1968) has investigated the body form of two chickadee species in a tension zone in Kansas, on which eight morphometric measurements were taken. It seems likely that interspecific mating, occasionally followed by back-crossing, was encountered, and field observations suggested that the hybrids were at some selective disadvantage during reproduction.

A somewhat similar situation was found in the Louisiana grackles, where putative subspecies showed a stepwise cline in colour but a continuous cline in the body conformation as assessed by discriminant functions based on four morphometric characters (Yang and Selander, 1968).

Some classical correlation studies on the inheritance of human body form can often be expressed to advantage in multivariate form, as de Groot and Li (1966) have shown.

One problem that faces the investigator, who is interested in rates of evolution, is the measurement of the amount of morphological change involved, and the techniques by which these changes should be assessed. These problems have been discussed by Lerman (1965) and Marcus (1969) both of whom have recommended the use of the generalized distances for measuring the amount of morphological change. Indeed, the analysis of continuous variation when this is supposed to follow Mendelian inheritance has been fully treated by Weber (1959, 1960a, b).

It should never be forgotten that the shape of an organism is the outward and visible manifestation of properties which must very frequently be associated with other inward and often invisible ones; for instance, Clark and Spuhler (1959) found that the fertility of human beings was different in the various body builds, and similar differences must await discovery in many polymorphic populations.

However, there is a deeper problem concerning the extent to which morphological change reflects genetic change; the relationship is far from proportionality, as any taxonomist knows who has to deal with groups that are similar in form but belong to distinct higher taxa, or groups that are radically distinct in form but belong to affine taxa. Some authors using multivariate methods, for instance Grewal (1962) and Brieger et al. (1963), have virtually taken a proportional relationship for granted, but Goodman (1969) has begun the long and complex task of considering the factors involved. He makes the point that comparably divergent forms will represent greater genetic divergence if derived from an ancestral population of low variability than from one of high variability. This is an arguable point, since the divergent forms might be derived from the most divergent members, genetically speaking, of the parent population, so that the variation which is, as it were, discounted, is variation inherent in the ancestral population whose translation to the progeny one might well wish to see retained in the measure of overall genetic variance. Goodman considers that Sokal's (1961) distance function is superior to the generalized distance as a measure of genetic divergence; the last word on this topic is unlikely to have been written, particularly as it forms part of a series of related issues central to the evolutionary interpretation of taxonomic practices. The situation in bacterial taxonomy is complicated by the fact that gene transfer may take place across wide phenetic taxonomic gaps (Jones and Sneath, 1970).

Oxnard (1969) reviewed the subject of shape and function and the necessity of the morphometric approach for an adequate appreciation of the subject. The relationship between evolutionary changes, assessed phenotypically, and the underlying genetic differences has been discussed for human populations by Cavalli-Sforza (1966) and by Cavalli-Sforza and Edwards (1964).

Williams *et al.* (1970) analysed four morphometric characters of the spermatozoa from nine strains of mice using both canonical and discriminant analyses. They found that adopting the multivariate approach improved the discrimination between the spermatozoa from the various strains, sometimes markedly so. Nevertheless, the full representation of the relationships between the nine strains required four dimensions, so that this example fails to illustrate that reduction of the effective dimensionality of a set of comparisons which has often been claimed as a major objective of multivariate studies, an objective which is, moreover, usually attained. However, it became clear that a full analysis of these relationships required a multivariate interpretation precisely because of the hypermultivariate nature of the underlying biological situation; for instance, earlier attempts at univariate analysis had shown no significant differences for some of the measurements even though to the experienced eye the spermatozoa were patently distinct. The multivariate approach showed that such univariate tests were maloriented.

Some dominance relationships were discovered when sperm from various hybrids between the strains of mice were measured and analysed. Williams *et al.* suggest that multivariate studies could help in investigations of the epigenetics of spermatozoa.

One way in which we can be reasonably sure that morphological change is being accompanied by genetic change is to measure the effects of genetic isolation in wild populations separated by a barrier to interbreeding, and compare these effects with those arising in populations separated by distances equivalent to those which separate the genetically isolated ones, but without any barrier to interbreeding. Rees (1970) has done this for white-tailed deer, by comparing populations on either side of the Straits of Mackinac with others as geogrpahically separated but without the particular, probably almost complete, barrier to mating that the Straits constitute. Rees found that when morphological change was assessed using ten cranial characters (for measuring changes in the skull) or seven dental and seven skeletal characters (for measuring changes in the mandibles) the differences attributable to the barrier of the Straits, which have been open for some 4000 years, exceed those which are to be found in populations separated by an equivalent distance but not by the Straits. The various characters used by Rees were combined into canonical variates, although these are called discriminant functions in his text.

There is, of course, a vast literature of genetical studies in which one character is measured or counted, or in which several characters are assessed separately as a consequence of some genetically oriented experiment. Whilst we admit freely that our self-imposed limitation to multivariate experiments is quite artificial in this context, the other choices before us were to omit all reference to the genetical aspects of morphometrics, or else to lengthen this chapter out of all proportion to its proper place in this book.

19
Co-ordinated Applications of Multivariate Analysis

There are so many ways of looking at multivariate problems that any one line of attack may be inadequate. When all calculations had to be done by hand there were obvious reasons for limiting the number of ways in which a given body of data was analysed; now that, for most ordinary purposes, computers can be called in to take over the computational work, there are advantages in the examination of as many aspects of the problem as seem reasonably likely to help the experimenter to gain new insights into his material. A commonly used combination of analyses involves the use of a principal components analysis to examine the variation within certain groups, followed by a canonical variate analysis to examine the variation between the groups.

If this particular combination is used there is some interest in knowing how principal components and canonical variates are related: Rouvier (1966) gives a most ingenious derivation of the expression for the canonical variate corresponding to the largest root of the appropriate determinantal equation in terms of the standardized principal components of the dispersion matrix pooled within groups.

It is most undesirable to make comparisons between methods of analysis without a reasonable exploration of the interpretative capacities of each; this is all the more true when the methods are closely related, amounting almost to rearrangements of the same calculation. It is by no means rare to see that one or other method is considered by a particular author to be in some respect superior to another when the capacities of the method considered inferior have been inadequately explored or understood. This is a transient phase of incomplete dissemination of the basic theory of the methods, comparable to that occurring a quarter of a century ago in univariate analysis, when one could find alleged comparisons of the effectiveness of the t-test and the one-way analysis of variance in the biological literature.

For reasons such as this we have included a series of coordinated multivariate analyses, using more than one analytical method, in the hope that these exercises will at least provide the reader with a starting point capable of adaptation to his particular interests.

A Belemnite Species of the Upper Cretaceous of the Soviet Union

The belemnite *Actinocamax verus* from nine localities in the Cretaceous of the Russian Platform was measured in respect of four characters, the length and breadth of the rostrum, the maximum width of the rostrum at right angles to the preceding measurement, and the asymmetry of the alveolar scar. Other characters were in fact measured but were, for various reasons, not used throughout the work. Some further simplifications of the original rather exhaustive analysis by Reyment and Naidin (1962) are made in this presentation.

Principal component analysis

Each of eight samples was subjected to a principal component analysis and it was found that the eight sets of latent vectors could be grouped into two sets; one set comprised the four samples from the Campanian horizons, and one from the Santonian, the other one the remaining samples from the Santonian horizons, one from the Upper Coniacian, and the pooled data for the entire set of eight samples. In both sets the first principal component was essentially a size vector, which for the pooled data was (0.96, 0.13, 0.20, 0.12) showing that most of the variation was in the largest length measurement.

The second principal component differed considerably between the two sets. In the first set it was a shape vector with most of the variation in the fourth (alveolar scar) measurement. In the second set it was also a shape vector but a distinct one with most of the variation in the third (width of rostrum) measurement. Reyment and Naidin interpret these differences as evidence for a measure of subspeciation during the period intervening between the older Upper Coniacian and the younger Campanian times. This interpretation now seems less plausible than a simpler one, namely that the second and third components of the first set are in fact the third and second components of the second set. This interpretation implies, simply, that although essentially the same changes of shape are occurring in both sets of samples, the vector corresponding to the second latent root in the first sample corresponds to the third latent root in the second, an interpretation made all the more plausible by the small amounts of variation taken up by these roots which, one feels, could easily interchange their order. The size vectors of both sets of samples are essentially parallel. It turns out that this switch is not uncommon for fossil material and has been observed among ostracods, for instance (Reyment, 1966a).

Generalized distance models

Generalized distances computed on the basis of four attributes gave a most unsatisfactory three-dimensional model in which some of the distances had to be "bent" drastically to make them fit. To some extent this undesirable feature is perhaps due to the four-dimensional nature of the "true" model but we always have to recall that such perturbations of the model may well be due to heterogeneity of the dispersion matrices in the original samples. When such heterogeneity was initially tested for, highly significant variation was found, the Upper Coniacian sample being particularly eccentric in this respect, although even after its removal there were still significant amounts of heterogeneity. However, the generalized distance model was still thoroughly unsatisfactory even when the two most aberrant samples had been removed and the heterogeneity had been reduced to a level a little below significance at the 5% level. This distortion of the matrix stems partly from the accumulation of small cumulative allometric changes almost inevitable in a palaeontological problem, and partly from secondary size sorting in the material owing to the effects of water transport. This sorting disturbs the evidence for a chronocline from Lower Coniacian to Lower Campanian times. An interesting byproduct of this study was the dual calculation of the generalized distances on a desk machine and electronically; quite substantial discrepancies were noticed and attributed to rounding-off errors on the desk machine which are liable to accumulate. In the worst cases the larger estimate was almost double the smaller one.

Genetic Clines in Body Shape of Fruit-flies

Five characters of *Drosophila subobscura,* namely, the length of the thorax; width of the head; wing length and width; and the length of the fore-tiba, were measured on samples from 12 localities ranging from Scotland to Israel. Misra (1966) comments that there will in general be a continuum of "normal" shape changes, such as Teissier

(1960) has discussed, assessable in terms of multivariate extension of the allometry equation, together with a number of dimensions of variation representing deviations from this "normal" continuum.

Misra uses a factor model (a principal components calculation applied to a correlation matrix with the unities of the leading diagonal replaced by communalities estimated from the averages of the elements of the corresponding columns) the merits of which are debatable. However, it seems unlikely that the size factor, which he isolated as the vector which corresponded to the largest root of the characteristic equation when so amended, is seriously different from the first latent vector.

The general size factor estimated in this way was remarkably close to a vector $(1,1,1,1,1)$ representing isometric growth in *Drosophila*. The two vectors make an angle of only $1°50'$. There is an apparently uniform north-south cline in the size of the fruit-flies expressed in the size factor, a population from Edinburgh having a mean score along the vector of 90.04 whereas a population from Qiryat Anavim, Israel, has a score of only 83.11, the remaining populations falling between these limits.

The data were then reanalysed using canonical variates; about 88% of the total variance was accounted for in terms of the first three variates. The variate corresponding to the largest root is something of a size vector, but it differs from that extracted from the modified correlation matrix in that it reflects relative, rather than absolute, size changes, with the thorax and head growing less rapidly than the other dimensions of the body. There is a general north-south cline reflected in this variate.

The next two variates do not show any clear-cut geographical trends. The second variate shows a contrast of wing-area and length of fore-tiba, with extremes which are geographically quite close (Lunz, Austria and Switzerland); the third variate illustrates an attenuation of the wings which again has geographically close populations showing extreme shapes; Pavia has the least, Formia the most, attenuation, both are in Italy. But in central Europe altitudes and associated climatic factors vary greatly, so that physical contrasts among the populations of fruit-flies might well be essentially geographical but not obviously so, unless examined with a knowledge of the altitude at which they were taken.

Misra compared the morphological changes in *Drosophila robusta*, published by Stalker and Carson (1947), with those found in *D. subobscura*. It appears that in the North American populations of *robusta*, size does not follow a north-south cline, but that the smallest individuals were bred in the warmest localities. There is thus every likelihood that the cline found in *subobscura* operates through the mediation of temperature.

Morphometric Changes in Heart-urchins of the English Chalk

The heart-urchins of the chalk in England form a virtually continuous series of measurable forms; the fact that there are no substantial breaks in the fossil record makes it all the more difficult to say when one species evolved into another. Nichols (1959) measured four characters (the length, breadth, height, and number of tube-feet) on eight samples of the genus *Micraster*. These samples he identified as *Micraster corbovis; M. cortestudinarum; M. coranguinum; M. senonensis;* intermediates between *coranguinum* and *senonensis: M. glyphus; M. stolleyi;* intermediates between *glyphus* and *stolleyi*: Reyment and Ramdén (1970) have recently reanalysed Nichol's data using principal components, principal coordinates, and canonical variates.

Principal components analysis

In all samples the first component, accounting for at least 89% of the total variance, consists of a general "size" component with all the coefficients positive.

Principal coordinates analysis

Reyment and Ramdén have also subjected the echinoid measurements to a principal coordinate analysis. The latent vectors showed a rapid falling off, being successively 42.96; 21.67; 6.69; and 3.56, although zero roots were not met with until the 87th. However, it is reasonable to consider the information as being concentrated in the first two components of variation.

By calculating the positions of each individual along the first two principal coordinate vectors, as is shown in Fig. 33, we can plot the positions of the measured individuals of *M. coranguinum, M. cortestudinarum* and *M. corbovis*. These three species evidently correspond to the three clusters of points on the chart, although there is a very slight overlap. A most interesting feature of the chart is that the individual points are arranged in a Y-shaped supercluster, which may be interpreted as evidence of phylogenetic connections between *corbovis* and *coranguinum* through *cortestudinarum*. It is significant that passage forms seem to occur in the form of points which straggle between the main clusters.

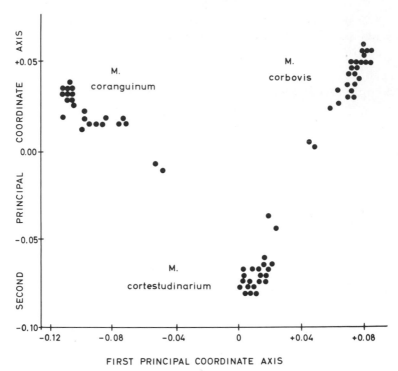

Fig. 33. Principal coordinates analysis of the genus *Micraster* showing the evolution of two species from an ancestral one. Data from Nichols (1959).

Canonical variate analysis

Figure 34 shows the arrangement of the eight samples on a chart whose axes are the first two canonical variates. The first variate, accounting for 75% of the total variance, is one in which the number of tube feet moves in one direction whereas the three other characters move in the other, in fact, this is a variate which makes an angle of only 60° with the "non-size" principal component. Here is fairly firm evidence of the hypothesis that principal components can act as signposts to the evolution of the group in question, given the difficulties of making comparisons between vectors in different spaces.

The second variate, accounting for 20% of the total variance, consists of a contrast between the breadth moving in one sense when the height and the number of tube-feet move in the other. There does not seem to be a pure size canonical variate.

When the means for the eight samples of heart-urchins are plotted along the two canonical variates of consequence, as shown in Fig. 34; there does not seem to be any relation between the burrowing habit and the position on the chart. In fact, alternate members of the arc formed by the means are representatives of deep burrowers and less deep, or non-burrowers. The most striking finding from this analysis is that those samples which are putatively intermediates between two

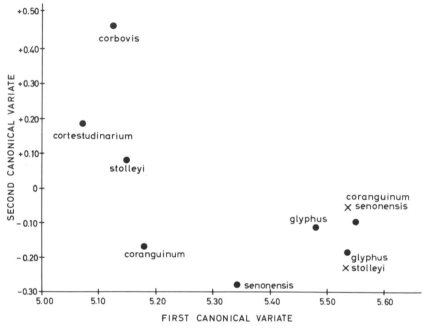

Fig. 34. Canonical variate analysis of the genus *Micraster* based on four morphometric characters. Data from Nichols (1959).

species do not occupy the positions that would be expected on this hypothesis. The *coranguinum* × *senonensis* intermediates are like neither of the "pure" species. The *glyphus* × *stolleyi* intermediates are very close to *glyphus* but not at all intermediate between that species and *stolleyi*. On the evidence of the canonical variate analysis, both samples of intermediates are virtually indistinguishable from *glyphus*. A possible explanation for this result is that there is a general tendency, according to Nichols, for the number of tube-feet to increase in the later geological strata: an increase in this character would tend to move the group which possessed it to the *glyphus* corner of the chart. It must be recalled that these "species" form part of a continuum of forms which can be separated only with difficulty on minute characters which are sometimes influenced by hybridization.

The investigations of Kermack (1954), who used an analysis of allometric growth to sort out the confused situation in these echinoids, have shown that only two main lineages can be recognized. Neither of these lineages can be regarded as completely separated from the ancestral stock. The first of these lineages occurs in the *latus* and *planus* zones of the chalk, terminating in a "dead-end" called *M. corbovis* which was contemporary with *M. cortestudinarum*, but separated from that species ecologically by its shallower burrowing habit. The second lineage appeared when *corbovis* died out; this is *M. (Isomicraster) senonenis* which to some extent occupied the ecological niche vacated by *corbovis*, *senonensis* being a shallow, perhaps partial, burrower. Although considered by some authors to be closely related to *corbovis*, *senonensis* is quite different from that species morphometrically, lying at the opposite corner of the chart.

20
Some Applications of Discriminatory Topology

Although there is now a very substantial literature on the applications of multivariate statistical methods of the kind that Rao (1952) has called Discriminatory Topology, this literature is often hard to come by. It is dispersed in a wide range of journals, and until comparatively recently has often been "buried" in obscure media of restricted circulation. Some useful "background" papers include Pearce (1965), Linder (1963), Rao (1964a, b), Boyce (1964), Blackith (1965), Reyment (1969a) and Yates (1950).

A complete bibliography of applications of multivariate analyses is almost unattainable because of the obscurity of some elements of the literature; however, to give the reader new to this field some idea of the breadth of application of these techniques we include highly condensed summaries of a range of investigations chosen principally to illustrate the field of application, and not, as in the case-histories, for their illustrations of the techniques; the two desiderata converge, however, and the summaries of the papers on applications should supplement the case-histories usefully.

The reader would be wise to bear in mind that earlier applications of factor analyses and some of principal components could often be reworked to advantage by less approximate techniques. Indeed, many early applications of factor analysis which we should have been glad to include for their biological interest have been reluctantly omitted because we cannot be sure to what extent the results obtained are artefacts of the approximations currently used at the time the paper was written.

Virtually no reference to multiple regression techniques, apart from some trend surface studies, has been made in these illustrations of morphometric analysis; there are two reasons for this omission, the first being simply that multiple regression techniques are, as Kendall (1957) puts it, multivariable but not multivariate. The second reason is that the regression approach to morphometric problems has almost always been unrewarding, since the technique is conceptually inappropriate and has been used simply because earlier computers were usually provided with multiple regression programmes but not often with truly multivariate ones.

Multivariate Studies of Nematodes

The rather featureless nature of nematodes makes the application of morphometric methods to nematological problems attractive in principle and difficult in practice. Moss and Webster (1969) have performed a numerical taxonomic study of some parasitic strongylates in which they classified their material on the basis of 62 characters, which were mainly morphological. There are two "classical", non-quantitative, attempts at constructing a taxonomy for the strongylates and that of Moss and Webster came down fairly definitely in favour of one rather than the other. The plant-parasitic and free-living nematodes are even more difficult than the animal parasitic ones, but Barraclough and Blackith (1962), who measured six characters on various members of the genus *Ditylenchus*, found only three underlying axes of variation of which one reflected the onset of maturation and sexual dimorphism, one size variation, and one specific differences of form.

Lima (1968) used 25 unweighted morphological characters on 76 populations of the free-living *Xiphinema* complex; these populations were clustered by a principal components analysis into seven putative species of which four were undescribed.

Sturrock (1963) used three responses of guinea pigs to infestations of *Trichostrongylus* and assessed the effects of different inoculations by calculation of the generalized distances between the batches of guinea pigs receiving the various inoculations. The resulting generalized distance chart (Fig. 35) serves instead of the more usual, but perhaps less effective, multiple regression calculation: it is effectively two-dimensional because the value of the generalized distance between groups I and IV, as determined from the chart constructed from the three distances I-II, II-III, and III-IV, together with the angles between the first and second vector ($0°$) and the second and third vector ($41°$) was 6.64, whereas direct calculation gave a figure of 5.58, in good agreement with the indirect estimate based on the assumption of two-dimensionality. The determination of the dimensionality of the chart seems easier using generalized distances than it does when using multiple regression techniques.

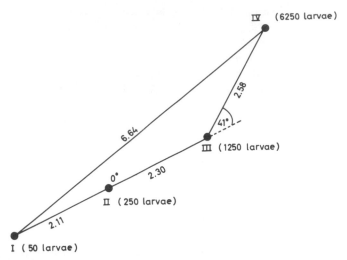

Fig. 35. Two-dimensional generalized distance chart of the responses of guinea-pigs to four graded inoculations of parasitic nematodes *(Trichostrongylus)* based on three responses. Redrawn from Sturrock (1963).

The relative efficacy of the three characters was also assessed by dropping each one in turn from the generalized distance between groups I and IV, and assessing the loss of discrimination involved: the order of decreasing efficacy was "Final weight of guinea-pig": "Egg count"; and "Worm burden".

Further multivariate studies on nematodes are reported by Moss and Webster (1970) together with a general assessment of the value of numerical work in the taxonomy of this group.

Archaeology (Stratigraphical)

The paper by Hodson *et al.* (1966), which forms the basis of this illustration, is important on two counts; firstly that it contains a discussion of the nature of the classificatory process which is realistic, a rare event, and secondly, that it includes an objective test of the validity not only of the multivariate morphometric techniques brought into play (proximity analysis, Shepherd, 1962; Kruskal, 1964a, b) but also of the conventional methods of assessment of archaeological artifacts. One of the most acute difficulties of any discussion of the relative merits of conventional and multivariate methods in classification is that there are no objective criteria against which the classifications can be judged. There is, therefore, a tendency for multivariate techniques to be condemned when they disagree with conventional methods, and regarded as superfluous when they agree. Similarly, when two or more conventional assessments are being compared, the absence of any agreed independent criteria of success places an undesirable emphasis on the personal authority of the protagonists, with the notorious sour relationships when agreement is not attained.

In the example discussed by Hodson *et al.*, both conventional and multivariate techniques of analysis can be evaluated against a chronology derived from the archaeological evidence ("horizontal stratification" in a cemetery) which is independent of that relating to the succession of brooches which forms the subject matter of this analysis.

Proximity analysis splays out the entities (brooches in this instance) in a space of some modest number of dimensions (one to four, say) using some measure of the likeness of the entities (a similarity index, see Chapter 22). The "strain" of this model, i.e. the extent to which the programme succeeds in reproducing the rank order of the inverse similarities by means of the rank order of the inter-point distances of the configuration, is calculated, and, in a new iterative cycle, the programme seeks to reduce the strain by rearranging the entities, and so on, in successive cycles, until a position of minimum strain is achieved. The rank-order of a classification is known to contain a high proportion of the total amount of information afforded by the classification (see Chapter 25).

The brooches were arranged in the following three ways:

(i) According to the judgement of four archaeologists working by the conventional methods customary in that field.
(ii) According to the judgement of an anatomist with no special knowledge of this branch of archaeology.
(iii) by a one- or two-dimensional proximity analysis.

The four archaeologists differed substantially among themselves in the arrangement of the brooches, only one of them producing an arrangement which conformed approximately to that defined by the chronology. The computer programme for proximity analysis, and the judgement of the anatomist, were also fairly well in accord with the chronology. Perhaps not the least of the conclusions that one can draw, however tentatively, from this trial is that within the framework of what, for want of a better phrase, we call conventional taxonomic practice, there is in fact a wide range of activities whose diversity is worth more attention than has been given to it in the past; there is also, within the framework of quantitative taxonomy, a much   wider   choice   of   techniques   than   many   conventional taxonomists suspect.

By the reasonably objective tests of this example, the computer programme, the anatomist, and one of the archaeologists all give roughly equivalent performances with three of the four archaeologists giving an apparently inferior performance.

Chromatographical and Other Chemical Data (Botanical)

Although biochemical aids to taxonomy are proving their worth in many   fields,   they   have   rarely   as   yet   been   combined   with morphometric techniques. Simon and Goodall (1968) have used Goodall's probabilistic similarity index to cluster the species of the leguminous plants of the genera *Medicago, Lotus, Trigonella,* and *Melilotus,* in respect of the numbers of shared phenolic substances revealed by two-dimensional chromatography. These authors point out that the ability to identify all the phenolic substances is not an essential prerequisite to using the chromatographic locations of the spots formed by these substances in taxonomic work. *Trigonella* and *Melilotus* species seemed to be closely clustered together with *Medicago radiata* in one main subdivision of the material. Two further such subdivisions were a group of seven species of *Medicago* and another group containing four species of the same genus.

There   was   only   a   limited   agreement   between   the   current conventional views on the taxonomy of these genera and the picture of their affinities which emerged from the biochemical study. The dangers of treating as distinct independent attributes the phenolic substances   which   might   have   been   synthesized   by   mechanisms generated by a common gene were stressed. It seems likely that the course of the differentiation of phenolic substances has followed a different   evolutionary   path   from   that   of   the   attributes   used   in conventional taxonomy.

Flake and Turner (1968) have constructed their own system for a numerical   taxonomic   classification   and   applied   it   to   some unspecified chemical data from the North American leguminous plant *Baptisia nuttaliana.* They claim that by this means they were able to detect geographical variation that had escaped the attention of an orthodox taxonomist working with this species.

Botany (A Study of Introgression)

A species-hybrid complex, in which the parental populations were the longleaf pine *(Pinus palustris)* and the loblolly pine *(P. taeda),* was assessed in terms of ten morphometric characters: needle length; fascicle shield diameter; interstomatal distance; bud scale length; bud scale width; cone length and width; seed length and width; and seedcoat thickness. Namkoong (1966) tried a variety of analytical techniques: Anderson's hybrid index (Anderson, 1953); an index composed of the elements of the first two principal components from the pooled within-groups correlation matrix; an index constructed from the canonical correlation between the five vegetative and the five reproductive characters choosing the most highly correlated vectors; a discriminant function; and lastly a distance analysis with the generalized distance from each parental species used as an axis of variation.

The indices prepared from the principal components and from the canonical correlations proved least useful. The discriminant function approach was most effective in distinguishing the populations, but was dependent on certain assumptions about the homogeneity of the dispersion matrices for the different groups, which could impair its usefulness. For this reason, Namkoong recommended the generalized distance approach in cases where these assumptions were either not valid or could not conveniently be tested.

Clifford and Binet (1954) also used the discriminant function to sort out the hybrids of two species of *Eucalyptus* in a hybrid swarm.

Botany (Plant Morphology)

Fifteen varieties of Indian tobacco *(Nicotiana rustica)* were measured by Murty *et al.* (1965) in respect of four characters, height, height exluding panicle; number of leaves; and leaf-area. A generalized distance analysis, followed by a supplementary canonical variate analysis, showed that the 15 varieties fell into four clusters on a plane defined by the two significant canonical variates

$$(+0.1303; -0.0158; +0.0063; +0.9911)$$
and
$$(+0.6186; +0.7052; +0.3372; -0.0763)$$

of which the first accounted for 61.9% of the total variation and the second for 30.7%. The first cluster constituted a group of tall varieties with long internodes and small leaves. The second cluster comprised plants that were intermediate in most respects. The third cluster contained plants of medium height but of small and few, leaves, whereas the fourth cluster contained but a single variety (Kharagpur) which had the largest leaf size. This distinct variety was grown near the coal-fields of Bihar, where the area is warmer than those areas used for growing the other varieties; and the microfauna and flora are different from those of other tobacco-growing regions.

The closeness of the agreement generally obtained when conventional (supposedly phylogenetically oriented) taxonomic studies are compared with purely phenetic ones is illustrated by Morishima (1969) who worked with strains of the rice plant. He compared 64 different strains using 24 characters, and clustered them into Asian, African, American and Oceanic groups. According to Morishima's interpretation of the cladistic relationships ascertained by Camin and Sokal's (1965) technique, the Asian group of strains must have advanced quite rapidly in the evolution of the group as a whole. There seems to be a general, and in Goodman and Paterniani's (1969) opinion well-founded view, that this agreement is closer when botanical morphometrics is performed with reproductive characters than with measurements made on vegetative parts of the plant.

Mastrogiuseppe *et al.* (1970) used a standard computer programme for stepwise discriminant analysis (based on the method of canonical variates) to study anatomical relationships in ten characters in Recent and fossil *Ginkgo* wood. They found that the morphometric analysis succeeded in bringing out significant differences not apparent through general microscopic investigation. The three characters— number of pits per millimetre of tracheid length on radial wall, ray height, and tangential diameter of late wood tracheids—accounted for most of the discriminatory power.

The green alga *Chlorodesmus,* which has been the subject of some confusion because taxa have been erected in the past on the basis of collections made at different stages of development and growing under different conditions, shows environmental variation (Ducker *et al.,* 1965). A canonical variate analysis based on only two characters, interdichotomal length and diameter, served to split the measured material into four groups which corresponded fairly closely with those finally arrived at by means of a cluster analysis using these two characters together with those of the dichotomy, branching habit, stipe, and presence or absence of crystals of calcium oxalate in the cytoplasm. Note that a mixture of quantitative and qualitative characters was used for this analysis. Watson *et al.* (1967) have continued this approach by combining 24 qualitative and 14 multistate characters to good effect in classifying the Ericales.

Whitehead (1955, 1956) used discriminant functions to sort out the confused taxonomic situation in the chick-weed genus *Cerastium* where attempts to identify the plants on the basis of single characters had led to difficulties. Applications of numerical taxonomic studies that do not find a place elsewhere in this book are those of Prance *et al.* (1969) dealing with the Chrysobalanaceae, and of Ornduff and Crovello (1968) dealing with Limnanthaceae.

Medical Applications of Multivariate Analyses

Apart from the special topic of somatotypy (see Chapter 1), multivariate techniques find a place in medical diagnosis mainly to combine tests or symptoms of inadequate diagnostic power into criteria which will allocate patients unequivocally into certain groups (Ledley and Lusted, 1959). Other uses are the prediction of the effects of Rh incompatibility (Williams, 1958) and the analysis of data from electrocardiograms (Pipberger, 1962). The field has been surveyed by Overall and Williams (1961a, b), and by Radhakrishna (1964) who points out that some measurements customarily made on patients seem to have very limited discriminatory value, and might well be dropped or replaced by better ones if they can be found. Even where the total discrimination is poor, the publication of such analyses, as Radhakrishna remarks, can stimulate and clarify further thought on the matter. Radhakrishna also stresses the value of abstract morphometrics in the surveys that form so vital a stage of preventive medicine and social medicine, spilling over into sociology, as in Freeman's (1970) work on performance during examinations.

More recently, Feldman *et al.* (1969) have compared successive screening of psychiatric patients with the use of discriminant functions for their classification into "excited" or "retarded" categories. Both techniques were highly effective, but the preference for successive screening expressed by these authors seems to be based on a somewhat naive understanding of multivariate methodology.

The outstanding success of Buck and Wicken's (1967) use of discriminant functions for specifying the risk of contracting lung disorders in various regions and occupations should dispel any lingering doubts about the usefulness of multivariate methods in medicine.

Immunology

Basford *et al.* (1968) have performed a principal components analysis on the immunological reactions of saline extracts of 20 species of beetles against rabbit antisera prepared from 14 such species.

This study is of particular interest because there are two quite divergent views concerning the higher taxonomy of the Coleoptera, of which the classification of Crowson (1955) which is avowedly a "natural" (to some extent polythetic) classification, and that of Leng (1920) which is more closely monothetic, may be taken as representative. In the event, Crowson's natural classification is supported by the immunological reactions.

Monothetic classifications (those based on splits defined by the presence or absence of selected characters thought to be of particular taxonomic value) are dear to the hearts of those conventional taxonomists who try to work from an *a priori* theoretical basis. As in the phytosociological context mentioned earlier, there is no reason why monothetic classifications should not be developed and tested. Until they have been tested, one might be justified in treating them with a certain scepticism, appropriate enough in the present case. It would be quite hard to construct a "natural" classification based on anything but a polythetic classificatory process, where the splits were made on some such criterion as maximizing the between-group variation relative to that within groups.

Mainardi (1959) has examined the generalized distances between gallinaceous birds based on their immunological reactions, for bird taxonomy (as opposed to bird identification) is confused because of the apparently close relationships involved.

Immunological reactions are only one kind of character that can be used for constructing discriminants of various sorts. In the case of bird variation just noted, multiple morphometric characters have been used to distinguish geographical forms by Storer (1950), and Power (1970).

Taxonomy and Biogeography (Entomological)

If there is one field of taxonomy where the application of conventional techniques should have resulted in something approaching finality it is that of the macro-Lepidoptera, where the classical approach has been intensively applied for more than a century. Recently Wilkinson (1967) has compared a vigorous revision using classical criteria with a numerical taxonomic study, and has applied multivariate techniques to show that (i) the unrevised classification can be shown to be unsatisfactory using principle coordinates (ii) the same multivariate technique clusters the material into three distinct groupings which the classical revision had suggested should have generic status, and that within these new genera there is further clustering into species-groups. In this case the results of the multivariate analysis accord well with those derived from an informed modern revision using classical techniques, and we may expect that this concordance will be a regular feature of good work by either approach. Wilkinson's later paper (1970) is interesting for its extensive use of various kinds of multidimensional representation, including the minimum spanning tree of Gower and Ross (1969) which links the most closely related species by distances inversely proportional to a similarity coefficient between them. Wilkinson's work is a model of forward-looking taxonomic practice.

Holloway (1969) has also applied numerical taxonomic techniques to the Lepidoptera but this time with a view to discerning the biogeography of the butterflies of India and relating this biogeography to the history of the subcontinent in terms of continental drift.

Acarology

Thomas (1968) has applied a principal components analysis to the similarity matrix formed from the assessment of six characters of larval ticks of the rabbit from 74 places throughout the U.S.A. The six tick characters were those selected by an earlier study as being effective discriminators. The first principal component, as so often happens, seems to represent size variation, at least partly associated with a north-south cline. There was also an east-west cline in respect of changes in the capitular appendages.

A factor analysis of 20 independent variables together with the six dependent variables was run, with rotation to simple structure. The overall size of the larvae was greatest in areas with low mean annual temperatures and total precipitation. The legs and capitular appendages were longest where the areas had high rainfall when the larvae were most abundant.

The numerical taxonomic approach has been tried by Sheals (1964) who assessed 23 qualitative and 39 quantitative characters on 22 species of mites. The similarity matrix was subjected to a principal coordinates analysis and the second and third vectors chosen, on the basis of empirical experience, as the axes of a chart on which the affinities of the species were displayed. A conventional dendrogram of affinities was also constructed, and the two methods of investigation found to differ appreciably. If the results of the orthodox taxonomic analyses are accepted as a basis for comparison, the dendrogram appears to recover less information about the affinities of the groups than does the principal coordinate approach. Earlier essays in the numerical taxonomy of the acarines have been reviewed by Funk (1963).

Pedology

Horton *et al.* (1968) have performed a canonical variate analysis on chemical data of soil from a gilgaied area under brigalow *(Acacia harpophylla)* in Queensland. The samples came from four depths of three positions on the microtopological features which are characteristic of gilgaied soils. Two significant canonical axes were found, one of which corresponds to variation in depth and the other to variation in position.

Hughes and Lindley (1955) measured exchangeable calcium, potassium, pH, and available phosphate in a group of six soils from Snowdonia. A canonical variate analysis showed that three vectors were needed adequately to describe the variation among the soils; these were

$$(+1.00 + 0.29 + 0.25 + 0.40)$$
$$(+1.00 - 1.98 - 0.60 + 6.31)$$
and
$$(+1.00 - 9.21 + 21.35 - 34.76)$$

One evident difference between the groups lay in the rainfall, which clearly distinguished, by a substantial generalized distance, all high rainfall soils from all medium rainfall soils irrespective of the underlying strata.

The numerical taxonomy of soils offers a tempting target for the exploitation of numerical methods, and has been taken up by Bidwell and Hole (1964) and later by Sarkar *et al.* (1966). These authors reduced an initial 61 soil characters to 22 by progressively eliminating those which had high correlations with other characters. Closely related soils remained clustered, even with the smaller number of characters, but the less closely affine soils drifted from group to group. Characters with correlations as low as 0.5 were eliminated, and it is arguable that it is just as deleterious to eliminate such characters as to retain them, and quite possibly more so. The object of the elimination was to reduce the apparently greater weight given to features measured by two or more correlated characters, but whether this objective is desirable is highly debatable.

An interesting byproduct of this work was the finding that the relation between the number of characters correlated at a particular level, and the level of correlation in question, is roughly linear.

Rayner (1966, 1969) has compared soils in Glamorganshire
(Wales) by means of dendrograms based on a numerical taxonomic
approach, and also by means of Gower's principal coordinates
method. A difficulty arises in such work, because the same type of
horizon does not always occur at the same depth, yet the horizons
cannot always be unambiguously classified from a knowledge of their
physical properties. However, the process of sorting horizons on the
basis of their similarities does not do this. For this reason there are
substantial differences between the dendrograms based on the
similarity of the most similar pairs of horizons, and on the average of
the similarities of the best matched pairs of horizons.

The principal coordinate approach gave a very clear interpretation
along two axes tentatively reified as reflecting acidity and redox
potential: the major soil types involved were distinctly separated
along these two axes of variation. Figure 36 shows these groups.
Although the picture given by the principal coordinates method was

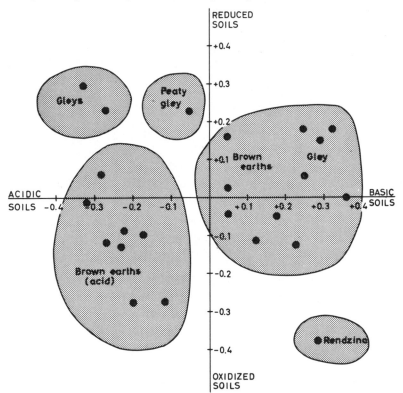

Fig. 36. A principal coordinate clustering of Glamorganshire soil profiles based on mean
horizon similarities: distances between the points approximately $\sqrt{2(1-S)}$ where S is the
measure of similarity. Redrawn from Rayner (1966).

rather like that given by the principal components technique, it did represent an improvement in terms of the ease with which the results could be interpreted. Multivariate pedology has come a long way since the pioneering work of Cox and Martin (1937). Much of this improvement stems from the recognition that orthogonal axes of variation, and in particular those arising from multivariate analyses of the data, give the experimenter the same intellectual and practical control over the experimental data that the analysis of variance provides for the univariate case, with the added advantage of affording multivariate vectorial representations of the processes at work in the soil. This approach comes out strongly from Holland's (1969) work on soil nutrients, in which he combined the variables into principal components and showed how their orthogonality was of conceptual value.

Marine Ecology

The relative abundances of 13 species of foraminifers from Christchurch Harbour, England, were assessed in terms of the physicochemical properties of the sediments in which they were found. These sediments and the bottom water were compared for dissolved oxygen, chlorinity, magnesium and calcium contents, temperature, pH, and depth of water, as well as the mud, silt, sand and gravel contents. Howarth and Murray (1969) then applied a variety of multivariate techniques to these data in order to understand the relationships involved.

One approach was to perform a principal components analysis, with subsequent rotation to a varimax solution in which the loadings on the factors are maximized. This part of the analysis used the environmental variables as "characters" of the fauna found in each locality, and is hence an R-mode analysis (see p. 209). The principal components analysis showed that the first four components accounted for more than half the total variation. The first factor extracted reflected changes in temperature, oxygen and pH: in all probability these features of the environment stem from the changes brought about by the river water, which tends to be colder and more acid, as well as slightly deficient in oxygen. Factors II and IV seemed to represent the changes brought about by incoming seawater, II concerned chlorinity and magnesium content, IV depended on the calcium content which was removed rapidly from the sea by the fauna, in contrast to the longer-term influences of chlorinity and magnesium content. Factor III represented the depth of water. The cluster analysis essentially confirmed these findings, and emphasized the similarity of the reactions of the camoebina and their dependence on the depth of water.

A *Q*-mode analysis, in which the sampling localities are grouped together according to the microfauna and physicochemical properties, was then tried. The first four components of this analysis took up 87% of the total variation, giving indirectly an indication of the existence of associations, or even of communities, of the organisms by reference to the localities having the most extreme composition of faunal elements. With the aid of some simpler techniques these multivariate methods added greatly to an understanding of the ecology of Christchurch Harbour.

Sailer and Flowers (1969) have investigated the shapes of lobsters by means of a linear discriminant function based on 16 out of an initial 24 morphometric characters. There seemed to be a distinction of shape between inshore and offshore specimens when due allowance had been made by matching for size and sex. Tagging experiments suggested that this distinction was accentuated, and perhaps brought about, by the homing tendency of the lobsters which, once displaced, tended to return to the area where they were reared: the effect was to create two populations differing morphometrically.

Linear discriminants, derived from a principal component analysis, have also been used to distinguish species within the Salmonidae by Ouellette and Qadri (1968), who used 43 morphometric measurements for the purpose. Fry (1970), who considers that sponges should be regarded as mixed populations of cells and cell products, used principal coordinates to distinguish them. Reyment (1969e) used the method of principal coordinates to analyse marine geochemical data in the form of trace-element determinations (Cd, Zn, Pb, V, Cu, Ni, Co, Mn, P, S) of sediments, and the factors of pH, Eh, O_2, $CaCO_3$, oxidizable organic matter and the median of the sediment grain sizes for five transects made in the Niger Delta, West Africa. Although the residual for two extracted coordinate vectors is quite small for data of this kind, being only 35.8%, no strong pattern is displayed by the plot of the first two coordinates. It was concluded that there is little differentiation in the sediment with respect to these variables, a reflection of its relatively homogeneous nature.

Meteorology

A field of growing importance for the use of multivariate methods is that of meteorology, where long-range weather forecasting is concerned with the problems of detecting relatively minor imbalances in the heat engine formed by the atmosphere which is carrying surplus heat from the tropics to the poles, where there is a heat deficit. The total amount of information available to the long-range forecaster in this highly "abstract" branch of morpho-metrics, in our sense of the term, is so great that some effective condensation is a prime requirement. An approach to such condensation has been made by using sets of empirical orthogonal functions (principal components) to represent the significant patterns of variation.

Taken at its face value, as Craddock (1965) points out in his review of the topic, this approach would involve the computation of the latent vectors of a matrix of order 1000, which is neither practicable nor, in all probability, even useful. However, the examination of regional networks of information can be made, and the principal components which represent large-scale patterns retained for comparison between the regions. An encouraging beginning was made when it was found that between 80 and 90% of the total variance of the dispersion matrix of 32 grid point values of the large-scale atmospheric temperature fields over Europe and the eastern North Atlantic, computed month by month, could be accounted for in terms of eight principal components, with smooth variation of these patterns of variation from one month to the next. Such economy in representation of the basic data was gratifying in itself, and also directed the attention of the forecasters to the most important large-scale features of the temperature distribution.

A principal components analysis of the mean monthly temperatures in Central England from 1680 to 1963 gave a series of components of which the first accounted for 92% of the total variance, and represents a combination of the first harmonic with part of the second, following in effect the annual variation of temperature. However, the second principal component proved to be the most interesting, although it accounted for only 2% of the total variance; this vector represented climatic changes in the mean annual temperature and showed, for the first time with any clarity, that since 1680 climatic variation in England involved warmer years of the maritime type and colder years of the continental type (i.e. years with and without an appreciable second harmonic in the second component). The third and fourth principal components, accounting for a little more and a little less than 1% of the total variance respectively, represented patterns of winter temperature, accommod-ating years in which the coldest part of the winter is either before or after the usual time.

21
The Relationships Between Morphometric Analyses and Numerical Taxonomy

It is as understandable as it is desirable that the ferment of new ideas concerning taxonomy should have thrown up an enormous range of topics at a rate which has made it difficult to integrate each one into the existing body of knowledge. Although the flow of new ideas continues, the time may have arrived to consider the relationship between the group of topics subsumed under the general heading of numerical taxonomy (Sokal, 1969) and the topics that fall into the more restricted field of morphometrics. It is quite possible to regard numerical taxonomy as a branch of morphometrics, and vice versa, but these practices leave us without a convenient designation for one or other activity. Perhaps the term "quantitative taxonomy" covers the two sets of ideas adequately. Another approach is to regard all these topics as branches of morphometrics and to describe multivariate changes of shape and size as falling into the province of discriminatory topology, since many of the techniques are based, directly or indirectly, on the discriminant function (Rao, 1952).

There is an undisguisable difference of emphasis between discriminatory topology and numerical taxonomy. In discriminatory topology, the methods used have been essentially those of mathematical statistics, although for practical reasons some compromise with the rather stringent requirements of the mathematical models has been necessary. In numerical taxonomic studies, on the contrary, there has been a much greater preoccupation with the underlying taxonomic concepts than there has with the underlying mathematical ones. Reyment (1969b) has remarked that modern numerical taxonomists have departed from the realm of mathematical statistics in developing their subject because of the special problems and difficulties that arise when a statistical approach is adopted, and in order to get round some of the less convenient properties of the basic data.

We will try to examine in what ways these two approaches to quantitative taxonomy have diverged. We have already seen that the distance between two populations of organisms (or of non-living material) is given by the equation

$$D^2 = \Sigma g_{ij} \mathbf{x}^i \mathbf{x}^j.$$

Now there is no difference of opinion about the $\mathbf{x}^i \mathbf{x}^j$ part of this expression, which is simply the vector of differences between the means of the groups compared in respect of the $\ldots ij \ldots p$ characters assessed. It follows that all the differences of opinion must be concentrated into the fundamental tensor g_{ij}, which describes the nature of the space in which these dissimilarities are to be arranged and compounded.

As we have also seen, the fundamental tensor is the inverse of the dispersion matrix of variances and covariances of the characters, which is to be equated to a Kronecker delta matrix of the same rank in order to eliminate the correlations between the characters.

Viewed in this light, the differences between numerical taxonomy and morphometrics may be considered to reduce to a consideration of the pre- and post-diagonal entries of the matrix which, once inverted, constitutes the fundamental tensor. In standardized terms, the matrix in discriminatory topology is

$$
\begin{bmatrix}
1 & r_{12} & r_{13} & r_{14} \cdots r_{1j} \\
 & 1 & r_{23} & r_{24} \cdots r_{2j} \\
 & & 1 & r_{34} \cdots r_{3j} \\
 & & & 1 \quad \cdots r_{ij} \\
 & & & \qquad 1
\end{bmatrix}
$$

where we do not need to write in the prediagonal entries since the matrix is axisymmetric (i.e. square symmetric). Kurczynski (1970) has extended the generalized distance to the case of discrete variables as used in numerical taxonomy.

When numerical taxonomy is practised using the "Pythagorean" distance between the groups, the matrix which is used implicitly in place of the correlation matrix for g^{ij} (g_{ij} prior to its inversion) is the unit matrix, $\mathbf{I}$,

$$
\begin{bmatrix}
1 & 0 & 0 & 0 \ldots 0 \\
 & 1 & 0 & 0 \ldots 0 \\
 & & 1 & 0 \ldots 0 \\
 & & & 1 \ldots 0 \\
 & & & \quad \ldots 1
\end{bmatrix}
$$

a matrix which specifies that all the correlations between the characters are zero. The fact that other measures of distance are employed in numerical taxonomic studies is hardly relevant, partly because they do nothing to improve this critical assessment of the correlations between the characters and partly because, for various reasons, they are falling into disuse.

The techniques of numerical taxonomy are essentially techniques for handling qualitative characters, counting the dissimilarities in such a way that the discriminant function linking two groups has coefficients of unity for all unshared characters and zero for all shared ones. These are, of course, *a posteriori* weights and their use does not infringe the principal of equal weighting *a priori*, but it is nevertheless true that some weighting system, implicit or explicit, is required by all taxonomic classificatory procedures, whether quantitative or conventional. In conventional monothetic classifications, the discriminant function consists of unity for the critical character, when it is shared, zero for the same character when it is not shared, and zero for all other characters.

When qualitative characters which do not vary within the groups are chosen, the correlations between the characters are not zero but indeterminate (Blackith, 1965) and the matrix then is

$$\begin{bmatrix} 1 & ? & ? & ?\ldots \\ & 1 & ? & ?\ldots \\ & & 1 & ?\ldots \\ & & & 1\ldots \end{bmatrix}$$

It is sometimes maintained that this matrix can be made determinate by substituting, arbitrarily, any constant such as zero for the ?'s, since the constant simply acts as a scaling factor. That this is not really a well thought-out position is shown by increasing the value of the constant from zero to unity, when the groups are being discriminated effectively on the basis of a single character completely correlated with all the others. Little though this situation may commend itself to taxonomists, it is questionably better than the assumption that correlations are zero throughout.

The rapid development of numerical taxonomy, despite (or possibly because of) its lack of attention to basic theory, is undoubtedly associated with its extreme simplicity. For many preliminary investigations it gives what is probably a sufficiently accurate general picture of the relationships between the entities. Nevertheless, the time is approaching when a critical look at the foundations of numerical taxonomy will be necessary, because various workers are beginning to consider such questions as the degree of concordance between various taxonomic techniques and even to doubt, perhaps correctly, whether any general concordance is possible. Such questions cannot be dealt with adequately on the basis of a method which can only be made to work if an important assumption is made for which there is no evidence. Fortunately, from the inception of the subject, its practitioners have been open-minded about all aspects of the methods employed, so that there can be every confidence that improvements will be made. We should not forget that numerical taxonomy has proved its worth and is now soundly established in the economically important fields of microbiology and food preservation (see, for instance, Oberzill, 1963; Gray, 1969; Pohja, 1960; and Gyllenberg *et al.,* 1963).

The advantages of working with oligovariate techniques are obvious; we have only to look at the vexed question of convergence (see, in particular, Chapter 26) to see how this topic assumes its present position solely because oligovariate techniques are not sufficiently well understood and applied.

Nevertheless, there is some practical force in Sokal and Sneath's (1963) comment that multivariate morphometrics is mainly useful at the taxonomic level of the species and below: at least that is the level at which it has received by far the most attention and it would be wrong to fail to discuss the difficulties of applying it to higher categories. At the species level, or thereabouts, the contrasts between taxa are often associated with shape and size differences and frequently need to be distinguished from various aspects of environmentally induced variation.

The acute sensitivity of morphometric analyses using quantitative data is at its most beneficial when there is a need to distinguish several interlocking patterns of growth. Once we move substantially above the genus, we are faced with categories in which the taxa differ by features which can often be expressed quantitatively only with some difficulty, and where changes of shape are not very important for the making of that particular distinction. The temptation to proceed by means of counts of qualitative dissimilarities is strong; and since much can be learnt from this way of progressing it should be encouraged at the same time as its limitations should be recognized. Fortunately, the prospects of working out a scheme of analysis much nearer to a truly oligovariate one seem to be good. The development of indices of similarity, which will cater for mixed qualitative, quantitative and even ranked data, is already with us

(Goodall, 1966b). Studies of the redundancy of discriminatory topological and of numerical techniques suggest that far fewer characters are needed in the former than the latter and fewer in either than has perhaps been thought necessary (p. 300).

Nevertheless, an investigation in which many different categories are involved is apt to be difficult because the effectiveness of each character is different when it is called upon to discriminate at various levels. In morphometric analyses this inequality is patent, the *a posteriori* weights reflect it. In numerical taxonomy, because the *a posteriori* weights can take only the two values 0 and 1, it is less obvious. However, one can see what is happening when, in the course of constructing a dendrogram, the higher taxa may differ by many character dissimilarities whereas the lower taxa may differ in few; indeed, one reason for choosing a number of characters, which may be substantially above that which a redundancy study would indicate as adequate, is that it is quite possible to have the pendant vertices of the dendrogram (see Chapter 24) failing to separate simply because there are no characters in the suite which happen to differentiate them. This sampling problem apparently remains unstudied.

We thus have three broadly distinguishable stages in the evolution of taxonomic practice:

Stage	*Nature of data*	*Treatment of data*
Conventional taxonomy	Qualitative	Qualitative
Numerical taxonomy	Qualitative	Quantitative
Discriminatory topology	Quantitative and qualitative	Quantitative

Although the term numerical taxonomy has sometimes been used to indicate both the last two stages, and morphometrics has been used more to indicate the last stage than any other, we broadly understand both the last stages in our usage of morphometrics, whilst recognizing that current practice in the description of quantitative taxonomy is neither clear nor consistent. Conventional taxonomists, especially palaeontologists, often resort to semi-quantitative treatment of their data, taken as the ratios of pairs of characters, but there is a broad distinction of the underlying philosophy which unites the last two stages mentioned above in contrast to the first stage. Kaesler (1967) has prepared a useful introduction to numerical taxonomy for palaeontologists, whose ideas on taxonomy and evolution do not always fit conveniently into the current spectrum of biological thought: Miller and Kahn (1962) give a general introduction to statistical methods, including morphometrics for geologists, and Burma (1949) was a pioneer of multivariate morphometric methods in palaeontology.

In addition to the distinctions mentioned above, there is a choice of hierarchical or non-hierarchical techniques. It seems likely that hierarchical techniques are almost always undesirable in theory, because even in biological classification the hierarchical structure can rarely be tested adequately and even random data sets will appear to have some hierarchical structure (Rohlf and Fisher, 1968). However, the consequences of using hierarchical techniques, when the structure of the experiment renders such a practice dubious, seem not to be very serious (McCammon, 1968b). Kaesler and McElroy (1966) used a hierarchical clustering of material from 80 wells in an Upper Cambrian sandstone, in respect of which 77 characters had been measured. The results made good geological sense despite the inappropriate model. McElroy and Kaesler (1965) published a non-hierarchical approach which allowed them to reify four underlying processes each of which generated a factor differentiating the material from the wells.

In all classificatory work, there are two approaches which complement one another. The material can be clustered into some taxonomic entities first, and then these entities can be arranged in some hierarchical system which, eventually, may be represented in one or two dimensions in a pattern reminiscent of the traditional phylogenetic tree. In a variation of this approach, the data are subdivided analytically instead of synthetically. In the other approach the hierarchical nature of the classification is not invoked, and a series of measurements of the distances between the various groups which can be discerned is used to construct a chart showing the degree of phenetic relationship not merely between groups at different levels.

This is the familiar classification—ordination controversy of phytosociologists (Chapter 16) and each approach has its own advantages and disadvantages, the important features of each illuminating the interpretations made by the other approach. Gower's (1967b) discussion of the use of distance techniques to assess the performance of different methods of cluster analysis is a good illustration of this complementarity.

22
Cluster Analysis

Cluster analysis is currently enjoying something of a boom in the world of systematics. The uninitiated might well ask what the term implies, for it gives *per se* no really clear indication of what it is supposed to do. Perusal of the voluminous literature on the subject very quickly brings one to the disquieting conclusion that there are, and can be, no hard-and-fast rules about what really determines a cluster (Rao, 1952). A break-down of the trends in ideas on the subject does seem to suggest that people, when they talk of cluster analysis, mean that they hope, by the use of some technique or other, to cause a grouping of points in multivariate space into disjoint sets, which sets are thought to correspond to pronounced characteristics of the sample.

Now, it is a sad fact, that different methods of cluster analysis may well bring about different geometrical distributions of the multivariate points. This may or may not be a bad thing. Several clustering methods have been proposed in the literature. In the following Chapter 23, some of the more interesting of these are reviewed.

It is perhaps worthwhile pointing out that clustering often imposes an hierarchical structure on the sample. This may be intuitively attractive to the taxonomist, but there is no great difficulty in causing the same taxonomic problem to take on quite different appearances just by switching some of the rules being applied (Goodall, 1968b). The hierarchic element may, however, be more a hindrance than a help in some generalized numerical-taxonomic studies.

There is now a large literature on methods of clustering, just as there is an even larger one on distance measurements (Rao, 1954). In fact, despite their superficial dissimilarity, a serious discussion of the one is almost impossible without mention of the other. Sokal and Sneath (1963) have given a detailed account of some of the older methods, together with a full critique. They show that there are 14 "indices of similarity" between taxa according to whether or not matched pairs of characters are equally weighted, or carry twice the weight of unmatched pairs, or unmatched pairs carry twice the weight of matched pairs, in the denominator, and whether negative matches are included in the numerator or excluded from it. These negative matches occur when an attribute is not available for study, because the character upon which it is measured is wanting in both groups. Whether or not these negative matches should be employed is a topic that does not lend itself to hard and fast rulings, because one is inevitably influenced by the nature of negative matches in the kind of material with which one is most familiar. The upshot of Sokal and Sneath's discussion is that common sense should be used.

A decision to exclude the negative matches in bacterial taxonomy may be based on the rational grounds that a whole series of negative matches may arise from the failure of a single enzyme at an early stage of a metabolic cycle, preventing other enzymes from expressing their presence. On the whole, however, Sokal and Sneath are in favour of the inclusion of negative matches in coefficients of similarity.

There is another aspect of the choice of characters in an index of similarity, which may give rise to serious differences of opinion; this is the question of rare or complicated characters from which, potentially, numerous attributes could be selected. A high proportion of what an orthodox taxonomist considers to be "good" characters fall into this category. Smirnov (1963, 1968, 1969) has developed a weighting system by which taxa sharing a rare character state (attribute) are considered to be more closely similar than two taxa which share a common attribute. Smirnov's index of similarity is thus probabilistic, with *a priori* weighting on the basis of the rarity of an attribute. Such a system would appear quite natural to a taxonomist who believed that his phenograms represented phylogenies. Sokal and Sneath feel that they cannot recommend Smirnov's index of similarity, principally on the grounds of its content of *a priori* weighting, though there are also some minor technical disadvantages. Sokal and Sneath's argument seems to be less than satisfactory in that they consider that the higher weighting will be "built-in" by the tendency to select several correlated attributes from the more complicated characters. But this tendency depends at least partly on the ease with which other attributes can be found; there is generally a distinct emphasis on the few complicated characters in organisms when few attributes can be found, and a reluctance to choose several from any one part when they can readily be found elsewhere. Moreover, just because the closely associated attributes tend to be most strongly positively correlated, the method of choice advocated by Sokal and Sneath will overweight the character. If one must have weighting of an effectively *a priori* kind, either by deliberate choice or by virtue of the technical inability of numerical taxonomy to allow for correlations, it seems preferable to have Smirnov's system, which is less arbitrary.

One can, of course, call the probabilistic weighting of characters into question by pointing out that such weighting will play down the similarity of organisms sharing many common attributes, whilst accentuating the similarity of organisms which may have few shared common attributes, but some shared uncommon ones. Carried to extremes, such a system could give highly undesirable results. The large suite of common attributes of the vertebrates, for instance, might be virtually weighted out of a comparison between a representative selection of vertebrates and invertebrates. If the vertebrates happened to include the exceptional fish that does not contain haemoglobin in the blood cells, and the invertebrates contained some of the dipterous insects whose larval stages happen to contain haemoglobin, one could imagine a situation arising where the high weight attached to the rare attribute (for animals as a whole) "haemoglobin in blood" would associate most of the vertebrates with the Chironomidae amongst the insects, and locate the exceptional fish with the remainder of the insects, in defiance of the many characters that contradict this association, which would have low weights. Everything hangs on the question of whether we can rest assured of the monophyletic nature of the rare characters: if, as in the illustration used above, we doubt it, probabilistic weighting could be disastrous. If we could be sure of monophyletic rare characters, then there would be much to be said for their use: we are back on the treadmill of phylogenetic speculation.

To some extent, the introduction of probabilistic ideas brings cluster analysis into the field of information theory. Orloci (1969) has made this comparison explicit, using information theory to cluster quadrats according to the trees found in them. Möller (1962) has outlined the methods for constructing probabilistic identification keys, necessarily on polythetic criteria.

Burnaby (1970) defined a similarity coefficient in which characters are weighted in proportion to the information they convey, and in inverse proportion to their overall association with other characters. For discrete characters, and under ideal conditions, it reduces to the ordinary Sokal-Mitchener coefficient. Under non-ideal conditions, Burnaby believed it permitted a more objective approach to numerical taxonomy by reducing the need for selection and scoring of characters. Gower (1970) commented on Burnaby's posthumously published work, as well as other weighted coefficients such as Goodall's and problems connected with their use. He pointed out that Burnaby's coefficient seeks to eliminate all effects of correlation irrespective of their origin, and that this point of view tends to characterize Goodall's (1966a) coefficient. He believes this often to be the wrong approach and concludes that inasmuch as Burnaby's method eliminates an unknown mixture of between- and within-class correlations it is inappropriate for many, if not all, problems.

There is, however, a real dilemma when the group of characters that one is considering is a part of an organ which may be lost or greatly modified in some organisms. If one of the rare or complicated characters, say a wing, is missing as in some apterous insects, but present in others which are to be included in the analysis, it is very hard to know whether one wants to adhere strictly to equal weighting of characters (*vide* Smirnov *loc. cit.*) and, if so, what steps should be taken to conserve equal weighting. For if each of ten attributes on a wing is equally weighted, but the primary character (the wing) is lost, the contrast between "wing present" and "wing absent" corresponds to ten unit increments of the "Pythagorean" distance between the two groups of insects. Proctor and Kendrick (1963) commend this practice of allowing each secondary character to have its own contribution to contrasts of this kind, and consider that the practice constitutes a modified weighting system. From the discussion of Smirnov's weighting system in Sokal and Sneath (1963) however, it seems likely that these authors would regard the practice as conforming to the equal weighting principle.

Burnaby (1966b) has drawn attention to the fact that the consequences of choosing one or other of the plethora of similarity indices are in many cases equivalent to the adoption of different schemes of character weighting. The concept of "equal weighting" is not as simple as it seems at first sight.

However, as Sokal and Sneath themselves hint, discriminant techniques might be an even better way of dealing with the problem, since there can be no question of *a priori* weighting and the effects of correlations between attributes are eliminated. An illustration is given on p. 315.

A development of significance for the long-term future is that of Goodall's (1966b) and Burnaby's (1970) similarity indices based on probability. These ideas share with the Smirnov index the capacity to represent rarity in a character by an enhanced weight if the taxonomist so desires. They can be used for all classes of character, and thus avoid the difficulties that arise when multistate or quantitative characters have to be included with qualitative characters in numerical taxonomy. With other similarity indices multistate characters have to be coded, and this coding (more than five spines on the femur, more than ten spines on the femur, etc.) into a sequence of dichotomous attributes overweights the character that is being included by giving it more than one attribute. Of most potential consequence, however, is the possibility of developing a procedure in numerical taxonomy in which the, at present, arbitrary levels at which the successive splits in the dendrograms are made will be controlled by specified significance levels. In this respect Goodall's work has affinities with Möller's (1962) construction of a probabilistic identification key. Because the similarity matrices for a cluster may differ from that for the population from which it was drawn, and subclusters may require their proper similarity matrices, a large amount of recalculation of these matrices at each step of the calculation is needed, which renders the aid of a computer almost essential. Actual applications of the new similarity index to bacterial testing experiments shows that it is somewhat more sensitive than the simple similarity index (Goodall, 1966a). Moreover, Klovan (1969) has used the probabilistic index to classify stromatoporoid coelenterates, which are particularly difficult subjects, with considerable success. He clustered individuals belonging to ill-defined species: this study was supplemented by a principal components analysis of the similarity matrix, allowing a classification to be built up which agreed well with generally accepted conventional assessments, despite the existence of disagreement between the conventional taxonomists engaged with this group.

Goodall (1967) has also provided a distribution for the simple matching coefficient introduced by Sokal and Michener (1958) which takes the value

$$M = \frac{a+d}{a+b+c+d}$$

where the positive and negative matches for two individuals A and B are given by the following contingency table:

		Individual A	
		+	−
Individual B	+	a	b
	−	c	d

To provide a distribution for this coefficient might, as Goodall suggests, facilitate the application of significance tests in numerical taxonomy. There is, nevertheless, considerable difference of opinion as to how desirable significance testing may be in this context.

Goodall also clears up a misunderstanding in the literature and shows that the expectation of M is not 0.5, as has been suggested, but that it lies between 0.5 and 1.0, so that 0.5 is the lower limit of the expectation, only realized when the joint probability of the alternatives f_{ij} is 0.5.

Balakrishnan and Sanghvi (1968) present a discussion of five more indices of similarity, of which one is much like the generalized distance of Mahalanobis, and all are, in fact, distance measures. A useful survey of many of the binary similarity coefficients has been given by Cheetham and Hazel (1969), who compare the properties of the various coefficients.

The highly subjective nature of the choice between clustering systems forces us to reappraise the whole question of clustering, since this is proving to be the Achilles' heel of numerical taxonomy. All clustering techniques can be divided into four main categories as follows (Orloci, 1967a):

(a) Monothetic processes, which define groups according to whether or not they possess a particular entity: for instance, phytosociologists divide sampling areas into those which possess indicator plants and those which do not.
(b) Polythetic processes, in which the criterion for the division of the ensemble into groups is not based on the possession or lack of some particular entity, but rather on some such criterion as the need to maximize the between-group variance relative to the within-group variance. We owe this important distinction to Sokal and Sneath (1963).
(c) Agglomerative techniques, in which the individual entities are grouped together on the basis of some working rule as to what constitutes a cluster in the given context.
(d) Divisive techniques, in which the ensemble of entities is divided into a series of sub-groups based on some criterion which may lead either to monothetic or polythetic classification.

Apart from the phytosociologists, few workers concerned with these multivariate problems employ monothetic processes, despite the fact that much classical taxonomy is effectively monothetic at any one level of the taxonomic hierarchy. We agree with Sokal and Sneath (1963), that polythetic classification is highly desirable, and with Pielou (1969) that monothetic methods of division are wasteful of information. Pielou also raises the question of whether clustering should be hierarchical or reticulate, that is to say whether the classificatory scheme should attempt to impose an evolutionary structure on the material or whether the affinities of the groups should be displayed without any imposition of the hierarchical structure. Any form of dendrogrammatic representation is necessarily hierarchical, whereas one advantage of principal components, or other truly multivariate representation, is that it is not necessarily, or even usually, hierarchical.

Monothetic methods of classification are, almost of necessity, divisive, and it may be that the doubts which many workers, not influenced by traditions peculiar to phytosociology, have about monothetic techniques have led to the very widespread use of agglomerative techniques. However, the subjectiveness of the choice of clustering procedures, a subjectiveness which stems essentially from the impossibility of defining a cluster in other than arbitrary terms (Rao, 1952), leads us to look again at the possibility of using divisive, polythetic, methods. For it is in this direction that we may hope to escape from the morass of clustering techniques about which it seems almost impossible to make sensible recommendations. This view seems to be broadly shared by Pielou (1969), but, as she remarks, whereas divisive-monothetic techniques are operational, divisive polythetic techniques suffer from severe problems at the level of their practical execution; such problems do much to detract from their theoretical superiority.

Let us look at these problems for a moment. Edwards and Cavalli-Sforza (1965) have put forward a scheme which divides the material to hand on the simple basis that the between-groups variance is maximized relative to the within-groups variance. There can be no doubt that for a modest number of entities (say, 10-13) this method works very well indeed, and can be performed by hand in an amount of time which makes the construction of a computer programme uneconomical, unless the method is likely to be required repeatedly (see, for instance, Blackith and Blackith, 1968). The difficulty arises when more extensive analyses are required, for the computer time increases dramatically with the addition of further entities to the ensemble. Gower (1967b) quotes a time of 54,000 years for an analysis of 41 entities by this means, on a computer with 5 μsec access time. However, as Pielou comments, it is usually practicable to impose certain restrictions on the partitions which have to be investigated, and these are particularly easy to apply when the calculation is done on a desk machine (Dagnelie, 1966).

Some advance on these lines seems to be imperative, for the division of opinion on agglomerative clustering techniques seems to be so deep and irreconcilable, and, what is worse, virtually untestable, that a divisive polythetic solution appears to be almost the only way out of the impasse. Agglomerative clustering even gives different results according to the number of entities clustered, irrespective of their nature (Crovello, 1968a, d) and changes of the suite of characters also modify the results (Crovello, 1969).

We should not forget, however, that clustering by means of ordination on charts whose axes are vectors (principal components, canonical variates) virtually meets all the requirements of a divisive polythetic technique. For the polythetic nature of the clustering is introduced by the truly multivariate nature of the comparisons, and the divisive nature of the distinctions between groups of entities is assured by the visual nature of the comparisons.

Among the authors who have put this point of view into practice, Orloci (1968a) has classified two species of heather and the intermediate hybrids from various localities by measuring the corolla, calyx, style, corolla lobe and pedicel lengths. A principal components analysis gave a clear-cut clustering into *Phyllodoce glanduliflora*, which occupied the negative quadrant of the principle component chart, between the first and second vectors: *P. empetriformis* which occupied the positive quadrant of the same chart but was also displaced in the direction of negative scores along the third vector, and finally a cluster of hybrid plants distinguished most clearly from the *empetriformis* parent in that they were displaced from the plane formed by the first two vectors in the direction of positive scores along the third principal component. Orloci makes the useful point that most clustering techniques will achieve some degree of sorting, even if there is no underlying structure to sort out. Greig-Smith (1964) has shown how the analysis of variance, applied to the component scores, may be used to distinguish real from arbitrary clustering.

The great strength, practical utility, and resilience of the many classical taxonomic processes surely stem from the fact that they are both divisive and polythetic, escaping the operational difficulties by arbitrary but, at least usually, wise restrictions on the possible splits. It is ironical but at present inescapable that the divisive, polythetic, system is the one which cannot in practice be duplicated on a computer except for fairly small numbers of entities to be classified. We must face the possibility that no agglomerative or monothetic divisive method can compete with this divisive, polythetic system, thus bringing numerical taxonomy to a period of quiescence. Recent, as yet unpublished, work by Barrs at Southampton University suggests that this impasse may be breached more rapidly than the formidable theoretical difficulties might suggest.

Machine Scanning of the Material

One of the most time-consuming tasks in numerical taxonomy is the examination of the material for suitable characters and the recording of the attributes of these characters. This is all the more severe when features of the internal anatomy of animals are to be included in the suite of characters. If an adequate classification could be made by machine scanning of the external form of the organisms, one aspect of taxonomic work would be greatly aided.

There are two rather distinct approaches to this topic. One is that of the specialized research biologist, anxious to use numerical taxonomic methods to get the best out of a large and laboriously acquired suite of carefully chosen attributes for some particular purpose, such as the revision of a sub-order of insects. A quite different approach, which seems nearer to much of the published writing of Sokal and Sneath (1963), is that of the worker without much specialized taxonomic knowledge, concerned to sort some large body of material into a hierarchy of taxa as expeditiously as possible. Both these approaches represent realistic attitudes on the part of a taxonomist under different circumstances.

Assuming that the second approach, that of sorting a mass of material, is uppermost in the taxonomist's mind, what can be done to place as large a part of the load as possible on to the computer? Rohlf and Sokal (1967) have explored the possibility of covering the outlines of some imaginary animals, "Caminalcules", which have been each hole in the card according to whether there is any part of the outline showing through the hole. Ledley (1964) has also explored this approach to the identification of chromosomes in a caryotype, a subject of increasing concern in medical genetics; the need to conduct routine examinations of caryotypes to detect the growing number of diseases whose aetiology has a genetic basis. Possible geological applications have been discussed by Dalke (1966).

Rohlf and Sokal's findings are most instructive; they show that the outlines of some imaginary animals, "Caminalcules", which have been invented by Dr. J. H. Camin to serve as test subjects for numerical taxonomic studies, could be scanned through the punched cards in both random and systematic fashions, and the data used to compute phenograms which agreed reasonably well with the classifications produced by more usual methods of selecting attributes, followed by analyses by either numerical taxonomic or "orthodox" methods. The size of the holes in the masking card must be kept small, otherwise the images assessed through the holes become unrecognizable. A surprising feature of these experiments in computer assessment of patterns is that some scanned patterns that appeared to be quite unrecognizable to the human eye, normally a powerful recognizer of patterns, still gave a reasonable phenogram when made the input of a computer programmed to produce phenograms.

It seems to be essential to standardize the size of the outlines of the organisms for machine scanning, and there are difficulties if the shapes differ to the extent that there are non-homologous parts showing through homologous holes in the masking cards. For instance, if the legs or other members of an animal project at different angles on two forms they will appear to be different even if they are in effect the same. Moreover, most organisms are three-dimensional, and although they could in principle be scanned from above, and perhaps below as well, the practical difficulties of avoiding the overlap of parts in real structures, as opposed to black and white figures, would be formidable.

Taking the results of Rohlf and Sokal's work at its face value, i.e. as a study in computer pattern recognition followed by the processing of the pattern into a phenogram, it is a remarkable success. As these authors comment, theirs is a "shotgun" method of recording taxonomic characters whereas the ordinary taxonomist prides himself on the care with which his characters are selected and assessed. Perhaps one contributory reason for the computer's success is that it had to construct the phenograms from a limited number of patterns all of much the same appearance; had it been faced, as it would in a mass sorting trial, with widely different forms, the performance might have been less good. However, the real if limited success of the experiment, even when wide holes were used, suggests that there is, as Rohlf and Sokal put it, substantial redundancy in the suite of attributes used in ordinary numerical taxonomic work, as well as in the information that the computer acquires from scanning the masking cards. To some extent the feeling that careful deliberation over the characters and attributes of a taxonomic study is necessary may be prejudice; in educational practice experienced examiners are generally distrustful of examinations in which the candidate marks squares on a card for "true or false" and thus opens the way to computer marking, but the evidence suggests that the judgements made by the computer are not so different from those made by examiners as to justify the much greater cost of formal, traditional, written examinations. So it may be with computer scanning of the outlines of taxa.

The question raised by Rohlf and Sokal as to whether there was redundancy in the input data opens up a discussion of the robustness of dendrogrammatic representation, which will be developed further in Chapter 25.

23
Comparison of the Clustering Techniques

The clustering process is essentially one in which, following some more or less arbitrary rules, individuals or groups which are more closely associated are brought together into a cluster which is then considered to be differentiated from other associations forming separate clusters. As Rao (1952) has cogently remarked, there is no clear definition of a cluster. Indeed, what may constitute a cluster for one worker may be of little consequence for another; in fact at different times one might be concerned to sort material into very broad categories initially, and at a later date to sort some fraction again for more detailed purposes. The operational definition of a cluster would probably have changed quite a lot as the investigator's interests became more concentrated.

The practical approach to *ab initio* clustering has been extensively discussed by Sokal and Sneath (1963) and does not require restating here. Of the various methods that these authors discuss, the weighted pair-group method seems to be at least as good as, and perhaps in some respects superior to, other techniques that build up clusters from individuals in an agglomerative way. Gower (1967b) has compared this weighted mean pair method with the divisive polythetic one published by Edwards and Cavalli-Sforza (1965) which splits the body of individuals or groups into smaller bodies in such a way that the sum of squares of distances between the groups so formed is maximized relative to that within the groups. This method requires the examination of the sums of squares between all the $2^{n-1}-1$ splits between n entities, and takes an inconveniently, and perhaps prohibitively, long time on a computer for $n > 20$. However, it has been used by Blackith and Blackith (1968), and has proved to be much less time-consuming than theory might predict. For fewer than 12 entities, some commonsense restriction can be placed on the range of splits to be examined. It is very hard to express such restrictions in a form suitable for programming but not nearly so hard when performing the computations by hand.

Yet another method suggested by Williams and Lambert (1959) divides the main body into groups according to whether they do or do not contain a particular variable (or character-state) giving a monothetic rather than a polythetic classification.

Following Gower (1967b), we may consider clustering methods as being either monothetic or polythetic, and either divisive or agglomerative. Sneath and Sokal's weighted pair-group method (and others discussed by them) is polythetic and agglomerative. Edwards and Cavalli-Sforza's method is polythetic and divisive. Williams and Lambert's method is monothetic and divisive. Gower tends to favour agglomerative methods, because there is no guarantee that divisive methods will not split entities arbitrarily. He also favours polythetic methods, partly because monothetic ones commit the investigator to divisive methods. Gower suggests that, although in general the weighted pair-group method, originally due to Sokal and Michener (1958), is perhaps the best for general purpose, classificatory work, one could also use it, with slight modifications, as an approximate algorithm to deal with cases where the Edwards-Cavalli-Sforza method can no longer be applied, because there are too many entities to be examined in respect of all possible splits.

Polythetic methods are to be preferred to monothetic ones, because entities may possess or lack the critical defining characteristic by chance, but this chance will determine the fraction of the dichotomization in which they will be confined. Generally, the special interests of the investigator will go a long way towards choosing his clustering techniques for him; the differences between the Sokal-Michener and the Edwards-Cavalli-Sforza methods, a choice between which will largely depend on the investigators' preference for divisive or agglomerative methods, is unlikely to be so great as to alter the conclusions to be drawn from the clustering, except at the lower taxonomic levels, but it is often at precisely these lower levels that crucial decisions have to be made.

However, within any general category of clustering techniques, there will be various options which the experimenter can follow; these include the possibility of standardizing the various character-states (attributes) to equalize the effects of differences in the units of measurement and the number of states either by means of multistate characters or by the use of logarithmic or other transformations in the analysis of continuous measurement. Here there is also a choice, so far as principal component analysis is concerned, of operating on the dispersion or correlation matrices. In hierarchical and ordination techniques one has the choice of distance or correlation measures.

The effects of making one or other of the various choices open to the experimenter are not always clear and may change substantially from one experiment to another. Many of the most cogent aspects have been discussed by Moss (1968), Michener (1963) and by Sokal and Michener (1967), who show the advantages of standardization (in numerical taxonomy). These authors discuss some of the vagaries of the methods which the experimenter may be forced to choose between, on a somewhat existentialist basis, for lack of any sufficient body of accumulated experience.

The choice of agglomerative (synthetic) techniques as opposed to divisive (analytical) ones is logical in the sense that it commits one to begin the classificatory process from the very beginning, and such an attitude may be appropriate for, say, an archaeologist faced with an assortment of artifacts. In relation to the problems of a taxonomist it seems less desirable; the implication that the whole body of classification in botany and zoology has to be worked over again is forbidding and unnecessary; there are great advantages to be obtained from a divisive technique which allows the experimenter to make use of his prior knowledge of the major groupings of the material where these have proved satisfactory. Nevertheless, there is much disagreement about the amount of reworking that is in fact desirable.

Wishart (1969) gives a series of computer programmes in Fortran II covering eight commonly-used clustering techniques.

Jardine *et al.* (1967) have constructed a logical model for taxonomic hierarchies and have shown how most average-link cluster analyses satisfy neither their "expansion" nor their "contraction" conditions for a satisfactory method: single-link clustering may group together at a relatively low level those taxa which happen to be united by chains of intermediates.

The choice between correlations and distance measures in clustering or ordination programmes is one in which, again, no finality is possible; nevertheless, we share the view of Boratynski and Davies (1971) that even if no serious differences are likely to arise from the use of one or other of these alternatives, the distance form is preferable. Jago (1969) found the distance analysis gave a clearer separation of grasshoppers of the genus *Cordillacris* than did correlations, and although correlations have their supporters in the literature, the balance of evidence seems to favour distances (Eades, 1965).

Methods for Handling Deviant Individuals;
Goodall's Deviant Index

In discriminatory analysis, one problem is to allocate an individual of unknown affinities to one or other of the clusters already recognized. In the more general case one may wish to assess the affinity of an individual to a particular cluster. There are three cases to be considered (Goodall, 1966c), in which the individual's affinities are assessed in terms of quantitative characters, qualitative characters and ranked characters. Where quantitative characters are measured, the measure of the centre of the cluster is taken as the mean. Where the characters are qualitative, the "norm" for the cluster is taken as the mode, and in the case of ranked characters the norm is taken to be the median.

One of the common difficulties of using keys for the identification of specimens is that, since keys are generally built on monothetic principles, the correct interpretation of every couplet in the key, up to the point at which the specimen is identified, is imperative. Goodall (1968a) comments that the need for a positive decision in each case often leaves the user almost as much in the dark as he would be if the key did not exist. Often, no doubt, the failure occurs through the inexperience of the user, who does not know how to interpret a couplet in the particular way envisaged by its constructor. But it has to be admitted that sometimes dichotomous keys, especially those dealing with rare material of which the constructor may have seen few specimens, or have been compelled to rely on earlier descriptions which themselves may have been of poor quality, are partly incorrect. In such cases the experienced worker has to know not only how to interpret those couplets which "work" but has to remember to circumvent the ones that do not. This circumvention, whether required through the failings of the key or the user, is difficult when the key is arranged on a monothetic pattern.

Goodall's computer key, with the programme in the computer language PL1, has several of the characteristics of multiple-entry keys, in which the choice of a subset of attributes and the order in which they are to be used may be at the discretion of the user. Such programmes obviate the situation, all too familiar to taxonomists, where the first couplet of the key cannot be cleared because some attribute is not visible, or otherwise not available. Since the identification proceeds by a series of interrogations between computer and user, time-sharing facilities are essential if the programme is to be economic. Artificial keys may now be generated by computers, and recent work in this field has been described by Pankhurst (1970).

The adoption of computer-based routines for identification should release the specialists in each group for more exciting and rewarding activities than routine identification, so that the current weight of routine work does not inhibit research. This step has been forced on palynologists because pollen presents a multiplicity of characters which do not necessarily fall into groups such that conventional taxonomic procedures could construct "natural" groupings (Walker *et al.,* 1968).

24
Graph Theory and Dendrograms

The theory of branching dendrograms (arborescences) is still in a relatively undeveloped state. The taxa in a dendrogram constitute the pendant vertices of the arborescence in the terminology of graph theory (Berge, 1962). If there are n such pendant vertices, the total number of distinct bifurcating arborescences that can be constructed is

$$\frac{1}{2n+1} \binom{2n+1}{n}$$

Lukasiewicz (in Berge, 1962) has suggested a neat way to represent dendrograms, of the kind much discussed by numerical taxonomists, without going to the trouble of printing a lineblock diagram. If we regard the dendrogram of Fig. 38, p. 299 as being generated by an algebraic monoid on an oriented (i.e. vector) plane, then by following strictly the rules that each bifurcation of the dendrogram is represented by the symbol 1, and each pendant vertex (terminal point) by 0, and also that the dendrogram is to be "read" from bottom (origin) to top (the currently existing taxa) and from left to right, we can specify the dendrogram of relationships within the orthopteroid insects shown below by the following operational symbols:

$$111100110001001010101010100$$

Evidently the symbolism, which contains all the information that can be extracted from the dendrogram, is far more concise and simple to publish. Moreover, it should be easy to programme computer outputs involving dendrograms so that the monoid symbolism is printed directly. One must always bear in mind that in these dendrogrammatic representations the non-pendant vertices must be regarded as swivelling points, so that there are many ways of representing the same dendrogram: for instance the monoid symbols:

$$1101010100111111000100100$$

generate a dendrogram identical with that shown above, after swivelling on its origin.

In general, the reader will want to know what names are to be attached to the pendant vertices, and if the named taxa are denoted by some suitably short symbol this can be printed in brackets after the zeros of the monoid. Thus, to replace a phenogram which occupies an entire page of a journal in print (Fig. 1 of Ehrlich and Ehrlich, 1967) we can write

$$1110(31)0(30)10(29)0(28)10(27)11110(26)0(25)10(23)-$$
$$110(24)0(22)0(21)0(20)10(19)110(18)0(17)110(15)-$$
$$0(14)0(16)10(13)10(12)0(11)1110(10)10(9)0(8)10(7)-$$
$$0(6)10(5)110(4)10(3)0(2)0(1)$$

where the figures in parentheses represent genera of butterflies. The one feature of the phenograms which is not represented by the monoid written out with its labels is the depth of the phenetic relationship.

Active research is in progress to try to codify the processes of at least some types of conventional taxonomy so as to represent these processes on hierarchical graphs. Rubin (1967) has reconstructed the theory of hierarchical classification to accommodate his "hill-climbing" algorithms for the optimal partitioning of the material. Estabrook (1968) has used the theory of partial orders to compute the shortest evolutionary paths which go to make up a dendrogram, given certain information about the evolutionary trends of the character states of the organisms in question. He has done this in the context of the Camin-Sokal model for cladograms, but one could conceive of an extension of this approach to the various other available models of cladograms, and even to its incorporation with the kind of evolutionary model suggested by Le Quesne's (1969) weeding out of those characters that have not followed the straight and narrow path of evolutionary progress.

It is ironical that the mathematical approach to hierarchical classification should be getting under way at a time when the desirability of hierarchies, without which trees of various kinds are irrelevant to the present situation, is being more and more questioned. Indeed, Cowan (1970) has baldly stated that one should not attempt to classify bacteria in what he regards as an unrealistic hierarchical structure, though workers with other groups might well feel that, even if we grant the proposition for bacteria, these organisms constitute something of a special case. The point, nevertheless, bears thinking about.

There is a disadvantage of dendrograms as compared with oligovariate charts, which stems precisely from the nested, hierarchical, nature of the dendrogram. Each taxon must be considered to fit into one of the clusters on the dendrogram, and cannot be represented as fitting into a zone of overlap between two such clusters. This problem is particularly severe when the dendrogram has been constructed on monothetic lines, since individuals, which are affine in respect of many characters other than that on which the split has been made, will perhaps be widely separated on the dendrogram. Polythetic dendrograms offer a substantial, though not complete, amelioration of this defect. In oligovariate representations, however, there is no reason why an individual should not be shown in its proper position intermediate between two clusters. Classification in linguistic analysis must permit an overlap between the classes if it is to be of any value (Needham, 1967) and one might think that any strait-jacket which forced data into an unsuitable mould could be dispensed with to advantage.

Graph theory has a part to play in releasing cluster analysis from the bane of nested, one-dimensional, hierarchies. An ingenious and realistic discussion of the classificatory process in these terms is given by Wirth *et al.* (1966) who produce a classification of orchids that is at once satisfying to a taxonomist in that it conforms to the thought processes that he would be inclined to employ, but quantifies the procedures involved and also provides a classification of the Oncodiinae which accords well with some conventional assessments of the subfamily.

Vinck (1962) has emphasized that taxonomic trees are rarely conceived as being more than two-dimensional, and often are effectively one-dimensional, to the great detriment of their representational powers. The Prim network approach of Cavalli-Sforza and Edwards (1967) is promising.

25
The Robustness of
Phenograms and
Distance Models

It is particularly rewarding to act as both a numerical and orthodox taxonomist faced with the practical problems of extracting material from, say, tropical rain-forests, and trying to find out not only the identity, but the affinity, of the new specimens.

One cannot work for long on a difficult group without pondering on the apparently miraculous emergence of a reasonable phenogram from a welter of inadequate information, lost types, poor descriptions, rare, or at least inaccessible, material, and the usual run of plain, all too human, errors that all of us make from time to time. And yet, despite all these handicaps, a high proportion of the conclusions of orthodox taxonomists will stand up to the accumulation of new information, although this proportion may seem to be less than it is because differences of opinion are generally more noticeable than agreements. How is it done? There seem to be two kinds of process at work. One kind relates to the somewhat authoritarian strain in some taxonomic writing, which leads to the publication of conclusions which were in fact arrived at by one set of mental processes but which are written up as if they were reached by quite other rationales. Cain (1959) calls this "cryptic taxonomy". Curiously, this habit of disguising the creative parts of an investigation is nowhere so marked as in mathematical papers, where, as Bickley and Gibson (1962) remark, "The fashion today seems to be to try to condition the minds of undergraduate mathematicians to the acceptance of ready-made abstractions . . . without reference to the process of abstraction and the array of facts whence the abstractions come. This makes mathematics difficult, even for mathematicians".

Some of the most "authoritarian" classification has been that of the phytosociologists, and it is perhaps retributive justice that in a phytosociological paper by the Braun-Blanquet school the classification arrived at by the authors is contradicted by the evidence they publish in the same paper (see Goodall, 1953).

Because of repeated claims from traditionally-minded taxonomists that classificatory dendrograms uninformed by phylogenetic thinking were of little value, Camin and Sokal (1965) have made a vigorous effort to construct "cladograms"; these are phenograms in which the characters are coded so that presumed "primitive" conditions are scored zero and presumed "advanced" conditions are scored unity. Further assumptions among the list given by Camin and Sokal are that the "primitive" state arose only once in the taxa being studied, and that evolution is irreversible so that "primitive" states cannot be regained after evolving to a more "advanced" state. These assumptions are, in fact, working hypotheses, which enable one to prepare a dendrogram in which the minimum number of "evolutionary changes" have taken place. One of several techniques is used to prepare a "procladogram", which is a kind of provisional cladogram; this provisional essay is then adjusted so as to invoke the minimum number of the "evolutionary" steps, by rearranging the arms of the cladogram until the number of evolutionary steps is minimal.

When the number of characters used in the preparation of these cladograms is small, the cladograms and the corresponding phenograms, prepared without the introduction of the "phylo-genetic" information which results in the recording of the attributes, will, as Camin and Sokal remark, be similar only when the similarities are due to recent divergence. However, this statement does not seem to be true when large numbers of characters are used in the calculations. Blackith and Blackith (1968) prepared a cladogram and a phenogram of some orthopteroid insects using 92 characters and found that it made little difference which taxonomic philosophy was adopted. Such similarity can hardly be explained away by remarks about recent divergence, since some of the material used in Blackith and Blackith's study is known on fossil evidence to have diverged in or before the Permo-Trias. The possibility always exists that for some reason the material they studied is exceptional, but the suite of characters was chosen with great care from the internal as well as from the external anatomy of the insects, and it is hard to see why any results obtained should be unrepresentative. The possibility exists that there is indeed an underlying relationship between the organisms used which will emerge from almost any unbiased attempt

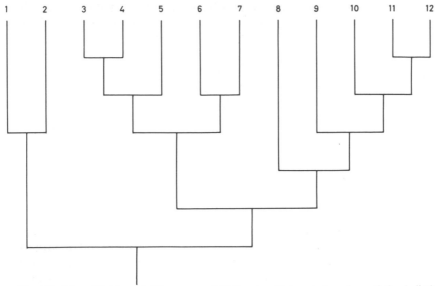

Fig. 37. The affinities of 12 groups of Orthopteroid insects based on their cladistic relationships as determined by the method of Camin and Sokal (1965). Data consisting of 92 characters. Redrawn from Blackith and Blackith (1968) so as to emphasize the similarity between Figs 37, 38 and 39 without violating the branching sequences for each. See Fig. 38 for labelling.

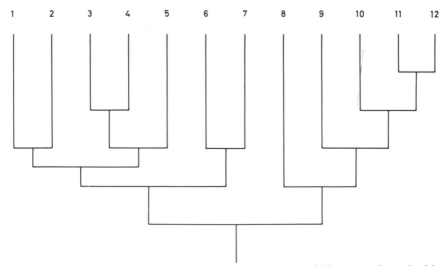

Fig. 38. The phenetic affinities of 12 groups of Orthopteroid insects as determined by the Edwards-Cavalli-Sforza (1965) technique using 92 characters. Redrawn as for Fig. 37. Numbers attached to pendant vertices of the dendrograms indicate: 1 Phasmatodea; 2 Dermaptera; 3 Gryllacrididae; 4 Tettigoniidae; 5 Grylloidea; 6 Mantodea; 7 Blattodea; 8 Eumastacoidea; 9 Tetrigoidea; 10 Pyrgomorphidae; 11 Acridinae; 12 Cyrtacanthacridinae.

to uncover it provided only that the number of attributes and their distribution are adequate. Taken at its face value, this remark may raise the spectre of Plato's ideal forms, but there are more rational, if more difficult, possible explanations. We have only to suppose that the form of the dendrogram, which in smaller experiments is determined by the sense of the supposed evolutionary steps, is in larger experiments more and more closely determined by the numbers of the dissimilarities rather than by their evolutionary sense.

An argument of this kind adverts to the ideas of Rohlf and Sokal (1967), already mentioned on p. 287, that there may be a substantial measure of redundancy in the data of a numerical taxonomic study. The effect of such redundancy will be to swamp information from the "phylogenetic" part of Camin and Sokal's model, because the "phenetic" information is overdetermining the pattern of relationships, to borrow a form of words from engineering practice. Let us look at the evidence for redundancy. Rohlf and Sokal found that scanning with coarsely punched masking cards did not degrade the computer's ability to construct a phenogram from the resulting scores of parts of animal outlines as much as they had expected. Blackith and Blackith repeated their analysis using a random subset of only 30 attributes which gave a phenogram differing in relatively trivial ways from that obtained using all 92 attributes. This result is positive evidence of heavy redundancy, at least near the "origin" of the phenogram. As one proceeds towards the smaller branches of the phenogram, the divisions of the taxa will be made on the basis of fewer and fewer dissimilarities, so that redundancy will not be uniform all over the phenogram.

Sneath (in discussion of Gray, 1969) notes that one can identify material using far fewer characters than are needed for classification.

Ehrlich and Ehrlich (1967) have also performed a large-scale numerical taxonomy of butterflies using 100 external characters and 96 internal ones. With these large numbers of characters they have been able to support a "weak" form of the non-specificity hypothesis, i.e. all the phenograms they computed, from internal or external characters, or those confined to the head, or thorax, or abdomen, bore a certain "family likeness" to one another. It does not seem likely that such differences as were found in the classification produced could be attributed to any deficiency in the numbers of characters employed; far more likely is it that, as these authors suggest, there is no single, unique, classification based upon overall resemblance, since there is no operational definition of what constitutes overall resemblance, though one is entitled to hope that with large numbers of characters the classifications will prove to be reasonably convergent.

There is nothing too unusual about delving deeply into a topic of both theoretical and practical importance, struggling to improve the logical basis as well as the practical conduct of the scientific approach, and suddenly finding oneself, as it were, emerging from the mists without ever having fully grasped the dimly seen shadows within. Some two decades ago, there was intense interest in the theory and practice of biological assay, with a strong emphasis on the investigation of dosage-response regression lines to try to find appropriate transformations which would the most nearly give rise to linear regressions, and sound estimates of the distances between such lines. Once a certain stage of the enquiry had passed, it became clear that the search for the "right" transformation was one in which no finality was possible even though much progress was, and needed to be, made. In attempts to discover the factors influencing graders' judgements of cheese curds, Harper (1952) also came upon some apparently irreduceable uncertainties in the methodology, leading him to complain that psychology had yet to formulate its own uncertainty principles.

The Ehrlichs' pessimistic conclusion from their investigation was that there is little need for numerical taxonomic studies because the effort would be greater than the likely reward. We should also recall that, since Ehrlich (1964) has expressed himself in deeply pessimistic terms about the very continuance of any sort of taxonomic work that a modern taxonomist might recognize as such, his later views on numerical methods are not likely to be optimistic; it is hardly logical to seek to improve an activity whose disappearance leaves one indifferent. The Ehrlichs' study was a monumental one, and there is ample evidence that, regarded strictly as an exercise in numerical taxonomy, far more work was done than is needed. The additional work has nevertheless greatly benefited our knowledge of the anatomy of butterflies. If we believe that there is a serious measure of redundancy in these large-scale investigations we should draw the appropriate conclusions, and cease to try to make the assessment converge on some scarcely attainable limiting form, but to attain a known degree of precision using methods of analysis which are as unbiased as we can reasonably make them.

There is strong evidence that when quantitative characters are used in a morphometric analysis, the generalized distances converge effectively for numbers of characters much smaller than those used in numerical taxonomy. For instance, the studies of Mahalanobis *et al.* (1949), which dealt with the anthropometry of groups in the United Provinces (now Uttar Pradesh), showed that a discriminatory topology based in 12 quantitative measurements was not appreciably different from an 8-measurement topology. Even as few as three characters were able to pick up all three components of variation in locust investigations (Albrecht and Blackith, 1957), despite certain misplaced theoretical reasons for doubting this ability.

In fact, there is not a great deal that theoretical analysis can do to resolve this question of redundancy, which is essentially a matter for experiment (Hotelling, 1951). However, Vinck (1962) has shown that redundancy may be calculated in the context of numerical taxonomy, and points out that even in the presence of a given amount of redundancy, some ambiguity may remain in the definition of the supposedly hierarchical structure of relationships. Garner and Carson (1960) have measured redundancy in multivariate terms.

Sheals (1964), anticipating Gower's publication (1966), noted that Gower's similarity index for qualitative characters, which consists of the simple sum of the similarities divided by the number of characters examined, is designed to be non-parametric. It is less obvious that, in the special but widely used case where the characters are scored on a 0,1 basis and all are qualitative, the Pythagorean squared distance between pairs of groups is virtually identical with Gower's expression, and is also non-parametric, so that the robustness of non-parametric methods as compared with parametric ones should emerge. Even where the methods are technically parametric ones, it seems intuitively likely that they will often behave as if virtually distribution-free, and to this property they may owe much of the robustness mentioned above.

An interesting question arises as to whether redundancy in one part of a classification can compensate for a deficiency in another. Intuitively one might expect that some compensation of this kind could arise, and indeed Crovello (1968b) has suggested that a phenetic analysis can tolerate a substantial amount of missing data (leading to asymmetry and hence to the possibility of compensation for the missing data by redundant information) and still give reasonably consistent results.

On the contrary, when there are errors of homology in the material under study, the effects of random mistakes are quite serious: Fisher and Rohlf (1969) suggest that a study composed of 74 characters could hardly tolerate more than some six errors of this kind but that distance methods were less sensitive to distortion than clustering based on correlations. There seems little doubt that this conclusion is confirmed by the difficulties which Kendrick and Weresub (1966) found in their attempts to perform a numerical taxonomy of the basidiomycetes: of the many critical features on which they comment, it is the failures of homology that seem to have contributed so much to the relative failure of their attempt. Of course, the basidiomycetes are an exceptional group, mercifully beyond the range of problem-groups with which most taxonomists have to deal. Nevertheless, as the endeavour to cover the higher taxonomic categories proceeds, problems of homology will inevitably loom larger and larger: we can see that this is so by contemplating (as has been suggested by Dr. F. Jeal of Trinity College, Dublin) a numerical taxonomy of the animal kingdom as a whole. Even if we are prepared to accept all the devices for simplifying the work, such as the use of exemplars, which are currently known, the formidable nature of the exercise remains daunting precisely at the level of homologies.

Once we are satisfied that redundancy exists, we must look at its likely influence on the phenograms, and also on the relationships between phenograms and other kinds of dendrograms such as cladograms. A useful starting point is the knowledge that the structural relationships between the various forms represented in the phenograms can be ascertained without any exact, direct, measurement of their shapes (by morphometric or numerical taxonomic techniques) but merely from a knowledge of the rank order of their relationships.

For instance, Kendall (1957) has shown that the first component of a principal component analysis is not greatly modified when numerical assessments are replaced by rankings. Similarly, Shepard (1962) has remarked that in the process of reducing a hypermultivariate situation to one of more modest dimensionality nearly all the essential structure of the relationships can be preserved using only rankings. Gower (1967a) has discussed this point in relation to multivariate analyses of the kind used in morphometric work.

Another highly relevant contribution is that of Williamson (1961), who compared principal component analyses and other vector techniques to link the abundances of 23 entities of the plankton content off the Scottish north-east coast to a suite of hydrographic assessments. Nearly all the information obtained was contained in the ranking of the data; rank correlation coefficients proved highly suitable for linking the one set of data with the other. Moreover, Burnaby (1966a) says that there appears to be no reason why quadratic discriminant functions should not be constructed using the ranks of the observed variates transformed to normal scores, although a certain amount of rescoring would be necessary for each population examined. We publish in Chapter 8 Burnaby's comparison of analyses along canonical variates using measurements and rankings, which shows that virtually all the information that the experiment was designed to retrieve is retained by the rankings.

Whatever differences of opinion there may be about the optimum solutions to the various statistical problems of morphometric analyses, it seems unlikely that the rank order of the relationships which are being studied will be disrupted by the use of suboptimal techniques of assessment, except in the occasional instance. We can, therefore, expect that almost all reasonable methods of assessment will lead to broadly the same general pattern of relationships, and that much of that part of the information, carried by the measures we obtain, but superfluous to the assessment of rank orders, will be redundant.

Naturally, a stable situation of the kind just described cannot be expected if only small experiments are carried out. Where, for instance, the total number of characters measured is small, there will probably be a big difference in a dendrogram according to whether it has been prepared as a cladogram or a phenogram. But where large numbers of characters are assessed, the cladograms will converge on the phenograms, because once the rank order of the groups is established there is little hope of making practical use of any other kinds of information. In a situation such as that described by Blackith and Blackith (1968), where three times as many attributes were used as were necessary to establish the broad pattern of relationships in the phenogram, it is perhaps hardly surprising that the putatively "phylogenetic" information incorporated into the cladogram had so little effect.

Considerations of this kind do something to resolve the dispute between "phylogeneticists" and "pheneticists". They show that this debate can only be prolonged if its ground is shifted somewhat from the positions currently held by each side. It is part of the phylogeneticists' case that information about the evolution of the forms should be fed into the cladograms, or their equivalent expression of relationships. Unless this information is to be swamped by that about the phenetic structure of the material, the total number of characters used must be kept low, probably considerably fewer than 30, and perhaps under 10. These few characters can no longer be considered as representative of the total structure of the organism, because it is precisely in the case where the total structure is adequately represented that the phylogenetic information is swamped. So the characters must, presumably, be selected on the strength of their ability to reflect evolutionary changes, i.e. they must be characters for which the experimenter is confident that he knows the evolutionary sequence, if they are to be used in a cladogram. In such an experiment the structure of the cladogram would be almost wholly determined in advance by the experimenter's views on the phylogeny of the groups he is dealing with.

The dispute is, then, both reduced in scope and sharpened in detail. It is reduced in scope because, for experiments of any substantial size in which many attributes are assessed, the same phenogram will emerge from the analysis with or without the "phylogenetic" component of the information. This is what Blackith and Blackith called "the irrelevance of phylogenetic speculation". However, the dispute is sharpened in detail because affinities can no longer be seen as measures of overall resemblance tempered by the injection of some phylogenetic speculation as persistently recommended by Mayr (1968). The attempt at compromise seems to be doomed. Either affinities are to be measured in terms of overall resemblance, or in terms of descent, but not both, at least not in the same diagram. Moreover, since the phenogram is so robust, no weighting system is likely to improve matters appreciably from the point of view of the phylogeneticist.

As early as 1963, Sokal was commenting that the various possible methods of estimating distance in taxonomic work were statistically robust in the sense that the results obtained by their use agreed reasonably well with one another (Sokal, 1963). We can now suggest that it is not so much the statistical robustness of the methods but the physical robustness of the relationships which are there to be discovered that is in question.

It would indeed be surprising if classifications were not robust, for, as Ouellette and Qadri (1968) comment, biological systems are highly integrated and their information content is highly redundant. This situation arises partly because of the need to keep all the functions of the organism in step during evolution, and partly because living organisms are not like machines whose worn or broken parts can be renewed: a failure of the lighting system will instantly make one aware of the value of the multiplicity of ways in which one can find one's way round a building by touch, sound, and memory: these are not the less valuable for being ordinarily redundant. Thus in addition to the redundancy built in to the system by the general genetic control of development, there are in some cases, as yet inadequately studied, "pure" redundancies for emergency use. There is, moreover, a clue to the nature of Rohlf's (1965) "uncertainty principle of taxonomy", formulated by an analogy with Heisenberg's uncertainty principle in physics. Rohlf is stating the undeniable proposition that different taxonomists, even when they are trying to use the same methods to analyse the same material, do not arrive at exactly the same arrangement of that material. This statement is as true of conventional as of numerical or other quantitative taxonomists. Even randomization of the order in which the original material is clustered gives disparate phenograms (Rohlf, 1965) but part of the disparity was made worse when correlations, rather than distances, were used to express relationships.

Theoretical taxonomists are in something of the position of the man in the street who, having given a series of almost incomprehensible directions to a passer-by seeking a particular place, ends up with the time-honoured phrase "You can't miss it". Almost any except the worst methods, handled by almost any except the worst taxonomists, will get to within a certain range of this framework of relationships between the groups being studied, although the dim outlines of the framework may well be clothed with a wide variety of labels.

Siehl (1970) reacts against the concept of pure numerical taxonomy and prefers to think along the lines of a numerical classification. He points out that this is actually embodied in the work of many of the biometrically working biologists and suggests that the techniques used in these studies may be suitably grouped under the neutral concept of numerical classification. He came to this opinion after detailed studies on fusulinids from Afghanistan.

26
Suggested Methods for Acquiring Phylogenetic Information

Once the necessary distinction between a phenogram (or other phenetic representation of relationships) and a cladogram or phylogram has been grasped, the question at once arises as to whether there is any way of recovering the information as to phylogeny which underlies the phenetic relationships. There seem in fact to be two broad approaches to this topic, neither of which has progressed very far, and both of which lead to difficulties in their theoretical development. Indeed, when one actually tries to think of ways of recovering the phyletic information inherent in a dendrogram, there is a striking contrast between the logical and practical difficulties which beset the work, and the classical view that the broad outlines, at least, of phyletic relationships were almost self-evident, at any rate to someone accustomed to handle the kind of biological material in question.

Minimum-evolutionary Criteria

One of the ways in which a reconstruction of supposed phylogenetic relationships has been undertaken is to make the assumption that evolution is parsimonious in the sense of minimizing the number of evolutionary steps required to attain the particular branching of the phenogram observed after a straightforward numerical taxonomic study. Several authors have embarked on such schemes of rearrangement, they include Camin and Sokal (1965), Edwards and Cavalli-Sforza (1964) Hendrickson (1968), Throckmorton (1965) and Wilson (1965). Cavalli-Sforza and Edwards (1967) have a more sophisticated method than Camin and Sokal (1965) for evolutionary cladistic relationships using Prim networks (Prim, 1958) but the input remains entirely phenetic.

Of these methods some are more developed than others, and have received more critical attention from other investigators. The Camin-Sokal approach becomes increasingly convergent on phenetic methods as the number of characters is increased (see Chapter 25) but long before this number has been attained the theoretical problems of determining the conditions for a minimum-length dendrogram become formidable, and even with four characters one runs into difficulties, partly, no doubt, because the theory of graphs, into which branch of mathematics the dendrogram manipulations fall, is at once far from easy and relatively undeveloped (Berge, 1962).

Wilson's approach has not escaped criticism. Colless (1967b) has drawn attention to some of the logical weaknesses in the argument. In general, although the pursuit of minimum-evolution cladograms is an activity to be welcomed, we have reserves about whether, in the long run, it will produce anything more constructive than a few fascinating problems in graph theory. One reason for thinking that such might be the case stems from a quite different approach to the extraction of phyletic information, adopted by Le Quesne (1969). Originally intended to serve as a basis for selecting characters for taxonomic studies Le Quesne's approach incidentally demonstrates that a substantial proportion of characters do not obey the basic postulates (that they should not have arisen more than once, and that they should not have undergone reversion to an ancestral (more-primitive) state). If, in fact, several reversions of this kind have occurred, then the search for a strictly parsimonious evolutionary pathway is likely to provide a misleading approximation to the actual course of evolution.

Le Quesne notes that if an organism has two characters with each one in the ancestral (primitive) condition, so that its suite of attributes can be coded (0,0) it may undergo a genetic rearrangement leading to the change of the first attribute from the ancestral to the advanced condition (1,0). The next stage of its evolution will *either* involve a change from (1,0) to (1,1) *or* a change from (0,0) to (0,1). Any other changes either imply that (1,1) has arisen twice (0,0 to 1,0 to 1,1) and (0,0 to 0,1 to 1,1) or else a reversal of an attribute (for instance, the formation of 0,1 *via* 0,0 to 1,0 to 1,1 to 0,1). Thus for any pair of characters we should, if they both obey the putative rules of taxonomic analysis, obtain mixtures of organisms with (0,0), and any two of the remaining three conditions, but not all four at the same time. We may note in passing that since Le Quesne's formulation of the problem (he uses another symbolism) and ours are completely symmetrical, it is quite unimportant whether one can or cannot specify the ancestral condition, since the argument works equally well the other way round.

Le Quesne goes on to sort through his characters two at a time, marking off those for which all four attributes are found. The number of marks is counted for each character, those characters having the highest numbers of marks are deleted, and the process repeated until all the marks have been removed. Those characters that are left are at least potentially those which have obeyed Camin and Sokal's "rules" as stated above. But even if all the characters were of this kind, one would not expect to find all four attributes present for every pair of characters; some combinations may have evolved and died out, others may simply not be represented in the material to hand, and other possibilities exist. Le Quesne, therefore, goes on to calculate the coefficient of character-state randomness; the number of pairs of characters for which all four attributes are actually found, divided by the sums of the probabilities of finding the four attribute combinations for each pair of characters. Evidently, the higher this ratio is, the fewer characters can be considered to have arisen only once.

In a pilot assessment based on data from the analysis of Wilkinson (1967), Le Quesne finds that 19 out of the 23 characters behaved as if they were uniquely derived, only four having to be eliminated. However, he states that in another genus the character-state randomness was much higher than in the pilot survey, and this was a very much larger genus.

We have to face the possibility that the proportion of characters which have not evolved according to the "rules" may be quite high. This is not a new possibility; Cain (1959) has commented that wherever taxonomists are looking at animals with eyes that are unclouded by traditional attitudes, there groups which are polyphyletic are being found, and the polyphyletic group is the living testimony to the fact that a character, or series of characters, has arisen at least twice. To the examples quoted by Cain, we can add Hinton's (1967) finding that plastron respiration has arisen not just twice but many times in insects. For example, plastron respiration has evolved independently in the beetles (twice), in the Tipulidae (Diptera) four times, and in each of seven dipterous families other than the Tipulidae once.

When applied to Blackith and Blackith's (1968) study, Le Quesne's criterion showed that only some 20 to 26 characters out of the original 92 survived to be included in his reassessment of the phenetic relationships among the orthopteroids (Le Quesne, 1971). This reassessment varies according to the precise technique by which unsuitable characters are eliminated, with varying numbers of characters retained in each technique. One of these reassessments is shown in Fig. 39, and this does not differ greatly from the purely phenetic interpretation resulting from the application of the

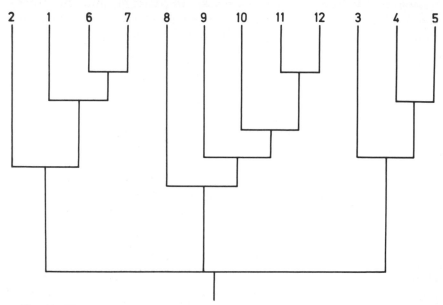

Fig. 39. The affinities of 12 groups of Orthopteroid insects based on 35 characters selected from the 92 of Figs 37 and 38 by Le Quesne's criterion of the uniquely derived character. Redrawn from Le Quesne (1971).

Edwards-Cavalli-Sforza analysis (cf. Fig. 38). Le Quesne deduces that there has been a considerable degree of repetition in the evolution of the orthopteran characters.

Comparisons with the Fossil Record

Ultimately, the proposition that a classification should be based on the phylogeny of the organisms to the exclusion of all other considerations, is an appeal to the fossil record. The relevance of this record is very variable; only for a limited number of vertebrates is there a sufficiently continuous record for it to be relevant to the problems of classifying living forms. Curiously, much less has been done with invertebrate fossils by way of numerical studies, in proportion to the generally much more extensive record of what, at least in many groups, has been a slow course of evolution. The vertebrate fossil evidence has been treated thoroughly by modern techniques of numerical analysis (Olson and Miller, 1958; Olson, 1964).

However, the relevance of the existing fossil record is often much less than would be inferred from the readiness with which phylogeneticists refer to it. Indeed, the criticism has sometimes been made that pheneticists are in some way "antievolutionary". Sokal and Sneath (1963) have done well to point out that it is not evolution which is criticized, so much as the readiness of some conventional taxonomists to pass speculation off as fact by referring glibly to a fossil record which may not be relevant to the problem in question. This situation is particularly irritating where phylogenetic speculation is never likely to be testable by reference to a fossil record; it is no accident that numerical taxonomy should have been nurtured in those fields, such as bacterial classification, where answers are required to classificatory questions as a matter of urgency, and where no fossil record of consequence is ever likely to be found.

Phenetic classification does not necessarily agree with phylogenetic classification even where the phylogeny is known: to expect it to do so is to misunderstand the nature of the classificatory process (Heiser *et al.*, 1965; Johnson and Holm, 1969).

There is at least one attempt to compare the fossil record with the results of a numerical taxonomic study, the thoughtful contribution of Bretsky (1969). In this work 42 species of bivalve molluscs of the family Lucinidae were assessed in respect of 42 characters of the shell. Among other treatments, the 42 x 42 matrix that resulted was subjected to a cluster analysis using as the index of similarity the crude distance between the taxa formed by summing the shared attributes. This distance is the complement of the Pythagorean distance.

As Bretsky comments, "the results of the study are not particularly encouraging to the contention that phenetic and phylogenetic classifications should be closely similar. At least in the Bivalvia, convergence appears to become troublesome at the generic level."

The reasons for these discrepancies are illuminating. Bretsky remarks that there are several cases of the independent origin of characters which could not be detected by phenetic means (and would thus give rise to the "troublesome convergence"). We have already commented on this kind of difficulty in the preceding pages; it is germane to note that although such methods could not be spotted by phenetic methods, they would in all probability not be spotted by conventional methods in the absence of the fossil record. But the fact that they have been discovered (by whatever means) raises the question, what is to be done about the classification as a consequence of the discovery? This question deserves more detailed discussion in the next section, anticipating which we conclude that if the phenogram is faithfully to mirror the course of evolution it is bound to include convergence, and thus differ from the evolutionary phylogram which is, necessarily, *not* a reflection of the course of evolution precisely because it excludes the symptoms of convergence. The unexpected conclusion from Bretsky's study should be that the phenogram and the phylogram differ because the phylogram itself does not mirror evolution adequately. Even if this were not so, Camin and Sokal (1965) have shown that phenetic and cladistic diagrams are quite distinct and will not necessarily agree.

Convergence and Divergence

If there is one proposition upon which conventional taxonomists agree it is that resemblance due to convergence is not to be regarded as "true" resemblance. At first sight the proposition seems plausible enough: one does not, indeed, want to put into a single division of the insects a praying mantis (Mantodea) an antlion (Mantispidae) and a bug (Phymatidae), because some members of each of these groups have developed raptorial forelegs, and look superficially similar. Nevertheless, the fear of convergence has become a phobia; almost all public discussion of numerical taxonomy seems to elicit the cry "What about convergence?". Some taxonomists, who pay lip-service to the ideal of not allowing convergence to influence their arrangements, have been by no means always successful in fulfilling their ideal. Dufour (1841) showed, in the early years of the nineteenth century, that the little pygmy mole-crickets (Tridactyloids), which burrow into sandy mud near water in southern Europe and the tropical regions of the world, are closely related to the grasshoppers and groundhoppers and not at all closely to the crickets. He did this by delicate dissections of the gut, on an insect a few millimetres long, carried out whilst on active service in the Peninsular war! However, for almost a century-and-a-half after Dufour's discovery the pygmy mole-crickets were classified with the crickets simply because they looked like them externally, having developed the outward form of small insects which burrow into sandy mud and live near water, including fossorial fore-legs and a shiny, hump-shaped, pronotum. Sokal and Sneath's (1963) discussion of the convergence problem is apposite and has not been superseded by subsequent discussions.

Nor have conventional taxonomists been much better at recognizing divergence, the converse of convergence, i.e. that animals which live quite different lives may come to look different even though they are fairly closely related. The blattids (cockroaches) live in warm, wet, places, are essentially nocturnal detritus feeders and are thigmotactic, that is to say they seek places where their backs can be pressed against some cover, for example in a slit or crack. The mantids are adapted to catching small arthropods in their forelegs, which are specially armed with spines for the purpose, they are essentially diurnal, live in the open, and have all-round vision with large eyes on heads which swivel freely. To look at, these two groups are quite distinct but, just as the internal anatomy betrayed the grasshopper-like structure of the pygmy mole-crickets to Dufour, so it has shown the relationship between blattids and mantids; it took a long time to do this because evidence from the internal anatomy is not the kind of thing that conventional taxonomists are used to reading about, and even today the two groups are classified as being considerably more distantly related than the anatomical evidence warrants.

We may, then begin to answer the question "what about convergence?" by saying that taxonomic methods which invoke the evidence, as far as possible, from a multiplicity of characters, drawn from as wide a range of the parts of the organism as possible, are rather less likely to be biased by convergence or divergence than are methods which are based on a small selection of characters, even though in most cases the characters will have been chosen, *inter alia,* because they are, in the taxonomist's judgement, less influenced by convergence than are others. This point was emphasized by Sokal and Sneath (1963).

But, we may also ask, is the influence of convergence the bane that it is so often made out to be? It is agreed that the evolutionary process involves both convergence and divergence, and some conventional taxonomists claim that phylogenetic methods produce a classificatory system more nearly reflecting the true course of evolution than phenetic methods can produce. One might expect, then, that conventional taxonomists would welcome the possibility of including evidence of convergence in their desire to arrive at a truly evolutionary classification. Convergence is in practice disregarded, divergence being the only direction of evolutionary change admitted, as in the lineage of some ancient nobility from which the bar sinister has quietly been dropped.

Provided that we do not seek to confine the evidence from numerical taxonomic exercises within a conventional dendrogram, there should be no difficulty in recognizing convergence and distinguishing it from other evolutionary relationships, always provided that we have the courage of our convictions and treat the vectors describing contrasts between the groups *as* vectors and do not disguise them as scalar quantities. Let us take a simple example, the relationship of the Tridactylidae to the Grylloidea, with which they were for long, and mistakenly, grouped, and two other groups, the Acrididae, short-horn grasshoppers and the Tettigoniidae, long-horn grasshoppers. As usual in numerical taxonomy we score each character as 1 when the group in question possesses the character and 0 when it does not. This simple example is given as an illustration of the principles at stake, not as a properly designed investigation of the relationships of these orthopteroid insects. It is, however, a real if shortened example which shows clear multivariate structure despite its dendrogrammatic form.

Character	Tridactylidae	Acrididae	Gryllidae	Tettigoniidae
less than 30				
antennal segments	1	1	0	0
six gastric caeca	1	1	0	0
burrowing habit	1	0	1	0
pronotum shield-like	1	0	1	0
cerci long	1	0	1	0
dorsoventrally				
flattened	1	0	1	0
hind legs saltatorial	1	1	0	0
no tympanum on				
forelegs	1	1	0	0

As Fig. 40 shows, the four groups can be arranged on a chart according to the generalized distances between them, with the vectors linking the terminal groups written out. This chart has a very clear "structure", with underlying contrasts of form as between the Tridactylidae and the Acrididae on the one hand (both of which belong to the suborder Caelifera) and the Tettigoniidae and the Gryllidae (both of which belong to the suborder Ensifera) on the other. At the same time, there is an evident contrast between the two

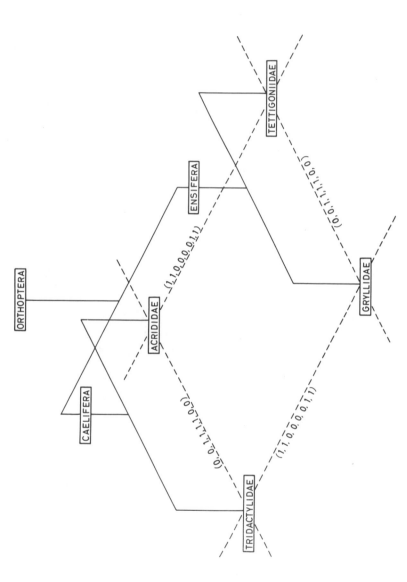

Fig. 40. Three-dimensional dendrograms of the phenetic affinities of four families of Orthopteran insects based on eight dichotomous characters. The diagram is arranged to show hierarchical structure as well as convergence and divergence. Dotted lines represent vectors (discriminant functions) between the families.

forms which, though not closely related, have adopted a semiburrowing mode of life, and those forms which are non-burrowing, with rare exceptions which do not concern us here.

There could be no mistaking the independent nature of adaptation to a burrowing habit and the "phylogenetic" association of the groups. Admittedly, in this simplified example, the two underlying contrasts of form are strictly orthogonal, making a right angle between them, and are particularly easy to discover; one could hardly hope for such an independent relationship in more complicated examples with many characters and groups. Moreover, in such cases the model might soon become so multidimensional (hypermultivariate) that its representation became difficult. However, any parts of a large model can be investigated in this way.

There is indeed little to inhibit one from treating the results of numerical taxonomic investigations as fully multivariate except the reluctance to recognize the vector properties of the links between the groups. Not to pursue this recognition is to saddle oneself with the problem of arranging the characters to prevent convergence from masking the other relationships, a difficult, and in many ways spurious, problem. Apart from practical difficulties of representation in three or more dimensions, any dendrogram can be converted into a generalized distance chart of the type of Fig. 40 using the positions of the groups on the chart as the pendant vertices of the dendrogram.

Closely allied to the contrast just drawn between hierarchical dendrograms and generalized distance charts is that between the analysis of a set of characters by the Edwards-Cavalli-Sforza divisive polythetic technique (see Chapter 25) and the ordination techniques based on multivariate axes of variation. Indeed, ideally and to a considerable extent practically, the ordination (generalized distance) and classification (hierarchical) approaches to the analysis of a set of data constitute the obverse and reverse of one and the same problem.

Is divergence any different in principle from convergence? Why, for instance, should it be considered in any way less harmful to exaggerate the differences between blattids and mantids than to underestimate the differences between crickets and tridactyloids? We are forced to the conclusion that convergence is just one cause of potentially poor estimation of resemblances just as divergence is another. In both cases the numerical taxonomist is in a stronger position than some conventional taxonomists, because of the basic principle of numerical taxonomy that the characters should be drawn from as wide a range of the organism's anatomy as possible. Conventional taxonomists differ widely on this issue among themselves.

This problem will not, in fact, ever be satisfactorily solved until truly multivariate methods of analysis are used; the dendrogram, whether purporting to be a phylogram or a phenogram, is too rigid a strait-jacket. The dimensions along which later convergence and divergence are to be assessed are most unlikely to be those along which the earlier evolution took place. In a sufficiently multidimensional representation, which can only be achieved by the use of multivariate methods, there will not be the problem of distinguishing closely affine from convergent groups except where the convergence has directly retraced the steps of an earlier, divergent, evolutionary process. And in such cases we are fully in accordance with Sokal and Sneath (1963), that these groups are to be regarded as closely related for all ordinary purposes. Although the general opinion seems to be that evolutionary steps are rarely retraced, the evidence seems inconclusive, and an extension of Le Quesne's (1969, 1971) criteria for discovering such deviations from the progressive course of evolution would be welcome.

The Importance of Higher Taxonomy

How does the existing unsatisfactory theoretical situation come about? We have a wide variety of dubious statements being made about taxonomy, some of which have been repeated for almost two centuries. There is doubt as to whether many of the existing allegedly monophyletic groups are in fact so. There is doubt as to whether the principles by which some conventional taxonomists declare that they work correspond in fact to the processes that they use. There is doubt about some of the objectives of a taxonomic study, and still more doubt as to whether the methods employed will lead to these objectives. There is debate about the principles on which characters are chosen, about how they should be used, and about the interpretation to be placed on the conclusions to which the consideration of the characters leads. And yet all of this uncertainty impinges on the vast majority of taxonomic investigations scarcely at all, and the practical impact of whatever theoretical views one has about taxonomy seems to be slight indeed when it comes to the actual conduct of a piece of taxonomic work.

The explanation we offer, with some reservations, is that there is a substantial difference between classification at the level of the exemplars (the material which actually comes to hand in a study) and that at the level of the notional groups into which these exemplars may be fitted. Sometimes this difference is expressed in the guise of a special role assumed by the species in taxonomy, often held to be a special role because of the biological species criterion, which in some form or other relates the capacity for interbreeding, or at least for gene flow, to the distinctiveness or otherwise of the species. We may take leave to doubt whether the biological species concept contributes very much to the topic, partly because it is so often effectively non-operational, but also because, on the rare occasions when the tremendous amount of work needed to examine the question has been undertaken, the results are too often far from conclusive. There are all degrees of interfertility between species, and it is hard to relate the survival of the progeny in the field to their survival in the laboratory; it is, after all, survival which counts as much as birth towards an ultimate gene-flow. Moreover, individuals from the ends of a cline may well be mutually sterile, yet we would be reluctant to consider them as distinct species, just as we decline to consider lions and tigers as belonging to a single species because they may hybridize in captivity.

There do not seem to be any adequate grounds for regarding any one level of organization as being different in kind from any other level. The only consistent distinction is that between an actual specimen and the notional higher categories into which it may be fitted. The theory of taxonomy applies of necessity with most force to the higher categories (higher that is, than the category of the exemplar, which is usually, though not necessarily, taken as an exemplar of a species). This is so because such higher categories are notional but at the practical level we become inured to theoretical difficulties and learn, if we are to survive as taxonomists, to gloss over them. Cain (1959) and Michener (1964) have both shown how unhelpful the binomial nomenclature may be, and how the problems that arise from its use impede practical taxonomy.

Theoretical taxonomy, then, is of most consequence in higher taxonomy which is the most neglected aspect of the subject today. Partly, no doubt, this situation has arisen because one needs to have a synoptic view of a group before its higher taxonomy can be revised; above all, this means ready access to a really comprehensive collection and to all the relevant literature, no mean requirements. Unless the revision is to be a local one, extensive travelling to museums and libraries, to say nothing of collecting trips, will be needed. We have estimated that the revision of a group of say, 300 species with a world-wide distribution may cost as much as £30,000 in salaries, travelling expenses, etc. Most such revisions would cost much less, partly because they would have to be done much less well than they should be because the costs cannot be recovered, partly because much of the real expense will be written off against other items.

For the practical purposes of identification of specimens, higher taxonomy forms a superstructure which, at a pinch, can be dispensed with. On the contrary, it does not seem to us that we can continue to teach higher taxonomy to students as if it were a topic of prime importance to biology, and simultaneously treat it as a research topic of so little consequence that only a very few workers are engaged therein. It cannot be claimed that all the problems of higher taxonomy are already solved, the exceptions are too obvious. More to the point might be the claim that any comprehensive attempt to investigate higher taxonomy on a broad front would be inconvenient, since workers in each compartmentalized subject would be compelled to revise their arrangements in the larger interests of uniformity in the subject rather than as an independent fragment in which considerations of classificatory levels can be made without reference to what is being done in other fields.

27
The Representation of the Results of Numerical Taxonomic Studies

Recently, the disadvantages of the phenogram as a representation of the results of a numerical taxonomic study have become so widely realized that other forms of representation are being sought. But these disadvantages are not entirely without remedy. The most serious is that a phenogram acts as a strait-jacket, forcing the data into a hierarchical system of representation, which is essentially one-dimensional, since the horizontal axis of a phenogram has almost no significance. Taxa which fall near to one another but are suspended from different branches of the tree do not necessarily have any close relationship, since each phenogram is to be considered as capable of rotation at all its "joints" as has been pointed out by Blackith (1965) and Gower (1967a). There is no reason, other than purely mechanical problems of construction, why phenograms should not be in three dimensions, and three-dimensional projections on to two-dimensional planes have been published (Blackith, 1965). For demonstration purposes phenograms with rotating joints, on the lines of "mobiles", would help to prevent too rigid an interpretation of some dendrograms.

Once there is a second dimension into which a phenogram moves, the question arises as to whether the angles between the horizontal parts of the dendrogram are arbitrary, or whether they too should be calculated. Each will generally bear some such interpretation as contrasts of form within a genus, or within a subfamily etc. These contrasts of form can be calculated using discriminant functions, and it is then a short step to using the generalized distance between the two groups at the ends of the arms to make the second (horizontal) dimension meaningful. But by this stage the similarity between a phenogram and a generalized distance chart will be so close that, logically, there is every reason to begin with such an end in view and calculate the positions of the groups on a canonical variate chart.

It has been argued that in a numerical taxonomic study one cannot use discriminatory techniques because the clustering process forms the essential first step without which there are no groups to be discriminated (Sokal and Sneath, 1963). This proposition is incorrect, since it ignores the basic distinction between ordination and hierarchical techniques. Moreover, personal experience suggests that it is not an accurate reflection of the real problems of many taxonomists. Even in an environment where one is familiar with very few of the organisms that are to be found, there is little difficulty in discovering groups of organisms; what is required is a knowledge of the mutual relationships and, ultimately, some kind of identification, of these groups. There is thus every reason to suppose that discriminatory techniques will serve the taxonomist, even at the "sorting" level adequately, as Du Praw's "Non-Linnean taxonomy" shows (see Chapter 11).

The construction of three-dimensional models of discriminant techniques applied to taxonomy enables an additional dimension to be employed, and also mobilizes the second dimension, which is generally ineffective in phenogram representations, to the full.

Charts based on the calculation of distance measurements are obvious candidates for three-dimensional models. One practical technique, amongst many that can be tried out, is the cutting of glass rods of a length proportional to the generalized distance or other measure of dissimilarity. These rods can then be inserted into small rubber balls, those made to imitate golf balls being of a convenient size. No glue is needed, since the ends of the rods, when lubricated with water to enable them to slide into a cut in the ball, will stick firmly when dry. The balls can then be painted or, if the model is to be photographed for publication, symbols may be drawn on the print once the photograph is available. Models made in this way have been published by Albrecht and Blackith (1957) and by Blackith (1957). Stiff wire "distances", advocated by Sokal and Sneath (1963) are easier to use and transport, but not quite so satisfactory aesthetically since the wire rarely remains straight during the fixing process.

Sometimes the model is under sufficient strain to make some representation of this strain desirable, in which case perspective drawings of the "bent" links between the groups may be shown (Reyment and Brännström, 1962).

If three or more latent vectors or canonical variates contain useful information about the contrasts of form among the material studied, there is the possibility of presenting this information in the form of a three-dimensional model. Almost all such models consist of a base-board, two of whose sides represent the first two axes of variation, the third axis being represented by the heights of the supports for whatever symbols have been chosen to represent the groups or entities displayed, which are then stuck into the base-board at the appropriate point. Such models allow the cye to pick out the salient features of the analysis quickly but they do present difficulties when the information comes to be published (Lysenko and Sneath, 1959).

The optimum angle from which to photograph the model, so as to display the underlying relationships in the analysis, may be such as to bring some of the symbols so far into the foreground as to obscure others. If the symbols are small, to minimize this problem, they may not look sufficiently distinct when the photograph is reduced for printing. It is often worthwhile making a few mock-up models, using only a small sample of the material if this is voluminous, before embarking on the final version, which needs to be prepared with some care lest defects of workmanship confuse or irritate the viewer.

As an alternative to the preparation of solid models, Rohlf (1968) has suggested the use of stereograms, in which two pictures of a model, drawn so that the points from which each is viewed subtend the same angle as would human eyes, are printed side by side. The theory and practice of stereoscopic models in multivariate statistics have been discussed by Fraser and Kovats (1966). Some readers seem to have considerably more difficulty than others in "reading" these stereograms as if they were in three dimensions, and Rohlf relates this difficulty to the possession of high visual acuity.

Another technique for representing three dimensions on paper is the SPAN diagram (Tryon and Bailey, 1966). This technique has been used to study two genera of plants, *Limnanthes* and *Salix*, by Crovello (1968a, b). In effect, the matrix of similarities between each of the taxa is analysed using the Key Communality Cluster analysis of Tryon and Bailey (see also Crovello, 1968c), whereby the number of dimensions needed to represent the relationships between the characters is reduced from the number of characters to some smaller number, which is then exhibited three at a time on circular diagrams in the reduced character space. These SPAN diagrams are plotted so that the position of each character is determined by its factor coefficients (the correlation of the character with an independent dimension in the reduced character-space).

Hope (1968) has illustrated the presentation of multivariate psychometric data using spherical maps. Carmichael and Sneath (1969) represent their classifications by means of taxometric maps.

Sneath (1967b) has declared that numerical taxonomy has a strong and justified leaning towards the hypermultivariate (i.e. with large numbers of variates) whereas the models for much multivariate analysis are ideally oligovariate (i.e. multivariate, but with only a few variates entering into the final representation of the relationships that are sought). This topic has a direct link with the problem of representing the results of a morphometric or numerical taxonomic study, because some ways of representing the results allow for more dimensions to be shown than do others. Cluster analyses leading to dendrograms are effectively one-dimensional, because the "horizontal" dimension cannot be considered as representing anything of biological consequence by virtue of the fact that any set of points on the same branch at the same level can be permuted (Gower, 1967a).

There must be some question, therefore, as to how far it is worth while to include a multiplicity of characters, leading to greater and greater dimensionality of the chart representing the relationships between the taxa studied, unless the published version of the chart is able to reflect this dimensionality. As Rohlf (1967) points out, one gets a better representation from models (in his case of principal component analysis) than from phenograms, although the phenograms are easier to construct. Part of the difficulty arises from the supposed zero correlations of the various characters used in a numerical taxonomic study; when, as in most morphometric analyses, the correlations between the characters are removed, there is much less difficulty in seeing how the representation of the relationships can be condensed to an oligovariate chart even though the original analysis was hypermultivariate. It is also easy to see that the large numbers of characters upon which some numerical taxonomists insist is supererogatory (cf. Chapter 25). For one reason, such an accumulation of characters is probably piling redundancy upon redundancy. For another, so long as, in a study which uses quantitative characters, there are only a few significant latent roots to the canonical equations which result, the chances are that more characters have been measured than is essential. Naturally, one wants to make sure that all the various patterns of growth or other contrasts of form have been recorded, so that some excess of characters is needed, but a large excess is evidently wasteful.

In any case, there is some doubt as to whether even the largest electronic computers will be able to take morphometric analyses using much more than 100 quantitative attributes. This estimate is of course flexible and related to the skill of the programmer. In a very clear introduction to the problems of morphometric analysis, Jeffers (1967a) has suggested that repeated analysis of the data will lead to a continuous process of adding new variables (characters), but deleting them as soon as it has been demonstrated that they add no new information to the description, so that only those of the new variables which are relatively uncorrelated with those already included will be retained. Jeffer's contribution is of particular value as a statement of the dynamic view of taxonomy, with new variables constantly being introduced and tested against the known pattern of variability. Blackith and Verdier (1960) for instance, introduced the dimensions of the fore-femur into acridology in this way. Sneath (1967b) has pointed out that the process of reducing, or attempting to reduce, the number of variables taking part in the erecting of the "framework of variability" is sometimes of illusory benefit because the very act of testing requires an amount of work comparable to that involved if the new characters are left, untested, in the initial suite of characters. However, this attitude is only justified if the analysis is only going to be used once. If the framework of reference is to be used several times, as for instance in organizations concerned with repeated assessments of the status of a limited range of variable material, a more dynamic approach to the taxonomic problems is economically justified.

28
Trend-surface Analyses of Transformation Grids

It is arguable that geology, where the development of statistical ideas has been later than in biology, has drawn more than it has as yet had an opportunity to give from the ever-changing pool of ideas in biometry. One promising contribution to this pool is the analysis of trend-surfaces, whose applicability to morphometric problems has been developed by Sneath (1967a). We have decided to include a short account of this topic despite our remark on multiple regression at the beginning of the book. An introduction for geologists is given by Krumbein and Graybill (1965). The underlying philosophy is closely dependent on the ideas of D'Arcy Wentworth Thompson already mentioned, which we can, with hindsight, see as much more difficult to express in suitable mathematical terms than Thompson himself seems to have thought. The use of transformation grids in relation to the study of growth and form has been discussed by Blackith *et al.* (1963). A cognate approach is that of the harmonic analysis of forms, using Fourier analyses. Bliss (1958) suggests the application of Fourier analyses to enable the problems of periodic regression to be studied. Lu (1965) has tried out three-cycle Fourier analysis to represent the frontal and lateral aspects of the human face, noting that in frontal view, the cosine terms measure the form of the symmetry and the sine terms the asymmetry. Lu also commends the Fourier coefficients as attributes in various types of discriminatory analyses for use in anthropometric work.

Trend-surface analysis uses the slightly different approach of fitting polynomials to surfaces. Evidently, for more complicated surfaces one is faced with the choice of a good fit using high-order polynomials and much computational labour, or a poorer fit with a simpler function. In its application to transformation grids, Sneath finds that third-order curves are adequate. Cooper (1965) brings quadratic discriminant functions to bear on the problems of pattern recognition.

There is an interesting dilemma for the analyst in that once the main trend of the surface has been removed one is left with the residuals, representing deviations from the fitted surface, partly because of errors of measurement and partly because in real analyses the fitted surface will not have taken into account some minor features of the surface which is being assessed.

The method has some merits as regards its being another way of viewing a set of data. Unfortunately, as usually practised, there are many formal statistical objections to the nature of the polynomials—for example, the arbitrary (often empirically based) deletion of some polynomial coefficients.

There is, nevertheless, no guarantee that these minor features do not contain significant information about the organism and some rather circumstantial evidence that they do. Quite simple deformations will convert the outline of a rabbit into that of a rhinoceros, as Thompson (1962) showed. One might think that nearly all the significant taxonomic information was contained in those parts of the two animals that have been smoothed out in his drawings. Sneath considers that increasing the taxonomic distance between two organisms does not, of necessity, modify the main trend surfaces but that it does lead to greatly increased waviness of the grid lines over short distances. It is interesting that a similar difficulty arises in the geological usage of trend-surface analysis since the presence of oilfields tends to be associated with the occurrence of high-order residuals after trends have been fitted to geological maps [Krumbein (1959), Merriam and Harbaugh (1963)]. Herein, one might think, lies the principal difficulty barring the way to machine recognition of shapes, partly because machines are not very good at such activities and partly because, in order to ensure that the machine picked up the minor features of consequence to the analyst, the number of points which it would require to memorize would cause serious congestion of the storage facilities.

Parks (1969) has presented multivariate facies maps in which areas are delimited by the local scores along successive principle components, each representing an element of the facies.

Lee (1969) has combined canonical correlation estimations with trend-surface analysis to good effect, after a preliminary principal component analysis had shown that the meaning, in his example, of the clastic ratio (sand + shale)/(carbonate + evaporite) was so ambiguous that it should not be included in the analysis. Canonical trend-surface analyses maximize the association between a set of geological variates (in this case, amounts or proportions of sand, shale, carbonate and evaporite) combined into a linear function, with linear functions of the X, Y coordinates of a map. It should test any hypothesis involving a general trend in all the variables, which are automatically weighted according to their error variances. Any number or type of geological variate can be incorporated although computational difficulties suggest that four variates may be about the optimum number, and the problems of interpretation may well impose similar limits. Highly correlated variables are undesirable because they render the correlation matrix almost singular and introduce computational problems.

Gittins (1968) has emphasized that trend-surface analysis can be considered as an extension of the ordination process for the identification (reification) of axes of variation whose nature has proved difficult to establish.

Fisher (1968) has made a most interesting use of trend-surface analysis in an investigation of the faunal resemblances of the 105 counties in the State of Kansas, in respect of the distributions of 64 species of mammals, 52 species of reptiles and 24 species of amphibia. The first step in the analysis was to calculate the similarities between the counties in respect of the species used as characters. A cluster analysis was then performed on the matrix. As a second major step, a factor analysis (Thurstone's complete centroid method; Thurstone, 1947) was carried out on the matrix of correlations between the species distributions. The scores for each county along each of five factors were then used as the raw material for a trend-surface analysis based on 6th degree equations.

The outcome of the analyses was that the trend-surface analysis of the factor scores represented the data more clearly and meaningfully than the dendrogram formed from the cluster analysis of the similarity matrix. Moreover, whereas the dendrogram forces the data into a straight-jacket, because elements which are quite closely related may find themselves quite widely separated on the dendrogram, the trend-surface analysis is able to accommodate the essential continuity of distributional data. As Fisher points out, the problem of an element which can be associated with two or more other groups of elements is not restricted to studies of geographical distributions, but occurs widely in taxonomy, and is to a large extent imposed on the data by certain habits of the human mind.

The trend-surface analysis was also superior, as a method of representation, to the display of the data along three orthogonal factors in the form of a stereogram. We may therefore hope to see this technique more widely used and refined for the handling of large bodies of data for which even three-dimensional charts are inadequate. We must, nevertheless, guard against abuse, for at the purely statistical level trend-surface analysis seems almost indefensible: there is, therefore a special responsibility on the user.

29
Some Applications of Morphometrics in Palaeontology

Biostratigraphy

Biostratigraphy is an important branch of applied palaeontology. It has as its main task the application of fossils, and hence of evolutionary sequences, to the subdivision of time sequences and their dating in geological terms. As practised in qualitative connections, the chronological subdivisions are made by identifying particular zonal fossils, which are normally limited to short time ranges. Quantitative biostratigraphy is a very new subject. It makes use of the evolutionary sequences represented by series of fossil species preserved in the sediment and studies, by morphometric means, changes in shape and size of critical forms over time.

The method of analysis uses a kind of time-series treatment of multivariate data, although, up to now, the procedures of true time-series analysis do not appear to have been applied. The practical application of morphometric biostratigraphy is clearly of importance in the petroleum industry in the detailed analysis of fossil sequences in boreholes. A well carried-out study of long-ranging forms and their morphologic variation (this contrasts with the aims of classical descriptive biostratigraphy), provides an excellent basis upon which to found a detailed fine-stratigraphic analysis of boreholes. Inasmuch as the fossil organisms, usually ostracods and foraminifers, will have been sensitive to environmental fluctuations, the possibilities of correlation between boreholes in different basins, or in widely differing parts of a sedimentary basin, are slight, but the method is valuable for work in restricted areas, such as a single sedimentary basin.

The technique of analysis consists of plotting the values of the first principal component of a matrix of morphologic variables as a function of time. The patterns obtained from these graphical representations are compared for each of the species considered and agreements noted. These chronologic patterns of variation are compared for the borehole sequences of the area. In one respect this method is not unlike the way in which pollen-analytical diagrams are used. A theoretical analysis of the method is given in Reyment (1971b).

Analysis of Morphologic Chronoclines in the Paleocene of Nigeria

Reyment (1966a) analysed the occurrences of several species of ostracods in the Nigerian Paleocene, obtained from borehole samples. The material was first studied by making graphs of each of the included carapace characters against location in the borehole. Then, by plotting the vector lengths represented by all of the variables considered simultaneously, it was found that many of the species, but not all of them, varied in the same way and it was possible to use the patterns of morphologic variation to characterize the time segments in the borehole from which the fossils were derived.

Morphometric Analysis of Sexual Dimorphism in Fossils

To the biologist, the idea of using statistical methods to separate the sexes may appear ludicrous, when all that is needed is a glance at the genitalia of the creatures in which one is interested. The palaeontologist is far less fortunate. In almost all cases in palaeozoology, at least, it is not possible to determine the sex of fossil remains by mere inspection. In order to arrive at some kind of working conclusion, the palaeontologist must make use of some kind of quantitative method, which one hopes may be sufficiently shrewdly chosen to bring out whatever morphologic differences of sexual origin there may be in the material. In the majority of published studies in which the determination of sexual dimorphism has played a prominent part, the method used has been a simple scatter diagram of two dimensions. The problem of the analysis of sexual dimorphism has been prominent in the study of fossil ostracods for a long time and it is in publications on this group that one finds the most examples. A scatter diagram of length and height of the ostracod carapace will usually be successful in disclosing not only the presence of sexual dimorphism in the material, but also the larval growth stages. In most situations, therefore, a scatter diagram of carapace dimensions is an effective way of unveiling the presence of sexual dimorphism in fossil crustaceans, and, in particular, those forms in which there is a relatively greater increase in the length dimension of males on passing from the final larval stage to the adult, than occurs in the females. In the majority of cases it is not possible to distinguish morphometrically between males and females of the larval stages. Histograms are also used, but these are only effective when size differences are strong.

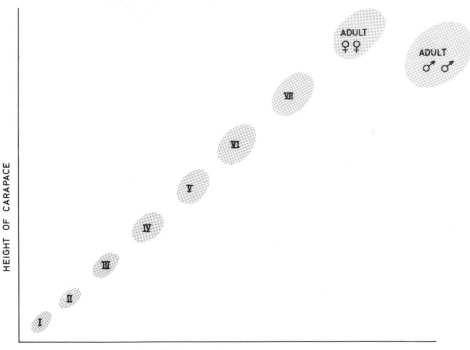

LENGTH OF CARAPACE
Fig. 41. Growth series for a Paleocene cytherid ostracod.

In quite a number of cases, the scatter diagram is unable to disclose the existence of sexual dimorphism in fossil material and an effective analysis requires the application of morphometric methods. Figure 41 shows a bivariate (length and height of carapace) plot for a Paleocene cytherid ostracod from Nigeria (Reyment, 1966a). The ellipses of bivariate scatter become successively more inflated as the stage of growth increases. This implies of course that the range of multivariate variability increases with each growth stage and that a comparison between, say, either of the adult categories and any of the middle instars is liable to be made more complicated, because of this heterogeneity in the variances and covariances.

The methods of most use in studying sexual dimorphism in fossils are the generalized statistical distance and principal components analysis. Figure 43 shows a graphical comparison of generalized distances for Nigerian Paleocene ostracods (Reyment, 1966a). Vertebrates are in some respects more difficult to deal with than are crustaceans and other arthropods, as there is usually no good way of knowing at what stage of development fossil bones of large animals were at the time of death. Certainly, quite good guesses can be and are being made, particularly with respect to human bones and some osteologists are quite skilled at distinguishing male bones from female remains. The level of skill is naturally very much less for fossils of other animals as regards the visual identification of the sexes on the basis of fossil bones.

The problem of carrying out generalized distance studies on vertebrates and other groups with continuously growing hard parts was tackled in a far-sighted contribution by Burnaby (1966a), followed up by a rigorous mathematical treatment of the subject by Rao (1966), who had encountered the problem through being a referee for Burnaby's paper. Burnaby's work was aimed at attempting the analysis of morphological measurements on continuously growing creatures by generalized distances with the introduction of the element of growth invariance, so that it would not matter that a sample of bones was made up of individuals which had died at quite different growth stages. This is a very important concept and one that is well worth practical exploration. Unfortunately, it requires, as a prerequisite, determinations of growth vectors and conclusions on environmental effects that in practice are very difficult to produce. Attempts have been made to use the vectors of principal components analysis as estimates of these vectors, but the results have mostly been poor.

Reyment (1969d) carried out some case studies on vertebrates, as well as other groups, by generalized distances and in particular, analyses of the dispersion matrices of males relative to females. The calculations required for doing this suite of analyses are given in the computer programme at the end of Chapter 7 on generalized distances where also the theory involved is briefly discussed.

Example of Dimorphism in a Trilobite

The Carboniferous trilobite genus *Archegonus* displays slight sexual dimorphism in size. A principal components analysis of the species *A. nitidus* was carried out, using the programme PNCOMP in order to investigate the structure of this dimorphism. The characters, length of pygidium, breadth of pygidium, breadth of rhachis and length of rhachis were measured on 30 individuals of the species, representing a range of growth stages. All variables were found to be very highly correlated (all coefficients of correlation in excess of 0.91) and the variances of the logarithmically transformed data were

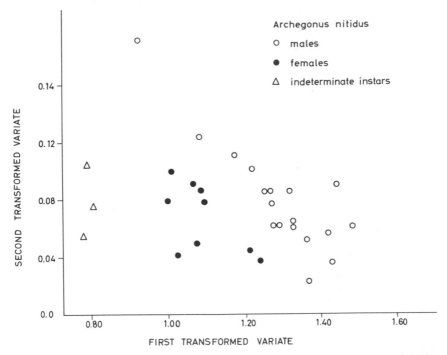

Fig. 42. Principal components analysis of a Carboniferous trilobite. The plot of the first and second transformed variates discloses sexual dimorphism in the pygidium.

observed to be roughly the same. The components analysis showed firstly that the last three roots of the dispersion matrix are almost isotropic and that there seems to be every reason to suspect that there is really only one principal axis. The plot of the transformed observations shows a clear delineation into the categories: males, females and larval individuals (Fig. 42).

Strength of Sexual Dimorphism in Paleogene Ostracods

Reyment (1966a) calculated the generalized distances between the sexes of 19 species of Nigerian Paleogene cytherid ostracods. He found the values to lie largely between 0.5 and 6.5, but occasional species were found to yield exceptionally high generalized distances in relation to the majority of the cytherids. The genus *Trachyleberis* occupies an exceptional position. In order to make the results readily comparable, the distances were displayed in an arbitrary graphical form (Fig. 43). The generalized statistical distance may thus be used as a quantitative expression of the strength of sexual dimorphism. The above investigation also showed that there is no clearcut ordering of the distances into generic groupings and it was suggested that the sexual-dimorphic generalized distance may be primarily a specific characteristic.

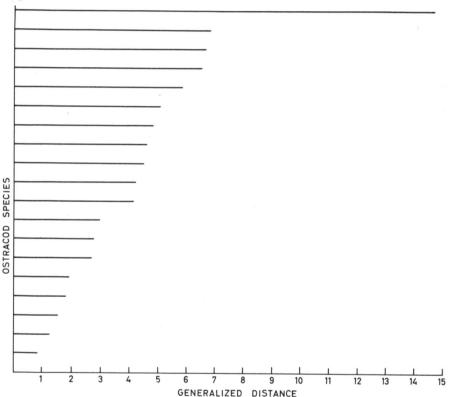

Fig. 43. Diagrammatic representation of the generalized distance between males and females for 19 species of Paleogene ostracods (after Reyment, 1966a).

Comparative Case Studies for Some Recent Organisms

Reyment (1969d) studied some living groups of organisms, using a palaeontological approach, in order to obtain information likely to be of use in the interpretation of fossils. The groups studied were frogs, turtles, crabs, brine shrimps and grasshoppers. None of the results obtained is in any way sensational from the zoological point of view, but they do provide the palaeontologist with useful comparative information. For the material of *Rana temporaria* analysed from Sweden, it was found that the ellipsoids of scatter for morphologic variables are differently inflated, but that the axes of the ellipsoids have the same orientation. This was taken to indicate that females have a wider range of morphometric dispersion but both sexes possess the same patterns of growth. The same analysis for *Rana esculenta* showed the dispersion ellipsoids to be both differently inflated and differently oriented. The mean vectors were not found to differ significantly from each other. It was suggested that males and females of this species show only weak dimorphism in the morphologic characters studied, but strong sexual dimorphism in their patterns of growth.

Material of the North American marten was found to possess strong sexual dimorphism in morphologic characters. The dispersion matrices were inflated to the same extent but there were some differences in the orientations of the principal axes of the ellipsoids of scatter, suggesting the possibility of differing patterns of growth.

Carcinus maenas, the common crab of tidal flats, was found to possess strong sexual dimorphism in eight morphocharacters and the scatter ellipsoids proved to be differently oriented and differently inflated. Similar results were found for the brine shrimp, although the generalized distance for these crustacea is not great.

A study of material of the grasshopper *Omocestus haemorrhoidalis* gave the result that sexual dimorphism in four morphocharacters is particularly strong. Comparisons of scatter ellipsoids for populations from the three isolated areas in Sweden in which the species occurs disclosed that interesting variations in inflation and orientation occur from place to place.

Dimorphism in a Group of Cretaceous Ammonites

The recognition of sexual dimorphism in ammonite cephalopods is a quite recent achievement and there is currently much interest in studies of this source of variation. The recently published symposium (Westermann, 1969) on sexual dimorphism in fossils contains numerous contributions on ammonites.

Reyment (1971a), in studying a group of species of the Cretaceous mammitid ammonite, *Benueites*, came to the conclusion

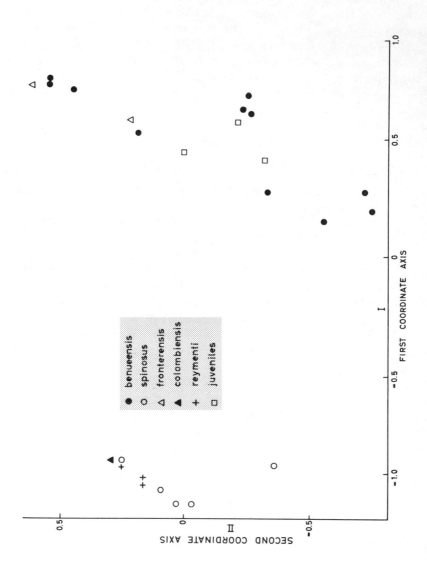

Fig. 44. Chart of the first two principal coordinates for species of the Cretaceous ammonite *Benueites* (redrawn from Reyment, 1971a).

that a type of dimorphism occurs in this group which differs from that previously described for the group. The known kind may be summarized in a few words as follows. The males are to a large extent miniatures of the females. There are numerous other differences, but the size criterion is the most striking one. The dimorphism in *Benueites* is not related to size or number of whorls in adults, but rather to constant differences in the ornament. It is not possible to say which of the morphs were males and which were females, as it is not easy to find a parallel with living cephalopods. Until this type of dimorphism has been documented for other ammonites, Reyment preferred to approach the taxonomic side of the study in a conservative manner. Hence, all morphological categories were given formal names. The principal coordinates analysis of 16 characters yielded a clear breakdown into the two morphologic categories, despite the fact that some of the individuals included in the analysis are transitional with respect to some of the features included in the study. The graph of the first two principal coordinates is shown in Fig. 44.

30
Some Closing Comments

We cannot too strongly emphasize that the ability to preserve and display what we have called "structure" in any morphometric analysis is the criterion by which we judge it. The reason for our recommending multivariate analyses, in preference to manipulations of ratios of characters and the more arbitrary techniques of factor analysis, is not that the latter are aesthetically less satisfying, though this is true for us, but that they simply do not do the job for which they are used as well as the more sophisticated methods. There will be situations where there is little or no "structure" in the charts or models, but experience suggests that such situations are indeed rare except in some "abstract" morphometric problems, particularly in ecology. Even when we have to deal with situations where the contrasts of form that are being studied are of much less than specific rank, i.e. are environmentally induced, or associated with genetic polymorphism, the "structure" is there and can be interpreted with not too much difficulty. Orloci (1968a, c) has discussed the definition of structure in phytosociology and in collections, but seems to have exaggerated the extent to which generalized distance and related techniques depend on multivariate normal samples for other than identification purposes: since Claringbould's (1958) work we may feel easier on this score.

There is now an enormous armoury of morphometric methods, amongst which there has been no sufficiently great pressure of "natural selection" to weed out those which are markedly suboptimal. One reason is that any one worker in the subject has to expend a considerable amount of time and energy in order to understand a new manipulative technique. The reluctance to abandon a technique which one feels one has mastered is quite understandable. Another difficulty is that techniques become associated with individual workers, who may feel, perhaps not wholly consciously, that to move on to another field may diminish a reputation, as an expert, that has been painfully accumulated over the years, and wish to see "their" technique handed on to their students. Because of the essential robustness of the structures which the methods are designed to uncover, these tendencies may be viewed indulgently, but some techniques do seem due for retirement. Our suggestions for this category include approximate methods based on matrices of correlations as already remarked by Eades (1965) and Colless (1967a), and those forms of factor analysis in which the communalities have to be guessed, since there is objective evidence that bad guesses cannot always be corrected and may vitiate the analysis (Seal, 1964).

It may be that, because of its early historical development, factor analysis has come to be used in some quarters as a method of first choice, instead of as a method to be chosen only when the nature of the problem calls for it. There does not seem to be any justification for factor analyses when the data are already clearly grouped into meaningful entities. We have yet to discover any illustration of the superiority of factor analysis over principal components analysis.

We are still in the phase of explosive evolution in morphometric analysis, and a great diversity of techniques, and, more important still, of mental approaches to the subject is essential, however much of a nuisance it may be to reach and to keep up with the flood. Any premature attempt to define an ideal way of dealing with multivariate situations is likely to reduce the present admittedly exasperating multiplicity of procedures into a few schools which may become almost non-communicating so far as the other schools are concerned. Such situations do great harm, especially when promoted on what is effectively a linguistic or national basis, and have arisen in the past in such subjects as phytosociology, philosophy and economics.

It seems necessary nevertheless to keep up the pressure of criticism on all forms of taxonomic endeavour to ensure that the pace of "natural selection" of the various techniques will be adequate. There is, as we have already explained, a tendency to stick with one group of techniques out of sheer inertia and pressure of work; we can sympathize with this tendency without condoning it.

One region in which patient understanding is more than usually necessary is in conventional "phylogenetic" taxonomy. Bretsky (1969) has remarked that it is much more difficult to explain how one sets up a phylogenetic scheme than to set out the basis for, say, numerical taxonomy. We are not wholly convinced that there is no case to be made out for creating phylogenies as special classifications, even though we share Colless' view that no adequate case has as yet been made out. One must remember that complex judgements are often all the more difficult to explain (in the sense of giving a coherent account of their basis) when in the past these judgements have been made in an atmosphere of uncritical acceptance. This is a well-known psychological mechanism (Scott Blair and Coppen, 1940). Craftsmen who can make sensible and consistent judgements in a situation where their judgements will be accepted almost without question become erratic when transformed to a situation where their judgement is actively being examined.

The reason for our occasional criticism of "authoritarian" judgements in taxonomy is not any desire to detract from the reputation of leading taxonomists, far from it, but the fact that authoritarian mental habits lead to "cryptic" taxonomy in the sense of Cain (1959). What one is not in the habit of explaining to others one is unlikely to explain clearly to oneself. It can be argued that, at a time when taxonomy is not attracting the support it needs to embrace that part of the world's fauna that is in peril of extinction, any criticism is unhelpful: but taxonomy is in a state of flux, apart from the impact of quantitative ideas, and cannot rest on its achievements, praiseworthy though they may be.

One of the reasons for the reluctance with which some "orthodox" taxonomists accept numerical ideas is that there are features of taxonomic practice which are needed but which are not taken care of, at any rate by the cruder forms of numerical taxonomy. There is evidence for the proposition, which a number of taxonomists believe to be essentially true, that the "key" critical attributes by which hierarchies can best be delineated evolve in a sequence one at a time (Crowson, 1953). These critical attributes are often complex morphological structures of the kind to which, for instance, Smirnov (1969) attaches much importance. They tend to be the attributes on which higher categories are distinguished, and it is disturbing to an orthodox taxonomist to see characters which he associates with the distinctions between orders or families being treated on the same level as those which, in his experience, distinguish species or infraspecific entities.

This reluctance to discard certain types of hard-won experience seems to us to be fully justified. Only in the cruder forms of numerical taxonomy is it necessary to treat all attributes on the same footing all the time. If a quantitative multivariate analysis, followed by an oligovariate representation, is used, the effectiveness of the critical attributes at each stage can be examined objectively. An attribute which distinguishes categories to which the rank of, say, order is thought appropriate will be heavily weighted on the discriminant linking such categories but lightly weighted on other axes of variation. Such *a posteriori* weights do not infringe any essential principle of quantitative taxonomy, but do afford valuable evidence which it is wasteful to ignore.

The effective alternative to *a posteriori* weighting is the use of probabilistic weighting, at least in agglomerative clustering. Probabilistic weighting, as we have seen (p. 280), leads to the dilemma of phylogenetic speculation, and we cannot as yet break into the circular arguments involved.

We have commented that the Achilles' heel of numerical taxonomy using agglomerative techniques seems to be the difficulty of giving any convincing reason for the choice of a clustering technique, other than personal preferences, or the convenient existence of computer programmes. This problem seems to be insoluble so long as agglomerative techniques of forming groups are employed: but it disappears as soon as divisive methods are introduced. Leaving monothetic division on one side, the only way of handling large numbers of entities by polythetic divisive techniques is by displaying them on charts, made up of various multivariate vectors (discriminant functions, principal components, principal coordinates and canonical variates). We have tried to demonstrate that multivariate ideas can be introduced into the classificatory process so that hierarchical as well as reticulate diagrams can be prepared, without recourse to agglomerative techniques, if the experimenter should feel that his problems require the imposition of a hierarchical structure.

This change, which we feel is likely to be forced on numerical taxonomy once its essentially exploratory phase comes to an end, will bring it within the general framework of multivariate studies, and for this reason we have done what we can to explore the relationships between the two approaches, as far as present knowledge permits. One sometimes finds in the literature statements that a fully multivariate approach is undesirable because of its supposed dependence on the hypothesis of multivariate normality, an hypothesis which is often difficult to verify in a particular context. We must at once concede that the distortions introduced by departures from multivariate normality have scarcely been investigated, so that on the face of it this criticism is justified. However, when the critics go on to use techniques which are even more drastic approximations (albeit disguised ones in some cases) one begins to wonder how far the criticism reflects a measure of prejudice. Claringbould (1958) has shown that several important methods work very well on dichotomous data, and Cooper (1965) has shown how several decision procedures are not closely dependent on the normal theory for which they were first evolved. It is for reasons such as this that we have tried to collect together the evidence that multivariate classification is mostly robust, and may be used with confidence.

Crovello (1968a) has commented that most numerical taxonomic studies of plants have been exploratory, and their authors have been reluctant to alter the taxonomy of the groups studied. Similar comments might be made of multivariate morphometric analyses, and of studies on animals as well as on plants. Caution in disturbing the existing system is highly desirable in any case; it is perhaps reinforced because so many of the workers who have pioneered quantitative taxonomic methods are not primarily taxonomists themselves, and several have come into the field from non-biological disciplines.

Obviously, while numerical techniques are still in an experimental stage, in the sense that their application to a particular field may have all sorts of unforeseeable snags, and every study is to some extent a pilot study, there is a strong case for restraint in interfering with the existing state of the taxonomy. However, when a reasonable body of experience has been built up, and the study is being conducted by or with someone who knows the conventional proceedings in that field, there seems no reason why substantial revisions should not be undertaken, all the more so because numerical methods force on the reviser a breadth of perspective which would be as desirable in the conventional taxonomies as it is imperative in the numerical. Wilkinson's (1970) revision of a butterfly genus is a case in point, and examples can also be given where numerical methods have decisively favoured one or other of the rival taxonomies in a particular field (e.g. Basford *et al.,* 1968; Sims, 1969), yet, as Boratynski and Davies (1971) have commented, a unique criterion for the excellence of a taxonomy is hard to come by, and may, indeed, not exist. The large investigations, such as those of Ehrlich and Ehrlich (1967) on butterflies, which are preceded by detailed surveys of the internal and external anatomies of the group, live up to the standards from which conventional methods sometimes fall short.

Numerical taxonomy in the restricted sense of the term as used by

Sokal and Sneath (1963) is at a cross-roads. In some ways the astringent critique of Crowson (1970) is justified. Numerical taxonomy has been proselytized with vigour, clarity and scope, and has bypassed the real mathematical problems with a disingenuity which may have been a short-term advantage in enabling many non-numerate biologists to grasp its broad outlines, but may prove to be a crippling handicap in the long run by weakening its theoretical basis to the point of collapse. Contrary to Crowson's implied view, numerical taxonomy is not the product of a sophisticated mathematical approach but of a naive, semiquantitative one. It has been distinguished for its vigorous application rather than its originality, since the basic hypotheses of Sokal and Sneath (1963) closely resemble the axioms of Mahalanobis (1937), as Rao and Varadarajan (1963) have remarked. In the development of numerical taxonomy, the vital role in any mathematical attempts at classification of efficient distance measurements based on Mahalanobis' Generalized Distance have been either forgotten or never understood. It is encouraging to observe that mathematical statisticians have begun to interest themselves in some of the aspects of numerical classification.

There seems to be no logically satisfactory way forward by using agglomerative clustering techniques: the most effective means of bringing polythetic divisive methods into use seems to be the use of multivariate methods for ordination and cluster analysis. Colless (1967a) voices a common misconception when he queries whether multivariate techniques can operate on material not already formed into discrete groups: this difficulty stems from the history of discriminant functions when used to allocate material to one of two or more groups. Principal components and principal coordinates are designed expressly for the purpose of ordinating supposedly homogeneous material.

Early claims that numerical taxonomy would bring stability to taxonomy look rather wan in the light of the inability of users of the agglomerative techniques to arrive at any preferred solution or any stable classification. Johnson and Holm (1969) and Heiser et al. (1965) have maintained that this uncertainty is a source of strength: that, in fact, the different classifications can be tailored to the needs of their users. However, there is no evidence that the changes arising from the use of different clustering methods, different numbers or kinds of characters, different stages of development, different sequences of feeding the material into the clustering process and so on, bear any useful relationships to the needs of potential users:

rather do they appear to be embarrassments to those who, a few years back, were welcoming numerical taxonomy as the harbinger of stability.

The much publicized objectivity of numerical taxonomic procedures is beginning to wilt under close scrutiny; it seems in fact not so different from that practised by the better conventional taxonomists, and to be constrained in practice for much the same reasons in each case, especially since the non-specificity hypothesis can no longer be relied upon, except as a first approximation to the truth (Michener and Sokal, 1966).

It remains true that, for many conventional taxonomists, the phylogenetic element in their classifications is thought to be desirable or even essential. Perhaps the most valuable function of the vigour with which numerical taxonomic studies have been prosecuted is that taxonomists have been forced to discuss this phylogenetic element instead of leaving it as an "acceptance" (Blackwelder, 1964) or statement with which there is wide agreement but for which there is little evidence. The more articulate conventional taxonomists have certainly made a most unconvincing job of defending the role of phylogeny and Colless (1967b) seems convinced that no such defence is possible; perhaps he is correct. Nevertheless, we have to remember that taxonomists are not used to explaining themselves (Mayr, 1968) and that even when they do find the time and energy to do so, the results are sometimes marred by an evident unfamiliarity with quantitative studies such as sap the reader's confidence in Hennig's (1965) and Crowson's (1970) deprecation of the quantitative approach in taxonomic research, despite the solid achievements of these authors in conventional taxonomy.

One of our objectives in this work has been to show how many isolated branches of quantitative biology and geology can be set within a general framework of multivariate morphometrics, concrete or abstract. As examples we have suggested phytosociology and zoogeography, palaeoecology, biostratigraphy and numerical taxonomy. It is no part of our intention to claim that such topics are always best regarded from such a quantitative standpoint, merely that the multivariate approach is so general that it illuminates problems once thought peculiar to individual disciplines. Our personal experience in biometrical consulting with palaeontologists, nematologists, acridologists and anthropologists shows that each group shares common morphometrical problems, when it comes to the interpretation of measurements, but is deeply reluctant to believe that progress made in one subject can be relevant to another.

It is our conviction that the substantial body of experience now available constitutes a discipline within which morphometrics can throw light on many such apparently individual problems and fields of study. For one thing, much of the literature on topics such as coefficients of similarity, clustering systems, ratios *versus* absolute measurements, and so on is redundant in the sense that workers reinvent the solutions to problems, and then publish and discuss them, not merely for each group of workers, but sometimes several times within each field of study (examples are given by Sokal and Sneath, 1963; Udvardy, 1969). We hope that we have performed a service similar to that of Sokal and Sneath in the more restricted field of numerical taxonomy by supplying an intellectual framework for morphometrics, so that research workers will feel convinced that there is indeed such a subject, or, rather, such a definite attitude of mind to research in this field: it will not suit all casts of mind but, for those whom it does suit, it is a powerful tool.

Our objectives were not wholly utilitarian, however; we feel that the power and generality of multivariate morphometrics is aesthetically rewarding, setting morphometric problems squarely in a context which includes relativity theory and elasticity theory and thus assimilating many aspects of quantitative biology into a single intellectual entity with many aspects of theoretical physics, and, as we showed in the introduction, of artistic endeavour.

We have given what we believe is a fair and reasonable exposition of the current state of what we term morphometrics. We hope, naturally, that our efforts will prove useful to quantitative biologists and earth scientists. We have experienced the process of writing the book as stimulating; it has greatly clarified our own thinking on the subject and has made us aware of the strengths of existing methodology and practice, but also of deficiencies, and weaknesses and gaps.

As the astute reader almost certainly will have deduced by now there are many problems to be commended to the urgent attention of biologically oriented mathematical statisticians. Among these we would mention the following. A fuller and more nuanced development of the subject of trend surfaces, a widening of the treatment of growth invariance and the problems relating to the estimation of the growth and ecological vectors (which could be usefully backed up by experimental work). There is an undoubted need for a deepening of the theoretical background of canonical variates analysis, particularly with respect to the robustness of this procedure in relation to heterogeneity in the dispersion matrices of the samples analysed.

Improvements in the applications of genetical theory to morphometric problems are clearly desirable. It is also important that in order to be able to study chronologic variation in morphometric characters, substantial work is needed on time-series studies on canonical variates and principal components.

GLOSSARY

Informal Explanations of Some Terms Used in the Text

Agglomerative Classification
A classification in which the body of material is clustered, generally according to somewhat arbitrary rules, into successively larger entities. Essentially a synthetic method of classification.

Allometry
Allometric growth means different growth rates of one organ relative to another, or to the entire body. Positive allometric growth of one organ in relation to another means that the first organ is growing relatively faster than the second organ. Negative allometric growth is the reverse of this situation. In addition to bivariate allometric growth, as envisaged above, it is also possible to think of a multivariate allometric extension.

Ammonite
An extinct group of the molluscan class of Cephalopoda having a chambered calcareous shell.

Association Matrix
A means of expressing similarity (affinity) in a sample of individuals (taxonomic entities). There is an element (coefficient) for every pair of individuals. The main diagonal is usually made to contain ones. The elements usually run from nought to one, the former for no association whatever and the latter for perfect association.

Behrens-Fisher Problem
The problem of determining the probability of drawing two random samples, whose means differ by k (k may equal 0), from normal populations the difference of whose means is known but the ratio of whose variances is not known.

Belemnites
An extinct group of dibranchiate ten-armed cephalopods with a massive internal shell of calcite.

Brachiopods
A phylum of bivalved invertebrates, the shell of which is marked by a bilateral symmetry. An important group during the Palaeozoic but of minor significance today.

Bryozoans
A group of colonial organisms. Usually build their "colonies" of calcareous substance.

Cenozoic
See Geological Time Scale.

Calcarenite
A moderately fine-grained limestone.

Canonical Form of a Matrix
The simplest form to which a square matrix can be reduced by a certain type of transformation. The canonical form of a matrix has non-zero elements only in the main diagonal and is therefore a diagonal matrix.
    If **S** is a sample covariance matrix, it can be diagonalized by pre- and post-multiplication with an orthogonal matrix **B**,

$$\mathbf{B'SB} = \Lambda$$

$$\begin{bmatrix} \lambda_1 & 0 & \ldots & 0 \\ . & \lambda_2 & \ldots & . \\ . & & & . \\ . & & & . \\ 0 & 0 & \ldots & \lambda_p \end{bmatrix}$$

Here p is the number of variables. The orthogonal matrix **B** is the matrix of latent vectors of **S** and Λ is its diagonal matrix of latent roots.

Centroid Method of Factor Extraction
An approximate means of extracting the principal factor of factor analysis. Widely used, before the advent of the electronic computer, as a common starting point for factor and principal component analyses.

Characteristic Equation
This topic of matrix algebra is one of manifold importance in morphometry and comes into much use in principal component analysis, factor analysis and principal coordinate analysis.
    The definition of the latent roots and vectors is in terms of the equation:

$$\mathbf{Au} = \lambda\mathbf{u}.$$

A is a square matrix, **u** is a vector and λ is a scalar. If the scalar and vector exist, then:

$$\mathbf{Au} - \lambda\mathbf{u} = 0$$
$$(\mathbf{A} - \lambda\mathbf{I})\mathbf{u} = 0.$$

This equation has the solution

$$|\mathbf{A} - \lambda\mathbf{I}| = 0.$$

There is a non-null solution as long as the above determinant is zero. This is called the *characteristic equation*.
    For a matrix **A** of order n, the characteristic equation is a polynomial in λ of degree n, with solutions $\lambda_1, \lambda_2, \ldots, \lambda_n$. For each of these there will be a vector, hence, there are latent vectors $\mathbf{u}_1, \mathbf{u}_2, \ldots, \mathbf{u}_n$ corresponding to the latent roots.

Example

The matrix

$$A = \begin{bmatrix} 1 & 4 \\ 9 & 1 \end{bmatrix}$$

has the characteristic equation

$$\left| \begin{bmatrix} 1 & 4 \\ 9 & 1 \end{bmatrix} - \begin{bmatrix} \lambda & 0 \\ 0 & \lambda \end{bmatrix} \right| = 0$$

i.e.

$$\begin{vmatrix} 1 - \lambda & 4 \\ 9 & 1 - \lambda \end{vmatrix} = 0.$$

The characteristic polynomial, found by expanding, is

$$(1 - \lambda)^2 - 36 = 0$$

Hence, $\lambda = -5$ or 7 are the latent roots.
Using the relationship

$$Au = \lambda u$$

we can see that

$$\begin{bmatrix} 1 & 4 \\ 9 & 1 \end{bmatrix} \begin{bmatrix} 2 \\ -3 \end{bmatrix} = -5 \begin{bmatrix} 2 \\ -3 \end{bmatrix}$$

and

$$\begin{bmatrix} 1 & 4 \\ 9 & 1 \end{bmatrix} \begin{bmatrix} 2 \\ 3 \end{bmatrix} = 7 \begin{bmatrix} 2 \\ 3 \end{bmatrix}$$

Hence, the latent vectors are

$$\begin{bmatrix} 2 \\ -3 \end{bmatrix} \quad \text{and} \quad \begin{bmatrix} 2 \\ 3 \end{bmatrix}$$

Finding the Latent Vectors
Consider the matrix:

$$A = \begin{bmatrix} 1 & 4 & 1 \\ 2 & 1 & 0 \\ -1 & 3 & 1 \end{bmatrix}$$

This has the latent roots $\lambda_i = 0, -1, 4$. To find the latent vectors, insert these values in the characteristic determinant.
Thus

$$\begin{bmatrix} 1 - \lambda_i & 4 & 1 \\ 2 & 1 - \lambda_i & 0 \\ -1 & 3 & 1 - \lambda_i \end{bmatrix} \begin{bmatrix} \alpha_i \\ \beta_i \\ \gamma_i \end{bmatrix} = 0.$$

For **X** = **O**, this becomes

$$\begin{aligned}
\alpha_1 + 4\beta_1 + \gamma_1 &= 0 \\
2\alpha_1 + \beta_1 &= 0 \\
-\alpha_1 + 3\beta_1 + \gamma_1 &= 0
\end{aligned}$$

Put α_1 arbitrarily = 1, which leads to the solutions $\beta_1 = -2$ and $\gamma_1 = 7$. Hence, the corresponding latent vector is

$$\mathbf{u}_1 = \begin{bmatrix} 1 \\ -2 \\ 7 \end{bmatrix}$$

Similarly for the other two latent roots. Finally we have the matrix of latent vectors:

$$\mathbf{U} = [\mathbf{u}_1 \ \mathbf{u}_2 \ \mathbf{u}_3] = \begin{bmatrix} 1 & 2 & 3 \\ -2 & -2 & 2 \\ 7 & 4 & 1 \end{bmatrix}$$

We note the relationships

$$\mathbf{A} = \mathbf{U}\mathbf{D}\mathbf{U}^{-1}$$

and

$$\mathbf{D} = \mathbf{U}^{-1}\mathbf{A}\mathbf{U}$$

where **D** is the diagonal matrix of latent roots. Returning to the foregoing,

$$|\mathbf{U}| = \tfrac{1}{20}$$

and the inverse thereof is:

$$\mathbf{U}^{-1} = \tfrac{1}{20}\begin{bmatrix} -5 & 5 & 5 \\ 8 & -10 & -4 \\ 3 & 5 & 1 \end{bmatrix}$$

$$\mathbf{U}^{-1}\mathbf{A}\mathbf{U} = \begin{bmatrix} 0 & 0 & 0 \\ 0 & -1 & 0 \\ 0 & 0 & 4 \end{bmatrix} = \begin{bmatrix} \lambda_1 & 0 & 0 \\ 0 & \lambda_2 & 0 \\ 0 & 0 & \lambda_3 \end{bmatrix} = \mathbf{D}$$

Cladistic Taxonomy
A taxonomy based on the sequence of evolutionary branchings which an organism is considered to have undergone during its divergence from an ancestral form. The dendrogram (*q.v.*) formed from such relationships is called a CLADOGRAM.

Correlation Matrix
A category of considerable importance is the correlation matrix, which may be regarded as a standardized dispersion matrix, in which all variances are standardized to 1. The *sample* correlation matrix is written:

$$\mathbf{R} = \begin{bmatrix} 1 & r_{12} & \cdots & r_{1p} \\ r_{21} & 1 & \cdots & r_{2p} \\ \cdot & \cdot & & \cdot \\ \cdot & \cdot & & \cdot \\ \cdot & \cdot & & \cdot \\ r_{p1} & r_{p2} & \cdots & 1 \end{bmatrix}$$

As is well known, the elements are found by finding the correlation between each pair of variables. For variables X and Y the correlation coefficient is given by

$$r = \frac{\Sigma(X-\bar{X}) \cdot (Y-\bar{Y})}{\sqrt{\Sigma(X-\bar{X})^2 \cdot \Sigma(Y-\bar{Y})^2}}$$

For most of the procedures used in morphometry, it is required that the variables are *multivariate normally distributed*. The general run of biological material is of this kind or can be made so by a logarithmic transformation applied to the independent variates. However deviations from normality are not uncommon in fossils.

Dendrogram

A tree-like representation of relationships of a hierarchical nature: subsumes phenograms, phylograms, and cladograms (*qq.v.*).

Determinant

Every square matrix has a number associated with it known as the determinant of the matrix. An example of a second order determinant is

$$\begin{vmatrix} a_1 & a_2 \\ b_1 & b_2 \end{vmatrix}$$

The number is found as:

$$\text{Determinant} = a_1 b_2 - b_2 a_1.$$

The convention for writing determinants is usually det **A** or $|\mathbf{A}|$, with vertical bars on either side of the matrix, here, **A**.

Determinantal Equation

An equation whose coefficients include determinants. See characteristic equation.

Dichotomous Data

Data of the "present-absent" kind: either an attribute is present (for example, an ostracod has a posterior spine) or it is absent (there is no posterior spine). The simplest kind of qualitative data is dichotomous but some qualitative data in morphometrics are multi-state.

Dispersion Matrix

This is a matrix composed of the variances and covariances of p variables. In the case of the *sample* dispersion matrix we write it as follows:

$$\mathbf{S} = \begin{bmatrix} s_{11} & s_{12} & \cdots & s_{1p} \\ s_{21} & s_{22} & \cdots & s_{2p} \\ & & \cdots & \\ s_{p1} & s_{p2} & \cdots & s_{pp} \end{bmatrix}$$

s_{11}, s_{22}, s_{33} are the variances of the p variables. Covariance s_{12} is the covariance between variable 1 and variable 2, while s_{2p} is the covariance between variables 2 and p. This matrix is central to all morphometric (multivariate) work.

Generalizing, we write the covariances as $s_{ij}(i \neq j)$.

The covariance is calculated as

$$s_{xy} = \frac{1}{N-1} \Sigma(X-\bar{X}) . (Y-\bar{Y});$$

and

$$s_x^2 = \frac{1}{N-1} \Sigma(X-\bar{X})^2 \text{ is the variance for } x.$$

Divisive Classification

A classification in which the body of material is split into groups based on some rule such as maximizing the ratio of intergroup to intragroup variance. Essentially an analytical classification.

Entropy

Definitions of entropy will vary greatly from one context to another, and may include:

(i) The definition in ecology. When two or more species occupy adjacent areas, the boundary, where they mix, has high entropy as defined by

$$S = \log\{N!/\Pi N_i!\}$$

where S is the entropy, $N!$ stands for factorial N, the total number of individuals in question, and $N_i!$ represents the factorial number of individuals of the ith species, Π stands for "the product of all such terms" (as $N_i!$).

(ii) The definition of entropy as a measure of uncertainty in biological systems. Thus in the above example the areas occupied by only one species would have low entropy, because there is a high probability of finding any given species there, whereas in transitional zones the entropy is high because there is uncertainty as to which species any given individual will belong. More generally, any shift from the equilibrium state of a system leads to a definition of the entropy change (ΔS) in terms directly analogous to the Riemannian metric, which is central to the theme of this book (in the form of the expression for the Generalized Distance)

$$\Delta S = -\frac{1}{2} \sum_{i,j=1}^{n} g_{ij} . a_i . a_j$$

where the fundamental tensor g_{ij} must be a positive definite matrix and a_i is the displacement along the ith variate (character, in morphometric work). Evidently, the concept of the generalized distance is closely related to that of the entropy if we think of the contrast between two forms as a distortion of the shape of one with reference to the other. The bigger the distortion the greater will be the generalized distance between the forms, and the entropy (in its sense of improbability) of the second, distorted, form.

Euclidean Space

"Ordinary" three-dimensional space, the reference axes of which rectilinear and orthogonal (q.v.).

Foraminifers

A class of protozoans, the members of which normally secrete a calcareous shell or build a shell by agglutinating minute grains of sand or limestone.

Generalized Distance
The generalized statistical distance, first introduced by P. C. Mahalanobis. Symbolized by D^2, it expresses the distance between two forms in a Riemannian space the axes of which are not orthogonal.

Geochemistry
The application of chemistry to geological problems: most usually, the study of the chemical constitution of the earth.

Geological Time Scale
See table on page 354.

Geosynclines
Sites of large-scale down-buckling in the crust of the earth generally accompanied by the deposition of thick sediments in the trough.

Homogeneity of Matrices
In morphometry, usually meant to imply homogeneity of pooled dispersion matrices. This is a generalization of the concept of homogeneity of variances so important in the univariate analysis of variance. Many statistical procedures require that the matrices of dispersion are statistically homogeneous.

Hypermultivariate Structure
In many analyses the effective dimensionality of the final representation is much less than that of the character-space; sometimes, however, no worthwhile reduction of dimensionality is achieved, usually because each character is contributing information about a distinct dimension of variation. The structure is then said to be hypermultivariate.

Identity Matrix
This is a square matrix with "ones" on the main diagonal and "zeros" elsewhere. It is designated by **I**.

$$\mathbf{I} = \begin{pmatrix} 1 & 0 & \dots & 0 \\ 0 & 1 & \dots & 0 \\ & & \dots & \\ 0 & 0 & \dots & 1 \end{pmatrix}$$

Also called the Unit Matrix, it is the analogue in matrix algebra of the number 1.

Information Theory
The branch of probability theory founded in 1948 by C. E. Shannon, that deals with the likelihood of transmission of messages, accurate within specified limits, when the "bits" of information comprising the messages are subject to probabilities of transmission failure.

Kronecker Delta
The function δ_j^i of two variables j and i defined by $\delta_j^i = 1$ if $i = j$ and $\delta_j^i = 0$ if $i \neq j$.

Geological column and time scale

Era	System or period	Epoch	Approximate age in millions of years from radioactivity
CENOZOIC *(recent life)*	QUATERNARY (An addition to the old tripartite 18th-century classification)	Neogene { RECENT *(most recent)* PLEISTOCENE *(very recent)* PLIOCENE *(moderately recent)* MIOCENE *(slightly recent)*	17 (Miocene?)
	TERTIARY (Third, from the 18th-century classification)	Paleogene { OLIGOCENE *(dawn of the recent)* EOCENE *(early dawn of the recent)* PALEOCENE	60 (Paleocene, New Jersey)
MESOZOIC *(middle life)*	CRETACEOUS (from the Latin for chalk)		62 (Late Cretaceous, New Jersey)
	JURASSIC (Jura Mts., Europe)		
	TRIASSIC (from tripartite division in Germany)		
PALAEOZOIC *(ancient life)*	PERMIAN (Perm, a province of Russia)		230 (end of Early Permian)
	CARBONIFEROUS (from abundant coal in rocks)		
	DEVONIAN (Devonshire, England)		270 (late Middle Devonian, Saskatchewan)
	SILURIAN (the Silures, an ancient British tribe)		350 (end of Ordovician)
	ORDOVICIAN (the Ordovices, an ancient British tribe)		440 (Late Cambrian, Sweden)
	CAMBRIAN (Roman name for Wales)		470 (late Early Cambrian, Alberta)
PRECAMBRIAN	Many local systems and series are recognized		600 (late Precambrian, Africa) 2500 (Precambrian, Manitoba. At least 2700, possibly 3300, Rhodesia)

The generalized Kronecker Delta has k superscripts and k subscripts. All of the Kronecker Deltas are numerical tensors, and serve as the fundamental tensors of Euclidean space, since by specifying that all the pre- and post-diagonal entries are zero we are, in effect, saying that all the correlations between characters used as reference axes are zero, thus defining a Euclidean space in which the distance metric is the Pythagorean one.

Latent Roots

The roots of the characteristic equation of a matrix are called the latent roots of the matrix. Also referred to as eigenvalues, proper values, characteristic roots, and characteristic values.

Latent Vectors

The vectors corresponding to the latent roots. Also referred to as eigenvectors. See also "characteristic equation".

Linear Transformations

Consider the matrix equation

$$\mathbf{y} = \mathbf{A}\mathbf{x}.$$

The elements of $\mathbf{y}$ are linear combinations of those of $\mathbf{x}$. Matrix $\mathbf{A}$ is said to be the *linear transformation* of $\mathbf{x}$ into $\mathbf{y}$.

Also, the general result:

if $\qquad\qquad \mathbf{y} = \mathbf{A}\mathbf{x} \quad$ and $\quad \mathbf{x} = \mathbf{B}\mathbf{w}$

then $\qquad\qquad \mathbf{y} = \mathbf{A}\mathbf{B}\mathbf{w}.$

Linear transformations are an important part of many of the calculations in morphometrics.

Macropterous Forms

Forms having the wings fully developed, or almost so.

Matrix

Consider data in the form of a vector: a vector of measurements

$$(x_1, x_2, \ldots, x_N) \quad \text{where } N = \text{number of variables}$$

and the measurements have been made on, say, k specimens. We can arrange the observation vectors into a $k \times N$ array, called a *matrix*. Thus

$$\begin{bmatrix} x_1^1, x_2^1, & \ldots, & x_N^1 \\ x_1^2, x_2^2, & \ldots, & x_N^2 \\ & \ldots & \\ x_1^k, x_2^k, & \ldots, & x_N^k \end{bmatrix}$$

The superscripts number the individual.

You will quickly notice that matrices and vectors are important things indeed in morphometrics. Although the subject of linear algebra, the mathematics of matrices and vectors, can become very complicated, it is not difficult to gain a basic knowledge about them and how to use them in morphometric studies.

We can start our discussion by asking what a matrix is. A matrix is a rectangular array of numbers. The array

$$\begin{bmatrix} 7 & 5 & -1 \\ 6 & 3 & 2 \end{bmatrix}$$

is a matrix. Another example of a matrix is

$$\begin{bmatrix} 44 & 8 \\ 1 & 19 \\ -2 & 1 \end{bmatrix}$$

Matrix Inversion

Division in its usual sense does not exist in matrix algebra. The concept of dividing is replaced by multiplying by a matrix called the inverse of A and written as A^{-1}.

Example:
Consider the equations

$$m + a = 14$$
$$m + d = 12$$
$$m - a = 6$$

In matrix form

$$\begin{bmatrix} 1 & 1 & 0 \\ 1 & 0 & 1 \\ 1 & -1 & 0 \end{bmatrix} \begin{bmatrix} m \\ a \\ d \end{bmatrix} = \begin{bmatrix} 14 \\ 12 \\ 6 \end{bmatrix}$$

This is in the form

$$Ax = b$$

The solution is found as

$$x = A^{-1}b.$$

For just three variables, the inverse can be found through the determinant. Consider next the matrix:

$$A = \begin{bmatrix} 1 & 2 & 3 \\ 4 & 5 & 6 \\ 7 & 8 & 10 \end{bmatrix}$$

The determinant of this matrix is $A = -3$. We also require the *adjugate* matrix of A, found from the cofactors. For the first element of the first column this is

$$\begin{vmatrix} 5 & 6 \\ 8 & 10 \end{vmatrix} = 50 - 48 = +2.$$

For the second element:

$$\begin{vmatrix} 2 & 3 \\ 8 & 10 \end{vmatrix} = -(20 - 24) = +4.$$

Similarly,

$$\begin{vmatrix} 2 & 3 \\ 5 & 6 \end{vmatrix} = 12 - 15 = -3.$$

This gives the adjugate matrix

$$\begin{bmatrix} 2 & 2 & -3 \\ 4 & -11 & 6 \\ -3 & 6 & -3 \end{bmatrix}$$

The inverse is then obtained as

$$A^{-1} = \frac{1}{|A|} \cdot \text{adj } A'$$

where adj A' denotes the transpose of the adjugate matrix.
Hence

$$\begin{bmatrix} \dfrac{-2}{3} & \dfrac{-4}{3} & +1 \\ \\ \dfrac{-2}{3} & \dfrac{+11}{3} & -2 \\ \\ +1 & -2 & +1 \end{bmatrix}$$

We note that

$$A^{-1}A = AA^{-1} = I = \begin{bmatrix} 1 & 0 & 0 \\ 0 & 1 & 0 \\ 0 & 0 & 1 \end{bmatrix}.$$

Some further interesting properties of the inverse matrix are as follows:
(1) $|A^{-1}| = 1/|A|$.
(2) The inverse matrix is non-singular.
(3) If A^{-1} and B^{-1} exist,

$$(AB)^{-1} = B^{-1}A^{-1}.$$

For morphometric work, the matrices to be inverted are usually large and require the help of a computer.

Matrix Multiplication
Adjacent subscripts (which must be equal for conformability) "cancel out", leaving the *first* and *last* subscripts as the order of the product.
For example

$$A_{r \times c} B_{c \times s} C_{s \times t} D_{t \times u}$$

yields the product E_{ru}, thus of order $(r \times u)$.
We also note that
(1) AB is not necessarily $= BA$.
(2) AB may exist, but not necessarily BA.
(3) A^2 only exists when A is square.
(4) AB is said as "A postmultiplied by B".

Matrix Transpose
This is an important process in dealing with matrices. It means the switching of

rows and columns so that the rows of the one matrix are the columns of the other. Thus the transpose of

$$
\mathbf{A} = \begin{bmatrix} 18 & 17 & 11 \\ 19 & 13 & 6 \\ 6 & 14 & 9 \\ 9 & 11 & 4 \end{bmatrix}
$$

is the matrix

$$
\mathbf{A}' = \begin{bmatrix} 18 & 19 & 6 & 9 \\ 17 & 13 & 14 & 11 \\ 11 & 6 & 9 & 4 \end{bmatrix}
$$

The second matrix is called the *transpose* of the first.

It is usually denoted by a dash: $\mathbf{A}'$ is the transpose of $\mathbf{A}$.

Mean Vector
The mean vector is a vector the elements of which are the univariate means of each morphometric character.

Micropterous Forms
Forms having the wings greatly reduced or vestigial even in the adults: in a number of insect species the macropterous and micropterous forms occur side by side.

Multiple Regression
A method of relating p independent variates to one dependent variate. A much used statistical procedure. Discriminant functions may be interpreted, if one so will, as a form of multiple regression in which dummy variates are used.

Multivariate Analysis of Variance
The multivariate generalization of the analysis of variance (ANOVA) for testing the equality of mean vectors of several universes. Often written as MANOVA.

Nomogram
A graph consisting of three lines or possibly curves (usually parallel) graduated for different variables in such a way that a straight edge cuting the three lines gives the related values of the three variables.

Non-parametric Tests
A non-parametric test is one which makes no hypothesis about the value of parameters in a statistical density function, and thus absolves the user from the responsibility of discovering the form of the function.

Normalizing of Vectors
The division of a vector by its length (see also vector).

Null Hypothesis
A statistical hypothesis usually specifying the universe from which a random sample is assumed to have been drawn and which is to be nullified if the evidence from the random sample is unfavourable to the hypothesis.

Numerical Taxonomy
In a general sense, any form of quantitative taxonomy: in a restricted sense, a classification based on measures of affinity compounded from large numbers of dichotomous or coded characters, treated as if they were uncorrelated.

Order of a Matrix
The ROWS of a matrix are the arrays of numbers that go across a page. The COLUMNS of a matrix are the arrays of numbers that go down the page.
   In the matrix

$$\begin{bmatrix} 7 & 3 \\ 5 & 2 \end{bmatrix}$$

the first row is 7 3 and the second row is 5 2. The first column is $\frac{7}{5}$ and the second column is $\frac{3}{2}$.
   A matrix that has m rows and n columns is called an $m \times n$ matrix (the number of rows is always placed first in this designation). Two matrices are said to be of the same order if they have the same numbers of rows and columns.
   The vector is a special kind of matrix. For example, consider the row vector

$$(1 \quad 1 \quad 2 \quad 3)$$

This is a 1 (row) by 4 (columns) matrix. Likewise, the vector

$$\begin{bmatrix} 2 \\ -1 \\ 3 \end{bmatrix}$$

is a column vector but by the same token also a 3 (rows) x 1 (column) matrix. Two matrices are said to be equal if they have the same order and if corresponding elements are equal.

Orthogonality
Used of the geometrical representation of vectors to indicate that the angle(s) between them are right angles. Used of the algebraic representation of vectors to indicate that the sum of cross-products of the coefficients of two vectors is zero.

Orthogonal Matrix
A matrix, **A**, is orthogonal if it produces the identity matrix when multiplied by its transpose:

$$\mathbf{AA'} = \mathbf{I}$$

Example:

$$\mathbf{A} = \begin{bmatrix} \cos\theta & -\sin\theta \\ \sin\theta & \cos\theta \end{bmatrix}$$

A matrix of latent vectors is an orthogonal matrix.

Ostracoda
A group of the bivalved Crustacea having, usually, a calcareous shell.

Paedomorphism
The maintenance of the body form commonly found in immature stages when the animal is adult, or substantially fully grown.

Paleogene
The Cenozoic is divided into two parts, the younger Neogene, and the older Paleogene. The Paleogene comprises the Paleocene, Eocene and Oligocene. See also Geological Time Scale.

Phenetic Taxonomy
A taxonomy based on the general similarity of the organisms studied, without consideration of evolutionary relationships.

Phenogram
A dendrogram representing relationships based on phenetic similarity rather than on common descent.

Phylogram
A dendrogram purporting to reflect the evolution of the organisms to which it relates.

PL1
A programming language developed by the IBM corporation.

Population
In order to avoid confusion, the term population has been reserved for use in its biological meaning. Statistical populations are referred to as universes.

Product of Matrices
The first stage is to consider a *matrix-vector* product.

$$\mathbf{Ax} = \begin{bmatrix} 58 & 26 & 8 \\ 32 & 58 & 12 \\ 1 & 3 & 9 \end{bmatrix} \begin{bmatrix} 0 \\ 1 \\ 2 \end{bmatrix}$$

$$= \begin{bmatrix} 58(0) & + & 26(1) & + & 8(2) \\ 32(0) & + & 58(1) & + & 12(2) \\ 1(0) & + & 3(1) & + & 9(2) \end{bmatrix} = \begin{bmatrix} 42 \\ 82 \\ 21 \end{bmatrix}$$

In algebraic terms:

$$\mathbf{A} = \begin{bmatrix} a_{11} & a_{12} & a_{13} \\ a_{21} & a_{22} & a_{23} \\ a_{31} & a_{32} & a_{33} \end{bmatrix} \text{ and } \mathbf{x} = \begin{bmatrix} x_1 \\ x_2 \\ x_3 \end{bmatrix}$$

You will have noted that the vector has to contain the same number of elements as the rows of the matrix.
Then,

$$\mathbf{Ax} = \begin{bmatrix} a_{11}x_1 & + & a_{12}x_2 & + & a_{13}x_3 \\ a_{21}x_1 & + & a_{22}x_2 & + & a_{23}x_3 \\ a_{31}x_1 & + & a_{32}x_2 & + & a_{33}x_3 \end{bmatrix}$$

This may also be written, using summation notation, as

$$
Ax = \begin{bmatrix} \sum_{k=1}^{3} a_{1k}x_k \\ \sum_{k=1}^{3} a_{2k}x_k \\ \sum_{k=1}^{3} a_{3k}x_k \end{bmatrix}
$$

Finally, we consider the product of two matrices. This is the logical extension of matrix-vector multiplication. As you will have guessed, the two matrices have to be conformable. Thus, there must be the same number of columns in one matrix as rows in the other. The matrices

$$
\begin{bmatrix} a & a & a \\ a & a & a \end{bmatrix} \text{ and } \begin{bmatrix} a & a \\ a & a \\ a & a \end{bmatrix} \text{ are conformable.}
$$

Consider the matrices

$$
A = \begin{bmatrix} 1 & 0 & 2 \\ 3 & 1 & 1 \\ 1 & 2 & 1 \\ -1 & 3 & 2 \end{bmatrix}, \text{ and } B = \begin{bmatrix} 1 & 2 \\ 0 & 1 \\ 0 & -1 \end{bmatrix} = [x \quad w]
$$

The product of A with each column of B is

$$
Ax = \begin{bmatrix} 1(1) & + & 0(0) & + & 2(0) \\ 3(1) & + & 1(0) & + & 1(0) \\ 1(1) & + & 2(0) & + & 1(0) \\ -1(1) & + & 3(0) & + & 2(0) \end{bmatrix} \begin{bmatrix} 1 \\ 3 \\ 1 \\ -1 \end{bmatrix}
$$

and similarly for Aw. The matrix product is thus

$$
AB = \begin{bmatrix} 1 & 0 \\ 3 & 6 \\ 1 & 3 \\ -1 & -1 \end{bmatrix}
$$

The above definition of matrix multiplication holds only if the jth column of matrix B has the same number of elements as does the ith row of matrix A. Formally, the product may be written as:

$$
AB = \left\{ \sum_{k=1}^{c} a_{ik}b_{kj} \right\} \qquad \begin{array}{l} \text{for } i = 1, 2, \ldots, r \\ j = 1, 2, \ldots, s \end{array}
$$

Quadratic Form

With vectors

$$\mathbf{x}' = [x_1 \; x_2 \; x_3] \quad \text{and} \quad \mathbf{y}' = [y_1 \; y_2 \; y_3]$$

we consider the product $\mathbf{x}'$ $\mathbf{Ay}$, where $\mathbf{A}$ is some 3×3 matrix.

$$\mathbf{x}'\mathbf{Ay} = [x_1 \; x_2 \; x_3] \begin{bmatrix} 1 & 2 & 3 \\ 4 & 7 & 6 \\ 2 & -2 & 5 \end{bmatrix} \begin{bmatrix} y_1 \\ y_2 \\ y_3 \end{bmatrix}$$

$$= [x_1 y_1 + 4x_2 y_1 + 2x_3 y_1 + 2x_3, \; 2x_1 + 7x_2 - 2x_3, \; 3x_1 + 6x_2 + 5x_3] \begin{bmatrix} y_1 \\ y_2 \\ y_3 \end{bmatrix}$$

This is a second degree function, of the first degree in each of the $x's$ and $y's$. It is called a *bilinear form*. If the $y's$ are replaced by $x's$, then the expression is called a *quadratic form*.

$$\mathbf{x}'\mathbf{Ax} = x_1^2 + 4x_2 x_1 + 2x_3 x_1 + 2x_1 x_2 + 7x_2^2 - 2x_3 x_2 + 3x_1 x_3 + 6x_2 x_3 + 5x_3^2.$$

In algebraic terms

$$\mathbf{x}'\mathbf{Ax} = \sum_i x_i^2 a_{ii} + \sum_{j \; > \; i} \sum x_i x_j (a_{ij} + a_{ji}).$$

Thus, $\mathbf{x}'\mathbf{Ax}$ is the sum of squares of the elements of $\mathbf{x}$, each square multiplied by the corresponding diagonal element of $\mathbf{A}$, plus the sum of products of the elements of $\mathbf{x}$, each product multiplied by the sum of the corresponding elements of $\mathbf{A}$.

  Positive definite quadratic form: if $\mathbf{x}'\mathbf{Ax}$ is positive for all values of $\mathbf{x}$ other than $\mathbf{x} = 0$, the matrix is termed *positive definite*.

Raptorial Fore-legs

Legs armed with spines which close like a gin-trap round the prey.

Reduced Major Axis

An approximate method of regression analysis in which only one "regression line" appears.

Reification

The identification of the biological or geological nature of a vector by examination of the material which it ordinates.

Riemannian Space

An n-dimensional coordinate manifold $(x_1, x_2, \ldots, x_n)$ whose element of arc length ds is given by a symmetric quadratic differential form

$$ds^2 = g_{ij}(x^1, \ldots, x^n) \, dx^i \, dx^j$$

the g_{ij} are the components of the tensor known as the "fundamental metric tensor".

Robustness
As applied to statistics, means the capability of a statistical procedure to produce a fair result despite deviations in the data from the premises upon which the procedure is based.

Square Symmetric Matrix

$$A = \begin{bmatrix} 3 & 2 & 1 \\ 2 & 1 & 2 \\ 1 & 2 & 2 \end{bmatrix}$$

The matrix is said to be square when it has as many rows as it has columns. It is said to be symmetric when element a_{ij} equals element a_{ji}. A symmetric matrix is unchanged by transposition. Hence, $A = A'$.

Stromatoporoids
A group of extinct colonial lime-secreting coelenterates, related to the corals.

Symbatic Vectors
Parallel or almost parallel vectors which, on reification (*q.v.*), prove to reflect quite different underlying processes.

Tensor
An abstract concept (an invariant multilinear direction function) having a definitely specified system of components in every coordinate system under consideration. Tensors are operators which transform a vector **u** into another vector **v** by linear modifications of the components of the vector **u**; in morphometric analysis, to convert the vector describing sexual dimorphism in a species into that describing geographic variation would require multiplication by a tensor.

Time Scale, Geological
See Geological Time Scale.

Trilobites
A group of extinct (Cambrian-Permian) arthropods.

Time-Series
Data taken at time intervals.

Topology
That branch of geometry which deals with the topological properties of figures; i.e. any property of a figure *A* that holds as well for every figure into which *A* may be transformed by a topological transformation. In morphometrics, encountered as "discriminatory topology".

Triangular Matrix
A matrix with zeros in the locations above *or* below the principal diagonal. An *upper* triangular matrix is a square matrix whose non-zero elements fill the

places on or above the main diagonal. An example of such a matrix is:

$$\begin{bmatrix} 1 & 4 & 1 \\ 0 & 2 & 2 \\ 0 & 0 & 3 \end{bmatrix}$$

A *lower* triangular matrix has the places above the main diagonal filled with noughts. The following matrix is an example of this situation

$$\begin{bmatrix} 2 & 0 & 0 \\ 4 & -1 & 0 \\ 2 & 2 & 4 \end{bmatrix}$$

You will meet triangular matrices, for example, in the section on principal coordinates.

Unit Matrix
See Identity Matrix.

Universe
In this book, the statistical population.

Vector
Geometrically, a vector consists of a line with which both magnitude and direction are associated: algebraically, it is defined by the coordinates of its terminal point in a space of as many dimensions as there are measured characters (assuming that the other end of the vector lies at the origin).

In terms of morphometric analysis, the geometrical representation of vectors leads to generalized distance charts in which forms are separated by lines (vectors) whose length reflects the generalized distance between the forms, which is the amount of change in the direction specified by the orientation of the vector. Distinguish carefully between the length of a vector u defined as the scalar quantity $l = g_{ij}u_iu_j$ and the extent to which u separates two forms, which is itself a vector quantity (the generalized distance). Scalars, vectors, and tensors form a family of concepts such that scalars can be regarded as tensors of zero rank, vectors as tensors of rank unity, whereas the kind of tensor important in morphometrics (the fundamental tensor of Riemannian space) is of the second rank. Tensors of rank greater than two are not immediately relevant to morphometric work.

Vector Product
The vector product usually met with in morphometrics is the *inner product* of two vectors, a very important concept in linear algebra.

The inner product is the sum of the products of the corresponding elements. This is designated as a. x, if a and x are the two vectors. The inner product of two vectors is a number.

Wilk's Criterion
A determinantal ratio arising in the calculations of MANOVA and upon which tests of significance are based.

Bibliography and Author Index

The page on which a reference occurs is given in square brackets.

Adanson, M. (1759). Histoire naturelle du Sénégal; Coquillages. Avec la relation abrégée d'un voyage en ce pays pendant les années 1749-1753, 461 pp. Bauche, Paris. [7]

Adanson, M. (1763). "Familles des Plantes," Vol. I, 515 pp. Vincent, Paris. [7]

Adke, S. R. (1958). A note on the distance between two populations. *Sankhyā* 19, 195-200. [11]

Albrecht, F. O., *see* Blackith, R. E.

Albrecht, F. O. and Blackith, R. E. (1957). Phase and moulting polymorphism in locusts. *Evolution, Lancaster, Pa.* 11, 166-177. [58, 91, 302, 321]

Amtmann, E. (1965). Biometrische Untersuchungen zur introgressiven Hybridization der Waldmaus (*Apodemus sylvaticus* L. 1785) und der Gelbhalsmaus (*Apodemus tauricus* Pallas 1811). *Z. zool. Syst. Evol.* 3, 103-156. [245]

Amtmann, E. (1966). Zur Systematik afrikanischer Streifenhörnchen der Gattung *Funisciurus*. Ein Beitrag zur Problematik klimaparalleler Variation und Phänatik. *Bonn. zool. Beitr.* 17, 1-44. [148]

Anand, I. J., *see* Murty, B. R.

Anderson, E. (1953). Introgressive hybridization. *Biol. Rev.* 28, 280-307. [262]

Anderson, T. W. (1958). "An Introduction to Multivariate Statistical Analysis". 374 pp. John Wiley & Sons, New York. [47, 49, 63, 135]

Anderson, T. W. (1963). Asymptotic theory for principal components analysis. *Ann. math. Statist.* 34, 122-148. [65, 189]

Anderson, T. W. and Bahadur, R. R. (1962). Two sample comparisons of dispersion matrices for alternatives of immediate specificity. *Ann. math. Statist.* 33, 420-431 [63]

Andrew, L. E., *see* White, M. J. D.

Antila, M., *see* Gyllenberg, H.

Arber, A. (1950). "The Natural Philosophy of Plant Form", 247 pp. Cambridge University Press, Cambridge. [5]

Arnoux, J., *see* Fraisse, R.

Arunachalam, V., *see* Murty, B. R.

Ashton, E. H., Healy, M. J. R. and Lipton, S. (1957). The descriptive use of discriminant functions in physical anthropology. *Proc. R. Soc.* B 146, 552-572. [103]

Ashton, E. H., Healy, M. J. R., Oxnard, C. E. and Spence, T. F. (1965). The combination of locomotor features of the primate shoulder girdle by canonical analysis. *J. Zool.* 147, 406-429. [100, 101]

Austin, M. P., *see* Greig-Smith, P.; Noy-Meir, I.

Baer, J. L. (1969). Paleoecology of cyclic sediments of the Lower Green River Formation, Central Utah. *Brigham Young Univ. Geol. Studies* 16 (1), 3-95. [167]

Bahadur, R. R., *see* Anderson, T. W.

Bailey, D. W. (1956). A comparison of genetic and environmental principal components of morphogenesis in mice. *Growth* 20, 63-74. [243]

Baker, G. A. (1954). Organoleptic ratings and analytical data for wines analysed into orthogonal factors. *Food Res.* 19, 575-580. [150]

Balakrishnan, V. and Sanghvi, L. D. (1968). Distance between populations on the basis of attribute data. *Biometrics* 24, 859-865. [283]

Barkham, J. P. and Norris, J. M. (1970). Multivariate procedures in an investigation of vegetation and soil relations of two beech woodlands, Cotswold Hills, England. *Ecology* 51, 630-639. [229]

Barraclough, R. M. and Blackith, R. E. (1962). Morphometric relationships in the genus *Ditylenchus*. *Nematologica* 8, 51-58. [28, 258]

Bartlett, M. S. (1965). Multivariate statistics. *In* "Theoretical and Mathematical Biology" (T. H. Waterman and H. J. Morowitz, eds), pp. 202-224. Blaisdell Publ. Co., New York. [35]

Barton, A. D., *see* Laird, A. K.

Basford, N. L., Butler, J. E., Leone, C. A. and Rohlf, F. J. (1968). Immunologic comparisons of selected Coleoptera with analyses of relationships using numerical taxonomic methods. *Syst. Zool.* 17, 388-406. [264, 342]

Beatty, R. A., *see* Williams, D. A.

Bennett, B. M. (1951). Note on a solution of the generalised Behrens-Fisher problem. *Ann. Inst. statist. Math., Tokyo* 2, 87-90. [63]

Berge, C. (1962). "The Theory of Graphs", 247 pp. Methuen, London. [294, 308]

Bergson, H. (1911). "Creative Evolution", 425 pp. Macmillan, London. [6]

Bickley, W. G. and Gibson, R. E. (1962). "Via Vector to Tensor; an Introduction to the Concepts and Techniques of the Vector and Tensor Calculus", 152 pp. English Univ. Press, London. [10, 297]

Bidwell, O. W., *see* Sarkar, P. K.

Bidwell, O. W. and Hole, F. D. (1964). An experiment in the numerical classification of Kansas soils. *Proc. Soil. Sci. Soc. Am.* 28, 263-268. [267]

Binet, F. E., *see* Clifford, H. T.

Blackith, R. E. (1957). Polymorphism in some Australian locusts and grasshoppers. *Biometrics* 13, 183-196. [17, 321]

Blackith, R. E. (1958). An analysis of polymorphism in social wasps. *Insectes Soc.* 5, 263-272. [36, 242]

Blackith, R. E. (1960). A synthesis of multivariate techniques to distinguish patterns of growth in grasshoppers. *Biometrics* 16, 28-40. [44, 141, 243]

Blackith, R. E. (1962). L'identité des manifestations phasaires chez les acridiens migrateurs. *Colloques int. Cent. natn. Rech. scient.* No. 114, 299-310. [41, 42, 211, 242]

Blackith, R. E. (1963). A multivariate analysis of Latin elegaic verse. *Language and Speech* 6, 196-205. [158]

Blackith, R. E. (1965). Morphometrics. *In* "Theoretical and Mathematical Biology" (T. H. Waterman and H. J. Morowitz, eds), pp. 225-249. Blaisdell Publishing Co., New York. [40, 103, 257, 320]

Blackith, R. E., *see* Albrecht, F. O., Barraclough, R. M., Blair, C. A., Eyles, A. C.

Blackith, R. E. and Albrecht, F. O. (1959). Morphometric differences between the eye-stripe polymorphs of the Red Locust. *Scient. J.R. Coll. Sci.* 27, 13-27. [97, 136]

Blackith, R. E. and Blackith, R. M. (1968). A numerical taxonomy of the Orthopteroid insects. *Aust. J. Zool.* 16, 111-131. [284, 289, 298, 299, 304, 305, 309]

Blackith, R. E. and Blackith, R. M. (1969). Variation of shape and of discrete anatomical characters in the morabine grasshoppers. *Aust. J. Zool.* 17, 697-718. [98]

Blackith, R. E. and Kevan, D. K. McE. (1967). A study of the genus *Chrotogonus* (Orthoptera). VIII. Patterns of variation in external morphology. *Evolution, Lancaster, Pa.* 21, 76-84. [95, 96, 97, 216, 243, 244]

Blackith, R. E. and Roberts, M. I. (1958). Farbenpolymorphismus bei einigen Feldheuschrecken. *Z. Vererb Lehre* 89, 328-337. [57, 212, 216, 243]

Blackith, R. E. and Verdier, M. (1960). Quelques nouvelles techniques utilisables en analyse morphométrique chez les acridiens. II. Utilisation du fémur antérieur pour diverses discriminations. *Bull. Soc. ent. Fr.* 65, 260-273. [324]

Blackith, R. E., Davies, R. G. and Moy, E. A. (1963). A biometric analysis of the development of *Dysdercus fasciatus* Sign. *Growth* 27, 317-334. [325]

Blackith, R. M., *see* Blackith, R. E.

Blackwelder, R. E. (1964). Phyletic and phenetic *versus* omnispective classification. *In* "Phenetic and Phylogenetic Classification" (V. H. Heywood and J. McNeill, eds), pp. 17-28. Systematics Assn. Publ. No. 6, London. [27, 344]

Blair, C. A., Blackith, R. E. and Boratynski, K. L. (1964). Variation in *Coccus hesperidum* L. (Homoptera; Coccidae). *Proc. R. ent. Soc. Lond.* (A) 39, 129-134. [150]

Bleibtrau, H. K., *see* Giles, E.

Bliss, C. I. (1958). Periodic regression in biology and climatology. *Bull. Conn. agric. Exp. Stn.* 615 pp. [325]

Bocquet, C. (1953). Recherches sur le polymorphisme naturel des *Jaera marina* (Fabr.) (*Isopodes asellotes*). Essai de systématique évolutive. *Archs Zool. éxp. gén.* 90, 107-450. [6]

Bocquet, C. and Solignac, M. (1969). Etude morphologique des hybrides expérimentaux entre *Jaera (albifrons) albifrons* et *Jaera (albifrons) praehirsuta* (Isopoda; Asellota). *Archs Zool.* éxp. gén. 110, 435-452. [245]

Bödvarsson, H. (1967). Icelandic Collembola; Material from Westman Islands and from Hornafjördur. *Opusc. ent.* 32, 255-270. [238]

Bonnet, L., Cassagnau, P. and de Izarra, D.-C. (1970). Etude écologique des collemboles muscicoles du Sidobre (Tarn). II. Modèle mathématique de la distribution des espèces sur un rocher. *Bull. Soc. Hist. nat. Toulouse* 106, 127-145. [238]

Boratynski, K., *see* Blair, C. A.

Boratynski, K. and Davies, R. G. (1971). The taxonomic value of male Coccoidea (Homoptera) with an application and evaluation of some numerical techniques. *Biol. J. Linn. Soc. Lond.* 3, 57-102. [151, 291, 342]

Boyce, A. J. (1964). The value of some methods of numerical taxonomy with reference to hominoid classification. *In* "Phylogenetic and Phenetic Classification" (V. H. Heywood and J. McNeill, eds), pp. 47-65. Systematics Assn. publ., No. 6, London. [257]

Brandwood, L., *see* Cox, D. R.

Brännström, B., *see* Reyment, R. A.

Branson, H. R. (1953). A definition of information from the thermodynamics of irreversible processes. *In* "Information Theory in Biology" (J. H. Quastler, ed.), pp. 25-40. Illinois Univ. Press, Illinois. [12]

Bretsky, S. S. (1969). Phenetic and phylogenetic classifications of the Lucinidae (Mollusca: Bivalvia). (See Reyment, 1970c). [311, 339]

Brieger, F. G., Vencovsky, R. and Paker, I. U. (1963). Distâncias filogenéticas no gênero *Cattleya. Ciênc. Cult.* 15, 187-188. [246]

Brožek, J. and Keys, A. (1951). Evaluation of leanness-fatness; norms and interrelationships. *Br. J. Nutr.* 5, 194-206. [15]

Buck, S. F. and Wicken, A. J. (1967). Models for use in investigating the risk of mortality from lung cancer and bronchitis. *Appl. Statist.* **16**, 185-210. [264]

Bühler, P., *see* Rempe, U.

Burgoyne, P. S., *see* Williams, D. A.

Burla, H. and Kälin, A. (1957). Biometrischer Vergleich zweier Populationen von *Drosophila obscura. Revue suisse Zool.* **64**, 246-252. [214]

Burma, B. (1949). Multivariate analysis—a new analytical tool for paleontology and geology. *J. Paleont.* **23**, 95-103. [277]

Burnaby, T. P. (1966a). Growth-invariant discriminant functions and generalised distances. *Biometrics* **22**, 96-110. [29, 54, 303, 332]

Burnaby, T. P. (1966b). Distribution-free quadratic discriminant functions in palaeontology. *In* "Computer Applications in the Earth Sciences", pp. 70-77. State Geol. Surv., Lawrence, Kansas. [31, 54, 281]

Burnaby, T. P. (1970). On a method for character-weighting a similarity coefficient, employing the concept of information. *Math. Geol.* **2**, 25-38. [280, 281]

Burt, C. (1941). "The Factors of the Mind", p. 509. Macmillan, New York. [201]

Burton, D. L., *see* Heiser, C. B.

Butler, J. E., *see* Basford, N. L.

Buzas, M. A. (1967). An application of canonical analysis as a method for comparing faunal areas. *J. Anim. Ecol.* **36**, 563-577. [218]

Cain, A. J. (1958). Logic and memory in Linnaeus' system of taxonomy. *Proc. Linn. Soc. Lond.* **170**, 185-217. [5, 340]

Cain, A. J. (1959). Taxonomic concepts. *Ibis* **101**, 302-318. [7, 297, 309, 319, 340]

Cain, A. J. (1962). The evolution of taxonomic principles. *In* "Microbial Classification" (G. C. Ainsworth and P. H. A. Sneath, eds), pp. 1-13. Cambridge Univ. Press, Cambridge. [5]

Camin, J. H. and Sokal, R. R. (1965). A method for deducing branching sequences in phylogeny. *Evolution* **19**, 311-326. [263, 298, 299, 307, 312]

Campbell, B. (1962). The systematics of man. *Nature, Lond.* **194**, 225-232. [59]

Carlberg, G., *see* Gyllenberg, H.

Carmichael, J. W. and Sneath, P. H. A. (1969). Taxometric maps. *Syst. Zool.* **18**, 402-415. [323]

Carson, D. H., *see* Garner, W. R.

Carson, H. L., *see* Stalker, H. D.

Casetti, E. (1964). Multiple discriminant functions. Tech. Rept. No. II, O.N.R. Geography Branch, Evanston, Ill. North-western Univ. 63 pp. [157]

Cassagnau, P. (1963). Les collemboles d'Afrique du nord avec une étude de quelques espèces du Nord-Constantinois. *Bull. Soc. Hist. nat. Toulouse* **98**, 197-206. [238]

Cassagnau, P., *see* Bonnet, L.

Cassagnau, P. and Matsakis, J. Th. (1966). Sur l'utilisation des comparaisons multiples en biocénotique édaphique. *Rev. Ecol. Biol. Sol.* **2**, 463-474. [233]

Cassie, R. M. (1962). Multivariate analysis in the interpretation of numerical plankton data. *N.Z. J. Sci.* **6**, 36-59. [219]

Cassie, R. M. (1967). Principal component analysis of the zoo-plankton of Lake Maggiore. *Mem. Ist. Ital. Idrobiol.* **21**, 129-144. [224]

Cassie, R. M. and Michael, A. D. (1968). Fauna and sediments of an intertidal mud flat: a multivariate analysis. *J. exp. mar. Biol. Ecol.* **2**, 1-23. [219]

Cattell, R. B. (1965a). Factor analysis: an introduction to essentials. I. The purpose and underlying models. *Biometrics* 21, 190-215. [201, 202]

Cattell, R. B. (1965b). Factor analysis: an introduction to essentials. II. The role of factor analysis in research. *Biometrics* 21, 405-435. [201, 202]

Cavalli-Sforza, L. L. (1966). Population structure and human evolution. *Proc. R. Soc. Lond.* (B) 164, 362-379. [247]

Cavalli-Sforza, L. L., *see* Edwards, A. W. F.

Cavalli-Sforza, L. L. and Edwards, A. W. F. (1964). Analysis of human evolution. *In* "Genetics Today", pp. 923-933. Oxford Univ. Press, Oxford. [247]

Cavalli-Sforza, L. L. and Edwards, A. W. F. (1967). Phylogenetic analysis: models and estimation procedures. *Evolution* 21, 550-570. [296, 307]

Cheetham, A. H. (1968). Morphology and systematics of the bryozoan genus *Metrarabdotos. Smithson. misc. Collns* 153(1), 1-121. [152]

Cheetham, A. H. and Hazel, J. E. (1969). Binary (presence-absence) similarity coefficients. *J. Paleont.* 43, 1130-1136. [283]

Choi, S. C., *see* Kraus, B. S.

Christensen, K. (1954). Ratios as a means of specific differentiation in Collembola. *Ent. News* 65, 176-177. [28, 108]

Claringbould, P. J. (1958). Multivariate quantal analysis. *J.R. statist. Soc.* (B) 20, 113-121. [99, 338, 342]

Clark, P. J. and Spuhler, J. N. (1959). Differential fertility in relation to body dimensions. *Human Biol.* 31, 121-137. [246]

Clifford, H. T. and Binet, F. E. (1954). A quantitative study of a presumed hybrid swarm between *Eucalyptus eleaphora* and *E. goniocalyx. Aust. J. Bot.* 2, 325-336. [262]

Clunies-Ross, C. W. (1969). Profiles and association: two exploratory techniques in the multivariate analysis of surveys. *Statistician* 19, 49-60. [230]

Cochran, W. G. (1962). On the performance of the linear discriminant function. *Bull. Inst. int. Statist.* 39, 435-447. [35]

Cochran, W. G. and Hopkins, C. E. (1961). Some classification problems with multivariate qualitative data. *Biometrics* 17, 10-32. [48]

Cock, A. G. (1966). Genetical aspects of metrical growth and form in animals. *Q. Rev. Biol.* 41, 131-190. [241]

Colless, D. H. (1967a). An examination of certain concepts in phenetic taxonomy. *Syst. Zool.* 16, 6-27. [7, 339, 343]

Colless, D. H. (1967b). The phylogenetic fallacy. *Syst. Zool.* 16, 289-295. [7, 308, 344]

Cooley, W. W. and Lohnes, P. R. (1962). "Multivariate Procedures for the Behavioural Sciences", 211 pp. John Wiley & Sons, New York. [77, 135]

Cooper, P. W. (1963). Statistical classification with quadratic forms. *Biometrika* 50, 439-448. [54]

Cooper, P. W. (1965). Quadratic discriminant functions in pattern recognition. *I. E.E.E. Trans. Inform. Theory* 11, 313-315. [31, 54, 325, 342]

Coppen, F. M. V., *see* Scott Blair, G. W.

Cousin, G. (1961). Analyse des équilibres morphogénétiques des types structureaux spécifiques et hybrides chez quelques gryllides. *Bull. Soc. ent. Fr.* 86, 500-521. [6, 241]

Cowan, S. T. (1970). A heretical taxonomy for microbiologists. *J. gen. Microbiol.* 61, 145-154. [295]

Cox, D. R. and Brandwood, L. (1959). On a discriminatory problem connected with the works of Plato. *J.R. Statist. Soc.* (B) 21, 195-200. [159]

Cox, G. M. and Martin, W. P. (1937). The discriminant function applied to the differentiation of soil types. *Iowa State Coll. J. Sci.* 11, 323-331. [269]

Craddock, J. M. (1965). A meteorological application of principal component analysis. *Statistician* 15, 143-156. [271]

Crovello, T. J. (1968a). A numerical taxonomic study of the genus *Salix*, Section Sitchenses. *Univ. Calif. Publs Bot.* 44, 1-66. [285, 342]

Crovello, T. J. (1968b). The effect of missing data and of two sources of character values on a phenetic study of the willows of California. *Madroño* 19, 301-315. [302]

Crovello, T. J. (1968c). Key communality cluster analysis as a taxonomic tool. *Taxon* 17, 241-258. [204, 322]

Crovello, T. J. (1968d). The effect of change of number of OTU's in a numerical taxonomic study. *Brittonia* 20, 346-367. [285]

Crovello, T. J. (1969). Effects of change of characters and of number of characters in numerical taxonomy. *Am. Midl. Nat.* 81, 68-86. [285]

Crovello, T. J., *see* Ornduff, R.

Crowson, R. A. (1953). A possible new principle in taxonomy, and its evolutionary implications. *Nature, Lond.* 171, 883. [340, 344]

Crowson, R. A. (1955). "The Natural Classification of the Families of the Coleoptera", 187 pp. Lloyd, London. [265]

Crowson, R. A. (1970). "Classification and Biology", 350 pp. Heinemann, London. [343]

Da Gama, M. M. (1964). "Colêmboles de Portugal Continental", 252 pp. Dissert. Univ. Coimbra. [238]

Dagnelie, P. (1965a). L'étude des communautés végétales par l'analyse statistique des liaisons entre les espèces et les variables écologiques; Principes fondamentaux. *Biometrics* 21, 345-361 [225]

Dagnelie, P. (1965b). L'étude des communautés végétales par l'analyse statistique des liaisons entre les espèces et les variables écologiques: un exemple. *Biometrics* 21, 890-907. [202, 225]

Dagnelie, P. (1966). A propos des différentes méthodes de classification numérique. *Revue Statist. appl.* 14, 55-75. [285]

Dalke, G. W. (1966). Implementation of pattern recognition techniques as applied to geoscience interpretation. *In* "Computer Applications in the Earth Sciences", pp. 24-29. State Geol. Survey, Lawrence, Kansas. [286]

Daly, H. V., *see* Sokal, R. R.

Davies, D. E., *see* Stower, W. J.

Davies, R. G., *see* Blackith, R. E., Boratynski, K.

Defrise-Gussenhoven, E. (1957). Mesure de divergence entre quelques fémurs fossiles et un ensemble de fémurs belges récents; étude biométrique. *Bull. Inst. Sci. nat. Belg.* 33, 13 pp. [59]

De Groot, M. H. and Li, C. C. (1966). Correlations between similar sets of measurements. *Biometrics* 22, 781-790. [246]

De Izarra, D.-C., *see* Bonnet, L.

Delany, M. J. and Healy, M. J. R. (1964). Variation in the long-tailed field-mouse (*Apodemus sylvaticus* L.) in north-west Scotland. II. Simultaneous examination of all the characters. *Proc. R. Soc. Lond.* B 161, 200-207. [215]

Delany, M. J. and Whittaker, H. M. (1969). Variation in the skull of the long-tailed field-mouse *Apodemus sylvaticus* in mainland Britain. *J. Zool.* 157, 147-157. [106, 216]

Dempster, A. P. (1963). Stepwise multivariate analysis of variance based on principal variables. *Biometrics* 19, 478-490. [53]

Dempster, A. P. (1964). Tests for the equality of two covariance matrices in relation to a best linear discriminator analysis. *Ann. math. Statist.* 35, 355-374. [53, 62]

Doran, J. E., *see* Hodson, F. R.

Draper, J. (1964). Some statistical problems in research and development. *Statistician* 14, 311-318. [160]

Ducker, S. C., Williams, W. T. and Lange, G. N. (1965). Numerical classification of the Pacific forms of *Chlorodesmus* (Chlorophyta). *Aust. J. Bot.* 13, 489-499. [263]

Dufour, L. (1841). Recherches anatomiques et physiologiques sur les Orthoptères, les Hymenoptères, et les Neuroptères. *Mém. Acad. Sci. Paris* 7, 265-647. [312]

Dunn, O. J., *see* Weiner, J. M.

Du Praw, E. J. (1964). Non-Linnean taxonomy. *Nature, Lond.* 202, 849-852. [143]

Du Praw, E. J. (1965a). Non-Linnean taxonomy and the systematics of honey-bees. *Syst. Zool.* 14, 1-24. [143]

Du Praw, E. J. (1965b). The recognition and handling of honey-bee specimens in non-Linnean taxonomy. *J. apicult. Res.* 4, 71-84. [144]

Dürer, A. (1613). Les quatres livres d'Albrecht Dürer de la proportion des parties et pourtraicts des corps humains. Arnheim. [7]

Dwyer, P. (1951). "Linear Computations", 344 pp. Chapman and Hall, London. [136]

Eades, D. C. (1965). The inappropriateness of the correlation coefficient as a measure of taxonomic resemblance. *Syst. Zool.* 14, 98-100. [291, 339]

Edwards, A. W. F., *see* Cavalli-Sforza, L. L.

Edwards, A. W. F. and Cavalli-Sforza, L. L. (1964). Reconstruction of evolutionary trees. *In* "Phenetic and Phylogenetic Classification" (V. H. Heywood and J. McNeill, eds), pp. 67-76. Syst. Ass. Publ. No. 6, London. [307]

Edwards, A. W. F. and Cavalli-Sforza, L. L. (1965). A method for cluster analysis. *Biometrics* 21, 362-375. [284, 289, 299]

Ehrenberg, A. S. C. (1962). Some questions about factor analysis. *Statistician* 12, 191-209. [201, 203]

Ehrenberg, A. S. C. (1963). Some queries to factor analysts. *Statistician* 13, 257-262. [203]

Ehrenberg, A. S. C. (1964). Replies to Dr. Warburton, Mr. Jeffers, and Dr. Lawley. *Statistician* 14, 51-61. [203]

Ehrlich, A. H., *see* Ehrlich, P. R.

Ehrlich, P. R. (1964). Some axioms of taxonomy. *Syst. Zool.* 13, 109-123. [301]

Ehrlich, P. R. and Ehrlich, A. H. (1967). The phenetic relationships of the butterflies. I. Adult taxonomy and the non-specificity hypothesis. *Syst. Zool.* 16, 301-317. [295, 300, 342]

Eickwort, K. (1969). Differential variation of males and females in *Polistes exclamans*. *Evolution* 23, 391-405. [242]

Einstein, A. (1950). "Out of My Later Years", 282 pp. Thames and Hudson, London. [13]

Eklund, E., *see* Gyllenberg, P.

Elliott, O., *see* Giles, E.

Estabrook, G. F. (1968). A general solution in partial orders for the Camin-Sokal model in phylogeny. *J. theoret. Biol.* 21, 421-438. [295]

Estabrook, G. F., *see* Wirth, M.

Eyles, A. C. and Blackith, R. E. (1965). Studies on hybridization in *Scolopostethus* Fieber (Heteroptera; Lygaeidae). *Evolution* 19, 465-479. [90, 92]

Feldman, S., Klein, D. F. and Honigfeld, G. (1969). A comparison of successive screening and discriminant function techniques in medical taxonomy. *Biometrics* 25, 725-734. [264]

Ferrari, T. J., Pijl, H. and Venekamp, J. T. N. (1957). Factor analysis in agricultural research. *Neth. J. agric. Sci.* 5, 211-221. [229]

Field, J. G. (1969). The use of the information statistic in the numerical classification of heterogeneous systems. *J. Ecol.* 57, 565-569. [232]

Fisher, D. R. (1968). A study of faunal resemblance using numerical taxonomy and factor analysis. *Syst. Zool.* 17, 48-63. [327]

Fisher, D. R., *see* Rohlf, F. J.

Fisher, D. R. and Rohlf, F. J. (1969). Robustness of numerical methods and errors of homology. *Syst. Zool.* 18, 33-36. [302]

Fisher, R. A. (1936). The use of multiple measurements in taxonomic problems. *Ann. Eugen. Lond.* 7, 179-188. [12]

Fitzsimons, P., *see* Moore, J. J.

Flake, R. H. and Turner, B. (1968). Numerical classification for taxonomic problems. *J. theoret. Biol.* 20, 260-270. [261]

Flowers, J. M., *see* Sailer, S. B.

Fraisse, R. and Arnoux, J. (1954). Les caractères biométriques du cocon chez *Bombyx mori* L. et leurs variations sous l'influence de l'alimentation. *Rev. Ver. Soie* 6, 43-62. [52]

Fraser, A. R. and Kovats, M. (1966). Stereoscopic models of multivariate statistical data. *Biometrics* 22, 358-367. [322]

Freeman, P. R. (1970). A multivariate study of students' performance under examination. *J.R. Statist. Soc.* A 133, 38-55. [264]

Fries, J. and Matérn, B. (1966). On the use of multivariate methods for the construction of tree-taper curves. *Res. notes, Dep. Forest Biometry, Skogshögskolan, Stockholm* 9, 85-117. [142, 148]

Fry, W. G. (1970). The sponge as a population: a biometric approach. *In* "The Biology of the Porifera" (W. G. Fry, ed.). *Symp. zool. Soc. Lond.* 25, 135-162. Academic Press Ltd., London. [270]

Fujii, K. (1969). Numerical taxonomy of ecological characteristics and the niche concept. *Syst. Zool.* 18, 151-153. [216]

Funk, R. C. (1963). The application of numerical taxonomy to acarology. *In* "Advances in Acarology", pp. 374-378. Cornell Univ. Press, Ithaca, New York. [267]

Gabriel, K. R. and Sokal, R. R. (1969). A new statistical approach to geographic variation analysis. *Syst. Zool.* 18, 259-278. [215]

Garner, W. R. and Carson, D. H. (1960). A multivariate solution of the redundancy of printed English. *Psychol. Rep.* 6, 123-141. [302]

Gatty, R. (1966). Multivariate analysis for marketing research: an evaluation. *Appl. Statist.* 15, 157-172. [150]

Gibson, R. E., *see* Bickley, W. G.

Gilchrist, B. M. (1960). Growth and form in the brine shrimp *Artemia salina* (L). *Proc. R. Soc. Lond.* 134, 221-235. [212]

Giles, E. and Bleibtrau, H. K. (1961). Cranial evidence in archaeological reconstruction: a trial of multivariate techniques for the South-west. *Am. Anthrop.* 63, 48-61. [59]

Giles, E. and Elliot, O. (1962). Race identification from cranial measurements. *J. Forensic Sci.* 7, 147-157. [59]

Giles, E. and Elliot, O. (1963). Sex determination by discriminant function analysis. *Am. J. phys. Anthrop.* 21, 53-68. [58]

Gittins, R. (1965a). Multivariate approaches to a limestone grassland community. II. A direct species ordination. *J. Ecol.* 53, 403-409. [226]

Gittins, R. (1965b). Multivariate approaches to a limestone grassland community. III. A comparative study of ordination and association-analysis. *J. Ecol.* 53, 411-425. [226]

Gittins, R. (1968). Trend-surface analysis of ecological data. *J. Ecol.* 56, 845-869. [327]

Gittins, R. (1969). The application of ordination techniques. *In* "Ecological Aspects of the Mineral Nutrition of Plants" (I. H. Rorison, ed.), pp. 37-66. Brit. Ecol. Soc. Symp. No. 9. [225, 229]

Glahn, H. R. (1968). Canonical correlation and its relationship to discriminant analysis and multiple regression. *J. Met.* 25, 23-31. [133]

Goodall, D. W. (1953). Objective methods for the classification of vegetation. II. Fidelity and indicator value. *Aust. J. Bot.* 1, 434-456. [298]

Goodall, D. W. (1954a). Objective methods for the classification of vegetation. III. An essay in the use of factor analysis. *Aust. J. Bot.* 2, 304-324. [226]

Goodall, D. W. (1954b). Vegetational classification and vegetational continua. *Angew. PflSoziol.* 1, 168-182. [226]

Goodall, D. W. (1966a). Numerical taxonomy of bacteria—some published data re-examined. *J. gen. Microbiol.* 42, 25-37. [281]

Goodall, D. W. (1966b). A new similarity index based on probability. *Biometrics* 22, 882-907. [276]

Goodall, D. W. (1966c). Deviant index: a new tool for numerical taxonomy. *Nature, Lond.* 210, 216. [292]

Goodall, D. W. (1967). The distribution of the matching coefficient. *Biometrics* 23, 647-656. [282]

Goodall, D. W. (1968a). Identification by computer. *Bioscience* 18, 485-488. [292]

Goodall, D. W. (1968b). Affinity between an individual and a cluster in numerical taxonomy. *Biométr.-Praxim.* 9, 1-20. [278]

Goodall, D. W., *see* Simon, J. P.

Goodman, M. M. (1969). Measuring evolutionary divergence. *Jap. J. Genet.* 44 (suppl. 1), 310-316. [246]

Goodman, M. M. and Paterniani, E. (1969). The races of maize. III. Choices of appropriate characters for racial classification. *Econ. Bot.* 23, 265-273. [263]

Gould, S. J. (1966). Allometry and size in ontogeny and phylogeny. *Biol. Rev.* 41, 587-640. [28]

Gould, S. J. (1967). Evolutionary patterns in pelycosaurian reptiles; a factor analytical study. *Evolution* 21, 385-401. [147, 202]

Gould, S. J. (1969). An evolutionary microcosm: Pleistocene and recent history of the Land Snail, *P.(Poecilozonites)* in Bermuda. *Bull. Mus. comp. Zool. Harv.* 138, 407-532. [205, 206]

Gower, J. C. (1966). Some distance properties of latent root and vector methods used in multivariate analysis. *Biometrika* 53, 325-338. [164, 202, 302]

Gower, J. C. (1967a). Multivariate analysis and multidimensional geometry. *Statistician* 17, 13-28. [142, 171, 174, 203, 303, 320, 323]

Gower, J. C. (1967b). A comparison of some methods of cluster analysis. *Biometrics* 23, 623-637. [172, 277, 284, 289, 290]

Gower, J. C. (1967c). Growth-free canonical variates and generalised inverses. Tech. Memorand. 67-1215, Bell Telephone labs. (unpublished). [172]

Gower, J. C. (1968). Adding a point to vector diagrams in multivariate analysis. *Biometrika* 55, 582-585. [166, 168]

Gower, J. C. (1969). The basis of numerical methods of classification. *In* "The Soil Ecosystem" (J. G. Sheals, ed.), No. 8, pp. 19-30. Syst. Assn. Publs., London. [144]

Gower, J. C. (1970). A note on Burnaby's character-weighted similarity coefficient. *Math. Geol.* 2, 39-45. [281]

Gower, J. C. and Ross, G. J. S. (1969). Minimum spanning trees and single linkage cluster analysis. *Appl. Statist.* 18, 54-64. [266]

Grafius, J. E. (1961). The complex trait as a geometric construct. *Heredity, Lond.* 16, 225-228. [6, 243]

Grafius, J. E. (1965). A geometry of plant breeding. *Mich. St. Univ. Res. Bull.* 7, 59 pp. [6, 243]

Gray, T. R. G. (1969). The identification of soil bacteria. *In* "The Soil Ecosystem" (J. G. Sheals, ed.), No. 8, pp. 73-82. Syst. Assn. Publ., London. [275, 300]

Graybill, F. A., *see* Krumbein, W. C.

Greig-Smith, P. (1964). "Quantitative Plant Ecology", 2nd Edn., 256 pp. Butterworths, London. [285]

Greig-Smith, P., Austin, M. P. and Whitmore, T. C. (1967). The application of quantitative methods to vegetation survey. I. Association analysis and principal component analysis. *J. Ecol.* 55, 483-503. [229]

Grewal, M. S. (1962). The rate of genetic divergence in the C57BL strain of mice. *Genet. Res.* 3, 226-237. [246]

Griffiths, J. C. (1966). Applications of discriminant functions as a classificatory tool in the geosciences. *In* "Computer Applications in the Earth Sciences", pp. 48-52. Lawrence, Kansas, State Geol. Surv. [56]

Groves, C. P. (1963). Results of a multivariate analysis on the skulls of Asiatic wild asses; with a note on the status of *Microhippus hemionus blanfordi* Pocock. *Ann. Mag. nat. Hist.*, ser. 13 6, 329-336. [51]

Guppy, J., *see* Walker, D.

Guthrie, W. K. C. (1962). "A History of Greek Philosophy", 539 pp. Cambridge Univ. Press, Cambridge. [2]

Gyllenberg, H., Eklund, E., Carlberg, G., Antila, M. and Vartio-Vaara, U. (1963). Contamination and deterioration of market milk. VI. Application of taxometrics in order to evaluate relationships between microbial characteristics and keeping quality of market milk. *Acta Agric. scand.* 13, 177-194. [275]

Hagler, H., *see* Kaufmann, H.

Haldane, J. B. S., *see* Kermack, K. A.

Harbaugh, J. W., *see* Merriam, D. F.

Harberd, D. J. (1962). Application of a multivariate technique to an ecological survey. *J. Ecol.* 50, 1-17. [226]

Harman, H. H. (1960). "Modern Factor Analysis", 474 pp. Chicago University Press, Chicago, Ill. [204]

Harper, R. (1952). Psychological and Psycho-physical studies of craftsmanship in dairying. *Br. J. Psychol., Monogr. Suppl.* 28, 1-63. [301]

Harper, R. and Baron, M. (1951). The application of factor analysis to tests on cheese. *Br. J. appl. Phys.* 2, 35-45. [150]

Hashiguchi, S. and Morishima, H. (1969). Estimation of genetic contribution of principal components to individual variates concerned. *Biometrics* 25, 9-15. [243]

Hazel, J. E., *see* Cheetham, A. H.

Hazel, L. N. (1943). The genetic basis for constructing selection indices. *Genetics, N.Y.* 28, 476-490. [245]

Healy, M. J. R. (1968). Multivariate normal plotting. *Appl. Stat.* 17, 157-161. [21]

Healy, M. J. R., *see* Ashton, E. H., Delany, M. J., Yates, F.

Heiser, C. B., Soria, J. and Burton, D. L. (1965). A numerical taxonomic study of *Solanum* species and hybrids. *Am. Nat.* 99, 471-488. [311, 343]

Hendrickson, J. A., Jr. (1968). Clustering in numerical cladistics; a minimum-length directed tree problem. *Math. Biosci.* 3, 371-381. [307]

Hennig, W. (1965). Phylogenetic systematics. *Ann. Rev. Entom.* 10, 97-116. [344]

Herne, H. (1967). How to cook relationships. *Statistician* 17, 357-370. [Preface]

Hinton, H. E. (1967). The structure of the plastron in *Lipsothrix* and the polyphyletic origin of plastron respiration in Tipulidae. *Proc. R. ent. Soc. Lond.* A 42, 35-38. [309]

Hocker, H. W. (1956). Certain aspects of climate as related to the distribution of Loblolly pine. *Ecology* 37, 824-834. [217]

Hodson, F. R., Sneath, P. H. A. and Doran, J. E. (1966). Some experiments on the numerical analysis of archaeological data. *Biometrika* 53, 311-324. [259, 260]

Hole, F. D., *see* Bidwell, O. W.

Holland, D. A. (1968a). Component analysis: an aid to the interpretation of data. *Expl. Agric.* 5, 151-164. [142, 161]

Holland, D. A. (1968b). The component analysis approach to the interpretation of plant analysis data from groundnuts and sugar-cane. *Expl. Agric.* 4, 179-185. [162]

Holland, D. A. (1969). Component analysis—an approach to the interpretation of soil data. *J. Sci. Fd Agric.* 20, 26-31. [269]

Holland, D. A., *see* Pearce, S. C.

Holloway, J. D. A. (1969). A numerical investigation of the biogeography of the butterfly fauna of India, and its relation to continental drift. *Biol. J. Linn. Soc. Lond.* 1, 373-385. [266]

Holm, R. W., *see* Johnson, M. P.

Honigfeld, G., *see* Feldman, S.

Hope, K. (1968). "Methods of Multivariate Analysis", 165 pp. University of London Press. [323]

Hopkins, C. E., *see* Cochran, W. G.

Hopkins, J. W. (1966). Some considerations in multivariate allometry. *Biometrics* 22, 747-760. [29]

Horton, I. F., Russell, J. S. and Moore, A. W. (1968). Multivariate covariance and canonical analysis: a method for selecting the most effective discriminators in a multivariate situation. *Biometrics* 24, 845-858. [267]

Hotelling, H. O. (1931). The generalisation of Student's ratio. *Ann. math. Statist.* 2, 360-378. [12]

Hotelling, H. O. (1935). The most predictable criterion. *J. educ. Psychol.* 26, 139-142. [134]

Hotelling, H. O. (1936). Relations between two sets of variates. *Biometrika* 28, 321-377. [20, 134]

Hotelling, H. O. (1951). A generalised T-test and measure of multivariate dispersion. 2nd Berkeley Symp. math. Stat. Prob. California Univ. Press. [302]

Hotelling, H. O. (1957). The relations of the newer multivariate statistical methods to factor analysis. *Br. J. Statist. Psychol.* 10, 69-79. [203]

Howarth, R. J. and Murray, J. W. (1969). The foraminiferida of Christchurch Harbour, England: a reappraisal using multivariate techniques. *J. Paleont.* 43, 660-675. [219, 269]

Hughes, R. E. and Lindley, D. V. (1955). Application of biometric methods to problems of classification in ecology. *Nature, Lond.* 175, 806-807. [267]

Huxley, J. S. and Teissier, G. (1936). Terminologie et notation dans la description de la croissance relative. *C.r. Séance Soc. Biol., Paris* 150, 934-936. [6]

Ivimey-Cook, R. B. (1969). Investigations into the phenetic relationships between species of *Ononis* L. *Watsonia* 7, 1-23. [152]

Ivimey-Cook, R. B. and Proctor, M. C. F. (1967). Factor analysis of data from an East Devon heath: a comparison of principal component and rotated solutions. *J. Ecol.* 55, 405-413. [227]

Jago, N. D. (1969). A revision of the systematics and taxonomy of certain North American gomphocerine grasshoppers (Gomphocerinae, Acrididae, Orthoptera). *Proc. nat. Acad. Sci. Philadelphia* 121, 229-335. [291]

Jardine, C. J., Jardine, N. and Sibson, R. (1967). The structure and construction of taxonomic hierarchies. *Math. Biosci.* 1, 173-179. [291]

Jardine, N. (1969). The observational and theoretical components of homology: a study based on the morphology of the dermal skull-roofs of rhipidistian fishes. *Biol. J. Linn. Soc. Lond.* 1, 327-361. [33]

Jardine, N., see Jardine, C. J.

Jeffers, J. N. R. (1962). Principal component analysis of designed experiment. *Statistician* 12, 230-242. [202]

Jeffers, J. N. R. (1964). A reply to Mr. Ehrenberg's questions. *Statistician* 14, 54-55. [202]

Jeffers, J. N. R. (1965). Correspondence. *Statistician* 15, 207-208. [160, 243]

Jeffers, J. N. R. (1967a). The study of variation in taxonomic research. *Statistician* 17, 29-43. [28, 324]

Jeffers, J. N. R. (1967b). Two case studies in the application of principal component analysis. *Appl. Statist.* 16, 225-236. [227]

Jenkins, G. I. (1967). Multivariate methods applied to product testing and specifications. *Statistician* 17, 141-155. [160]

Johnson, M. P. and Holm, R. W. (1969). Numerical taxonomic studies in the genus *Sarcostemma* R. Br. (Asclepiadaceae). *In* "Modern Methods in Plant

Taxonomy", pp. 199-217. Academic Press, London and New York. [311, 343]

Jolicoeur, P. (1959). Multivariate geographical variation in the wolf, *Canis lupus* L. *Evolution* 13, 283-299. [144, 215]

Jolicoeur, P. (1963a). The degree of generality of robustness in *Martes americana*. *Growth* 27, 1-27. [44, 244]

Jolicoeur, P. (1963b). Bilateral symmetry and asymmetry in limb bones of *Martes americana* and man. *Rev. can. Biol.* 22, 409-432. [44]

Jolicoeur, P. (1963c). Les combinaisons multidimensionelles de caractères anatomiques quantitifs. *Proc. XVI Int. Congr. Zool.* I, 183. [215]

Jolicoeur, P. (1963d). The multivariate generalisation of the allometry equation. *Biometrics* 19, 497-499. [6, 28]

Jolicoeur, P. (1968). Interval estimation of the slope of the major axis of a bivariate normal distribution in the case of a small sample. *Biometrics* 24, 679-682. [28]

Jolicoeur, P. and Mosimann, J. E. (1960). Size and shape variation in the painted turtle, a principal component analysis. *Growth* 24, 339-354. [65, 148]

Jolicoeur, P. and Mosimann, J. E. (1968). Intervalles de confiance pour la pente de l'axe majeur d'une distribution normale bidimensionelle. *Biométr.-Praxim.* 9, 121-140. [28, 65, 148]

Jones, I. B., *see* Stower, W. J.

Jones, D. and Sneath, P. H. A. (1970). Genetic transfer and bacterial taxonomy. *Bact. Rev.* 34, 40-81. [246]

Jöreskog, K. G. (1963). "Statistical Estimation in Factor Analysis", 145 pp. Almqvist and Wiksell, Stockholm. [202, 204]

Jux, U. and Strauch, F. (1965). Die "Hians"—Schille aus dem Mittel-devon der Bergisch Gladbach—Paffrather Mulde. *Fortschr. Geol. Rheinld. Westf.* 9, 51-86. [111]

Jux, U. and Strauch, F. (1966). Die mitteldevonische Brachiopoden gattung *Uncites* de France 1825. *Palaeontographica, Abt. A.* 125, 176-227. [111]

Kaesler, R. L. (1967). Numerical taxonomy in invertebrate paleontology. *Univ. Kansas Geol. Dept. spec. Publn.* No. 2, 63-81. [276]

Kaesler, R. L. and McElroy, M. N. (1966). Classification of subsurface localities of the Reagan sandstone (Upper Cambrian) of central and northwest Kansas. Computer applications in the Earth Sciences, pp. 42-47. Lawrence, Kansas, State Geol. Survey. [277]

Kaesler, R. L., *see* McElroy, M. N.

Kahn, J. S., *see* Miller, R. L.

Kälin, A., *see* Burla, H.

Kaufmann, H., Hagler, K. and Lang, R. (1958). Analyse anthropologique et statistique des Walsers orientaux de Romanches de l'Oberhalbstein. *Archs suisses Anthrop. gén.* 23, 1-328. [56]

Kendall, M. G. (1957). "A Course in Multivariate Analysis", 185 pp. Griffin, London. [23, 136, 257, 303]

Kendall, M. G. and Stuart, A. (1966). "The Advanced Theory of Statistics", Vol. III, 552 pp. Griffin, London. [49, 50, 210]

Kendrick, W. B., *see* Proctor, J. R.

Kendrick, W. B. and Weresub, L. K. (1966). Attempting neo-Adansonian computer taxonomy at the ordinal level in the basidiomycetes. *Syst. Zool.* 15, 307-329. [33, 303]

Kermack, K. A. (1954). A biometrical study of *Micraster coranguinum* and *M. (Isomicraster) senonensis*. *Phil. Trans. R. Soc. Lond.* B 237, 375-428. [254, 255, 256]

Kermack, K. A. and Haldane, J. B. S. (1950). Organic correlation and allometry. *Biometrika* 37, 30-41. [28]

Kevan, D. K. McE., *see* Blackith, R. E.

Key, K. H. L. (1967). Operational homology. *Syst. Zool.* 16, 275-276. [33]

Keys, A., *see* Brožek, J.

Klein, D. F., *see* Feldman, S.

Klen, L., *see* Smith, J. E. K.

Klovan, J. E. (1970). Numerical classification of *Stictostroma* Parks from the Devonian of southern Ontario, Canada (*see* Reyment, 1970c). [282]

Kokawa, S. (1958). Some tentative methods for the age-estimation by means of morphometry of *Menyanthes* remains. *J. Inst. Poly. Osaka* D 9, 111-118. [28]

Kovats, M., *see* Fraser, A. R.

Kraus, B. S. and Choi, S. C. (1958). A factorial analysis of the prenatal growth of the human skeleton. *Growth* 22, 231-242. [241]

Krumbein, W. C. (1959). Trend-surface analysis of contour-type maps with irregular control-point spacing. *J. geophys. Res.* 64, 823-824. [326]

Krumbein, W. C. and Graybill, F. A. (1965). "An Introduction to Statistical Models in Geology", 465 pp. McGraw-Hill, New York. [325]

Kruskal, J. B. (1964a). Non-metric multidimensional scaling; a numerical method. *Psychometrika* 29, 115-129. [259]

Kruskal, J. B. (1964b). Multidimensional scaling by optimising goodness of fit to a non-metric hypothesis. *Psychometrika* 29, 1-27. [259]

Kullback, S. (1959). "Information Theory and Statistics", 395 pp. John Wiley & Sons, New York. [64, 113]

Kurczynski, T. W. (1970). Generalized distance and discrete variables. *Biometrics* 26, 525-534. [172, 236, 273]

Lachenbruch, H. P. A. (1968). On expected probabilities of mis-classification in discriminant analysis, necessary sample size, and a relation with the multiple correlation coefficient. *Biometrics* 24, 823-834. [54]

Laird, A. K., Tyler, S. A. and Barton, A. D. (1965). Dynamics of normal growth. *Growth* 29, 233-248. [28]

Lambe, E., *see* Moore, J. J.

Lambert, J. M., *see* Williams, W. T.

Lambert, J. M. and Williams, W. T. (1962). Multivariate methods in plant ecology. IV. Nodal analysis. *J. Ecol.* 50, 775-802. [210, 226]

Lance, G. N., *see* Ducker, S. C., Watson, L., Webb, L. J.

Lang, R., *see* Kaufmann, H.

Lawley, D. N. (1940). The estimation of factor loadings by the method of maximum likelihood. *Proc. R. Soc. Edinb.* 60, 64-82. [202]

Lawley, D. N. (1958). Estimation in factor analysis under various initial assumptions. *Br. J. statist. Psychol.* 11, 1-12. [202]

Lawley, D. N. (1960). Approximate methods in factor analysis. *Br. J. statist. Psychol.* 13, 11-17. [202]

Lawley, D. N. and Maxwell, A. E. (1963). "Factor Analysis as a Statistical Method", 117 pp. Butterworths, London. [202, 243]

Lawrence, P. N. (1963). A preliminary survey of the collembola of Morocco. *Bull. Soc. Sci. nat., Phys., Maroc.* 43, 29-34. [238]

Lecher, P. (1967). Cytogénétique de l'hybridisation expérimentale et naturelle chez l'Isopode *Jaera (albifrons) syei* Bocquet. *Archs Zool. éxp. gén.* 108, 633-698. [245]

Ledley, R. S. (1964). High-speed automatic analysis of biomedical pictures. *Science, N.Y.* 146, 216-223. [286]

Ledley, R. S. and Lusted, L. B. (1959). Reasoned foundations of medical diagnosis. *Science, N.Y.* 130, 9-21. [264]

Lee, P. J. (1969). The theory and application of canonical trend surfaces. *J. Geol.* 77, 303-318. [133, 326]

Lefèbvre, J. (1966). Étude, a l'aide de mensuration, de la conformation et de la croissance des bovins normands, 124 pp. Thèse, Fac. des Sciences, Univ. Caen. [241, 244]

Leng, C. W. (1920). Catalogue of the Coleoptera of America, North of Mexico. Sherman, New York. [265]

Leone, C. A., *see* Basford, N. L.

Le Quesne, W. J. (1969). A method of selection of characters in numerical taxonomy. *Syst. Zool.* 18, 201-205. [295, 308, 317]

Le Quesne, W. J. (1971). Further studies based on the unique character concept. *Syst. Zool.* 20, in press. [310, 317]

Lerman, A. (1965). On rates of evolution of unit characters and character complexes. *Evolution* 19, 16-25. [54, 246]

Levy, I. (1926). "Récherches sur les Sources de la Légende de Pythagore", 149 pp. Laroux, Paris. [2]

Li, C. C., *see* De Groot, M. H.

Lima, M. B. (1968). A numerical approach to the *Xiphonema americanum* complex. *C.r. int. Symp. Nematol., Antibes, 1965*, 30 pp. Brill. Leiden. [258]

Linder, A. (1963). Anschauliche Deutung und Begrundung des Trennverfahrens. *Method. Inf. méd.* 2, 30-33. [257]

Lindley, D. V. (1962). Factor analysis. *Statistician* 12, 169-171. [202]

Lindley, D. V. (1964). Factor analysis, a summary of discussion. *Statistician* 14, 47-49. [202]

Lindley, D. V., *see* Hughes, R. E.

Lipton, S., *see* Ashton, E. H.

Lohnes, P. R., *see* Cooley, W. W.

Lu, K. H. (1965). Harmonic analysis of the human face. *Biometrics* 21, 491-505. [325]

Lusted, L. B., *see* Ledley, R. S.

Lysenko, O. and Sneath, P. H. A. (1959). The use of models in bacterial classification. *J. gen. Microbiol.* 20, 284-290. [322]

Lyubischev, A. A. (1959). The application of biometrics in taxonomy. *Vestnik Leningradskogo Univ.* 14, 128-136. [50]

McCammon, H. M. (1970). Variation in recent brachiopod populations (*see* Reyment, 1970c). [211]

McCammon, R. B. (1966). Principal component analysis and its application in large-scale correlation studies. *J. Geol.* 74, 721-733. [219]

McCammon, R. B. (1968a). Multiple component analysis and its application in classification of environments. *Bull. Am. Ass. Petrol. Geol.* 52, 2178-2196. [220]

McCammon, R. B. (1968b). The dendrograph: a new tool for correlation. *Bull. Geol. Soc. Am.* 79, 1663-1670. [277]

McElroy, M. N. and Kaesler, R. L. (1965). Application of factor analysis to the Upper Cambrian Reagan sandstone of central and northwest Kansas. *The Compass* 42, 188-201. [277]

McElroy, M. N., *see* Kaesler, R. L.

Mahalanobis, P. C. (1928). A statistical study of the Chinese head. *Man in India* 8, 107-122. [59]

Mahalanobis, P. C. (1936). On the generalised distance in statistics. *Proc. natn. Inst. Sci. India* 2, 49-55. [17]

Mahalanobis, P. C. (1937). Normalisation of statistical variates and the use of rectangular coordinates in the theory of sampling distributions. *Sankhyā* 3, 35-40. [343]

Mahalanobis, P. C., Majumdar, D. N. and Rao, C. R. (1949). Anthropometric survey of the United Provinces, 1941; a statistical study. *Sankhyā* 9, 89-324. [301, 302]

Mainardi, D. (1959). Immunological distances among some Gallinaceous birds. *Nature, Lond.* 184, 913-914. [265]

Majumdar, M. N., *see* Mahalanobis, P. C.

Malmgren, B. (1970). Morphometric analysis of two species of *Floridina* (Chilostomata, Bryozoa). *Stockh. Contr. Geol.* 23, 73-89. [215]

Mantel, N. and Valand, R. S. (1970). A technique of non-parametric multivariate analysis. *Biometrics* 26, 547-558. [236]

Marcus, L. F. (1969). Measurement of selection using distance statistics in the prehistoric orang-utan *Pongo pygmaeus palaeosumatrensis*. *Evolution* 23, 301-307. [246]

Marcus, L. F., *see* Sarkar, P. K.

Martin, E. S. (1936). A study of an Egyptian series of mandibles with special reference to mathematical methods of sexing. *Biometrika* 28, 148-178. [59]

Martin, L. (1960). Homométrie, allométrie et cograduation en biométrie générale. *Biometrische Zeitschr.* 2, 73-97. [28]

Martin, W. P., *see* Cox, G. M.

Mastrogiuseppe, T. D., Cridland, A. A. and Bogyo, T. P. (1970). Multivariate comparison of fossil and recent ginkgo wood. *Lethaia* 3, 271-277. [263]

Matérn, B., *see* Fries, J.

Mather, K. (1949). "Biometrical Genetics", 162 pp. Methuen, London. [53]

Matsakis, J. T. (1964). "Comparaisons Multiples en Biologie", 201 pp. Imp. du Sud-Ouest, Toulouse. [233]

Matsakis, J. T., *see* Cassagnau, P.

Maxwell, A. E. (1961). Canonical variate analysis when the variables are dichotomous. *Educ. psychol. Measur.* 21, 259-271. [99, 100]

Maxwell, A. E., *see* Lawley, D. N.

Mayr, E. (1968). Theory of biological classification. *Nature, Lond.* 220, 545-548. [305, 344]

Merriam, D. F. and Harbaugh, J. W. (1963). Computer helps map oil structures. *Oil Gas J.* 61, 158-159, 161-163. [326]

Michael, A. D., *see* Cassie, R. M.

Michener, C. D. (1963). Some future developments in taxonomy. *Syst. Zool.* 12, 151-172. [318]

Michener, C. D. (1964). The possible use of uninomial nomenclature to increase the stability of names in biology. *Syst. Zool.* 13, 182-190. [291, 318]

Michener, C. D., *see* Sokal, R. R.

Michener, C. D. and Sokal, R. R. (1966). Two tests of the hypothesis of non-specificity in the *Hoplitis* complex (Hymenoptera, Magachilidae). *Ann. ent. Soc. Am.* 59, 1211-1217. [242, 344]

Middleton, G. V. (1962). A multivariate statistical technique applied to the study of sandstone composition. *Trans. R. Soc. Can. Ser. III* 56, 119-126. [30]

Middleton, G. V. (1964). Statistical studies on scapolites. *Can. J. Earth Sci.* 1, 23-34. [204]

Miller, R. L., *see* Olson, E. C.

Miller, R. L. and Kahn, J. S. (1962). "Statistical Analysis in the Geological Sciences", 483 pp. John Wiley & Sons, New York. [276]

Milne, P., *see* Walker, D.

Misra, R. K. (1966). Vectorial analysis for genetic clines in body dimensions in populations of *Drosophila subobscura* Coll. and a comparison with those of *D. robusta* Sturt. *Biometrics* 22, 469-487. [251]

Möller, F. (1962). Quantitative methods in the systematics of actinomycetales. IV. The theory and application of a probabilistic identification key. *Giorn. Microbiol.* 10, 29-47. [280, 282]

Moore, A. W., *see* Horton, I. F.

Moore, C. S. (1965). Inter-relations of growth and cropping in apple trees studied by the method of component analysis. *J. hort. Sci.* 40, 133-149, [142]

Moore, C. S. (1968). Response of growth and cropping of apples to treatments studied by the method of component analysis. *J. hort. Sci.* 43, 249-262. [142]

Moore, J. J., Fitzsimons, P., Lambe, E. and White, J. (1970). A comparison and evaluation of some phytosociological techniques. *Vegetatio* 20, 1-20. [229]

Morishima, H. (1969). Phenetic similarity and phylogenetic relationships among strains of *Oryza perennis* estimated by methods of numerical taxonomy. *Evolution* 23, 429-443. [263]

Morishima, H., *see* Hashiguchi, S.

Moser, C. A. and Scott, W. (1961). "British Towns: a Statistical Study of Their Social and Economic Differences", 169 pp. Oliver and Boyd, London. [156]

Mosimann, J. E., *see* Jolicoeur, P.

Moss, W. W. (1968). Experiments with various techniques of numerical taxonomy. *Syst. Zool.* 17, 31-47. [291]

Moss, W. W. and Webster, W. A. (1969). A numerical taxonomic study of a group of selected strongylates (Nematoda). *Syst. Zool.* 18, 423-443. [258]

Moss, W. W. and Webster, W. A. (1970). Phenetics and numerical taxonomy applied to systematic nematology. *J. Nemat.* 2, 16-25. [259]

Mound, L. A. (1963). Host correlated variation in *Bemisia tabaci* (Gennadius) (Homoptera: Aleyrodidae). *Proc. R. ent. Soc. Lond.* A 38, 171-180. [52]

Moy, E. A., *see* Blackith, R. E.

Mukherjee, R., Rao, C. R. and Trevor, J. C. (1955). "The Ancient Inhabitants of the Jebel Moya", 123 pp. Cambridge University Press, Cambridge. [37, 54, 55]

Mulhall, H., *see* Talbot, P. A.

Murdie, G. (1969). Some causes of size variation in the pea aphid, *Acyrtosiphon pisum* Harris. *Trans. R. ent. Soc. Lond.* 121, 423-442. [149]

Murray, J. W., *see* Howarth, R. J.

Murty, B. R., Anand, I. J. and Arunachalam, V. (1965). Subspecific differentiation in *Nicotiana rustica* L. *Indian J. Genet. Pl. Breed.* 25, 217-223. [262]

Naidin, D. P., *see* Reyment, R. A.

Namkoong, G. (1966). Statistical analysis of introgression. *Biometrics* 22, 488-502. [262]

Needham, R. M. (1967). Automatic classification in linguistics. *Statistician* 17, 45-54. [296]

Nichols, F. (1959). Changes in the chalk heart-urchin *Micraster* interpreted in relation to living forms. *Phil. Trans. R. Soc. Lond.* B 242, 347-437. [253]

Norris, J. M., *see* Barkham, J. P.

Noy-Meir, I. and Austin, M. P. (1970). Principal coordinate ordination and simulated vegetational data. *Ecology* 51, 551-552. [229]

Oberzill, W. (1963). Beiträge der mikrobiologische Grundlagen-forschung zu industriellen Verfahrensfragen. *Ost. Chem. Zeit.* 64, 65-70. [275]

Olson, E. C. (1964). Morphological integration and the meaning of characters in classification systems. *In* "Phenetic and Phylogenetic Classification" (V. H. Heywood and J. McNeill, eds), pp. 123-156. Syst. Assn. Publ. No. 6, London. [311]

Chicago Univ. Press, Chicago, Ill. [103, 243, 311]

Orloci, L. (1966). Geometric models in ecology. I. The theory and application of some ordination methods. *J. Ecol.* 54, 193-215. [227, 228]

Orloci, L. (1967a). An agglomerative method for classification of plant communities. *J. Ecol.* 55, 193-205. [229, 283]

Orloci, L. (1967b). Data centering: a review and evaluation with reference to component analysis. *Syst. Zool.* 16, 208-212. [209]

Orloci, L. (1968a). Definitions of structure in multivariate phytosociological samples. *Vegetatio* 15, 281-291. [285, 338]

Orloci, L. (1968b). Information analysis in phytosociology. *J. theoret. Biol.* 20, 271-284. [225]

Orloci, L. (1968c). A model for the analysis of structure in taxonomic collections. *Can. J. Bot.* 46, 1093-1097. [338]

Orloci, L. (1969). Information analysis of structure in biological collections. *Nature, Lond.* 223, 483-484. [280]

Ornduff, R. and Crovello, T. J. (1968). Numerical taxonomy of the Limnanthaceae. *Am. J. Sci.* 55, 173-182. [264]

Osborne, R. H. (1969). The American Upper Ordovician standard. XI. Multivariate classification of typical Cincinnatian calcarenites. *J. Sedim. Petrol.* 39, 769-776. [163, 167]

Ouellette, R. P. and Qadri, S. U. (1968). The discriminatory power of taxonomic characteristics in separating salmonoid fishes. *Syst. Zool.* 17, 70-75. [270, 305]

Overall, J. E. and Williams, C. M. (1961a). Models for medical diagnosis: Factor analysis. I. Theoretical. *Method. Inf. Med.* 5, 51-56. [264]

Overall, J. E. and Williams, C. M. (1961b). Models for medical diagnosis: Factor analysis. II. Experimental. *Method. Inf. Med.* 5, 78-80. [264]

Oxnard, C. E. (1967). The functional morphology of the primate shoulder as revealed by comparative anatomical, osteometric, and discriminant function techniques. *Am. J. phys. Anthrop.* 26, 219-240. [100]

Oxnard, C. E. (1968). The architecture of the shoulder in some mammals. *J. Morph.* 126, 249-290. [100, 103, 104]

Oxnard, C. E. (1969). Mathematics, shape and function: a study in primate anatomy. *Am. Scient.* 57, 75-96. [247]

Oxnard, C. E., *see* Ashton, E. H.

Paker, I. U., *see* Brieger, F. G.

Pankhurst, R. J. (1970). Key generation by computer. *Nature, Lond.* 227, 1269-1270. [293]

Parks, J. M. (1969). Multivariate fascies maps. *In* "Symp. Computer Applications in Petroleum Exploration", pp. 6-18. State Geol. Survey, Lawrence, Kansas. [326]

Paterniani, E., *see* Goodman, M. M.

Pearce, S. C. (1959). Some recent applications of multivariate analysis to data from fruit trees. *Ann. Rep. E. Malling Res. Stat.* (1958) 73-76. [22, 140]

Pearce, S. C. (1965). The measurement of a living organism. *Biométr.-Praxim.* 6, 143-152. [13, 257]

Pearce, S. C. and Holland, D. A. (1960). Some applications of multivariate analysis in botany. *Appl. Statist.* 9, 1-7. [161]

Pearce, S. C. and Holland, D. A. (1961). Analyse des composantes, outil en recherche biométrique. *Biométr.-Praxim.* 2, 159-177. [44, 161]

Pelto, C. R. (1954). Mapping of multicomponent systems. *J. Geol.* 62, 501-511. [219]

Penrose, L. S. (1954). Distance, size and shape. *Ann. Eugen., Lond.* 18, 337-343. [29]

Perkal, J. and Szczotka, F. (1960). Eine neue Methode der Analyse eines Kollektivs von Merkmalen. *Biometr. Z.* 2, 108-116. [243]

Pielou, E. C. (1969). "An Introduction to Mathematical Ecology", 286 pp. John Wiley & Sons, New York. [284]

Pijl, H., *see* Ferrari, T. J.

Pimentel, R. A. (1958). Taxonomic methods, their bearing on subspeciation. *Syst. Zool.* 7, 139-156. [214]

Pimentel, R. A. (1959). Mendelian infra-specific divergence levels and their analysis. *Syst. Zool.* 8, 139-159. [214]

Pipberger, H. V. (1962). Analysis of electrocardiogram by digital computer. *Method. Inf. Med.* 1, 69-71. [264]

Pitcher, M. (1966). A factor-analytic scheme for grouping and separating types of fossils. Computer applications in the Earth Sciences, pp. 30-41. State Geol. Survey, Lawrence, Kansas. [204, 205]

Pohja, M. S. (1960). Micrococci in fermented meat products; classification and description of 171 strains. *Suom. Maatal. Seur. Julk.* 96, 1-80. [275]

Pons, J. (1955). The sexual diagnosis of isolated bones of the skeleton. *Hum. Biol.* 27, 12-21. [59]

Power, D. M. (1970). Geographic variation of red-winged blackbirds in Central North America. *Univ. Kansas Publn.* 19, 1-83. [265]

Prance, G. T., Rogers, D. J. and White, F. (1969). A taxometric study of an angiosperm family; generic delimitation in the Chrysobalanaceae. *New Phytol.* 68, 1203-1234. [264]

Prim, R. C. (1958). Shortest connection networks and some generalisations. *Bell. Syst. tech. J.* 36, 1389-1401. [307]

Proctor, J. R. and Kendrick, W. B. (1963). Unequal weighting in numerical taxonomy. *Nature, Lond.* 197, 716-717. [281]

Proctor, M. C. F. (1967). The distribution of British liverworts; a statistical analysis. *J. Ecol.* 55, 119-136. [229]

Proctor, M. C. F., *see* Ivimey-Cook, R. B.

Qadri, S. U., *see* Ouellette, R. P.

Quenouille, M. H. (1952). "Associate Measurements", 242 pp. Butterworths, London. [136]

Radhakrishna, S. (1964). Discrimination analysis in medicine. *Statistician* 14, 147-167. [264]

Ramdén, H.-A., *see* Reyment, R. A.

Rao, C. R. (1948). The utilisation of multiple measurements in problems of biological classification. *J. R. statist. Soc.* B 10, 159-203. [144]

Rao, C. R. (1952). "Advanced Statistical Methods in Biometric Research", 292 pp. John Wiley & Sons, New York. [18, 47, 143, 257, 272, 278, 284, 289]

Rao, C. R. (1953). Discriminant functions for genetic differentiation and selection. *Sankhyā* 12, 229-246. [245]

Rao, C. R. (1960). Multivariate analysis: an indispensable tool in applied research. *Sankhyā* 22, 317-338. [13]

Rao, C. R. (1961). Some observations on multivariate statistical methods in anthropological research. *Bull. intn. statist. Inst.* 33, 99-109. [59]

Rao, C. R. (1964a). Sir Ronald Aylmer Fisher—the architect of multivariate analysis. *Biometrics* 20, 286-300. [257]

Rao, C. R. (1964b). The use and interpretation of principal components analysis in applied research. *Sankhyā* 26, 329-358. [148, 170, 171, 203, 257]

Rao, C. R. (1966). Discriminant function between composite hypotheses and related problems. *Biometrika* 53, 339-345. [332]

Rao, C. R., *see* Mahalanobis, P. C., Mukherjee, R.

Rao, C. R. and Slater, P. (1949). Multivariate analysis applied to differences between neurotic groups. *Br. J. Psychol. statist. Section* 2, 17-29. [99]

Rao, C. R. and Varadarajan, V. S. (1963). Discrimination of Gaussian processes. *Sankhyā* 25, 303-330. [343]

Rasch, D. (1962). Die Factoranalyse und ihre Anwendung in der Tierzucht. *Biometr. Z.* 4, 15-39. [202]

Rayner, J. H. (1966). Classification of soils by numerical methods. *J. Soil Sci.* 17, 79-92. [268]

Rayner, J. H. (1969). The numerical approach to soil systematics. *In* "The Soil Ecosystem" (J. G. Sheals, ed.), pp. 31-39. Syst. Assn. Publ. No. 8, London. [268]

Rees, J. W. (1970). A multivariate morphometric analysis of divergence in skull morphology among geographically contiguous groups of white-tailed deer (*Odocoileus virginianus*) in Michigan. *Evolution* 24, 220-229. [247]

Reeve, E. C. R. (1941). A statistical analysis of taxonomic differences within the genus *Tamandua* Gray (Xenarthra). *Proc. Zool. Soc. Lond.* A 111, 279-302. [108]

Rempe, U. (1965). Lassen sich bei Säugertieren Introgression mit multivariaten Verfahren nachweisen? *Z. zool. Syst. Evol.* 3, 388-412. [245]

Rempe, U. and Bühler, P. (1969). Zum Einfluss der geographischen und altersbedingten Variabilität bei der Bestimmung von *Neomys*—Mandibeln mit Hilfe der Diskriminanzanalyse. *Ztg. Säugetierk.* 34, 148-164. [216]

Reyment, R. A. (1962). Observations on homogeneity of covariance matrices in paleontologic biometry. *Biometrics* 18, 1-11. [64]

Reyment, R. A. (1963a). Studies on Nigerian Upper Cretaceous and Lower Tertiary Ostracoda. II. Danian, Palaeocene and Eocene Ostracoda. *Stockh. Contr. Geol.* 10, 286 pp. [43, 59]

Reyment, R. A. (1963b). Multivariate analytical treatment of quantitative species associations: an example from palaeo-ecology. *J. anim. Ecol.* 32, 535-547. [150, 217]

Reyment, R. A. (1966a). Studies on Nigerian Upper Cretaceous and Lower Tertiary Ostracoda. III. Stratigraphical, Palaeoecological, and Biometrical conclusions. *Stockh. Contr. Geol.* 14, 151 pp. [43, 53, 212, 213, 251, 330, 331, 332, 333, 334]

Reyment, R. A. (1966b). *Afrobolivina africana* (Graham, de Klasz, Rérat): quantitative Untersuchung der Variabilität einer palaeozänen Foraminifera. *Ec. Geol. Helv.* 59, 319-338. [148, 149]

Reyment, R. A. (1968). Multivariate statistical analysis in geology. *Publn. Palaeont. Inst. Univ. Uppsala No. 74*, 18 pp. [53, 138]

Reyment, R. A. (1969a). Biometrical techniques in systematics. *In* "Systematic Biology", pp. 542-587. National Academy of Science, Washington, D.C. [257]

Reyment, R. A. (1969b). *Textilina mexicana* (Cushman) from the western Niger delta. *Bull. Geol. Instn. Univ. Uppsala, n. ser.* 1, 75-81. [148, 272]

Reyment, R. A. (1969c). Upper Sinemurian (Lias) at Gantofta, Skåne. *Geol. För. Stockh. Förh.* 91, 208-216. [98, 148]

Reyment, R. A. (1969d). Some case studies of the statistical analysis of sexual dimorphism. *Bull. geol. Instn. Univ. Uppsala, n. ser.* 1, 97-119. [41, 332, 334]

Reyment, R. A. (1969e). Interstitial ecology of the Niger delta; an actuopaleoecological study. *Bull. geol. Instn. Univ. Uppsala, n. ser.* 1, 121-159. [167, 236, 270]

Reyment, R. A. (1969f). Covariance structure and morphometric analysis. *Math. Geol.* 1(2), 185-197.

Reyment, R. A. (1969g). A multivariate paleontological growth problem. *Biometrics* 25, 1-8. [65]

Reyment, R. A. (1970a). Problems and pitfalls in computer statistics. State Geol. Survey, Lawrence, Kansas. 118 pp., unpublished.

Reyment, R. A. (1970b). Eigen-theory in numerical taxonomy. *Bull. geol. Instn. Univ. Uppsala, n. ser.* 2(8), 67-72. [166]

Reyment, R. A. (ed.) (1970c). Symposium on biometry in palaeontology. *Bull. geol. Instn. Univ. Uppsala, n. ser.* 2, 1-89. [330]

Reyment, R. A. (1971a). Vermuteter Dimorphismus bei der Ammoniten-Gattung *Benueites. Bull. geol. Instn. Univ. Uppsala, n. ser.* 3(1), 1-18. [335]

Reyment, R. A. (1971b). Spectral breakdown of morphometric data. *Math. Geol.* 4, in press. [217]

Reyment, R. A. and Brännström, B. (1962). Certain aspects of the physiology of *Cypridopsis* (Ostracoda: Crustacea). *Stockh. Contr. Geol.* 9, 207-242. [214, 321]

Reyment, R. A. and Naidin, D. P. (1962). Biometric study of *Actinocamax verus* S.L. from the Upper Cretaceous of the Russian platform. *Stockh. Contr. Geol.* 9, 147-206. [59, 250]

Reyment, R. A. and Ramdén, H.-Å. (1970). Fortran IV program for canonical variates analysis for the CDC 3600 computer. Computer contr. 47, 39 pp. State Geol. Survey, Lawrence, Kansas. [89, 109, 111, 253]

Reyment, R. A. and Sandberg, P. (1963). Biometric study of *Barramites subdifficilis* (Karakasch). *Palaeontology* 6, 727-730. [44, 148]

Reyment, R. A. and van Valen, L. (1969). *Buntonia olokundudui* sp. nov. (Ostracoda, Crustacea): a study of meristic variation in Paleocene and recent ostracods. *Bull. geol. Instn. Univ. Uppsala, n. ser.* 1, 83-94. [166, 243]

Reyment, R. A., Ramden, H.-Å. and Wahlstedt, W. J. (1969). Fortran IV program for the generalized statistical distance and analysis of covariance matrices for the CDC 3600 computer. Computer contr. 39, 42 pp. State Geol. Survey, Lawrence, Kansas, 42 pp. [62]

Ricard, F. H., *see* Rouvier, R.

Rinkel, R. C., *see* Sokal, R. R.

Rising, J. D. (1968). A multivariate assessment of interbreeding between the chickadees, *Parus atricapillus* and *P. caroliensis. Syst. Zool.* 17, 160-169. [245]

Roberts, M. I., *see* Blackith, R. E.

Rogers, D. J., *see* Prance, G. T., Wirth, M.

Rohlf, F. J. (1965). A randomisation test of the non-specificity hypothesis in numerical taxonomy. *Taxon* 14, 262-267. [306]

Rohlf, F. J. (1967). Correlated characters in numerical taxonomy. *Syst. Zool.* 16, 109-126. [35, 323]

Rohlf, F. J. (1968). Stereograms in numerical taxonomy. *Syst. Zool.* 17, 246-255. [322]

Rohlf, F. J., *see* Basford, N. L., Fisher, D. R., Sokal, R. R.

Rohlf, F. J. and Fisher, D. R. (1968). Tests for hierarchical structure in random data sets. *Syst. Zool.* 17, 407-412. [277]

Rohlf, F. J. and Sokal, R. R. (1967). Taxonomic structure from randomly and systematically scanned biological images. *Syst. Zool.* 16, 246-260. [300]

Ross, G. J. S., *see* Gower, J. C.

Rouvier, R. R. (1966). L'analyse en composantes principales; son utilisation en génétique et ses rapports avec l'analyse discriminatoire. *Biometrics* 22, 343-357. [249]

Rouvier, R. R. (1969). Pondération des valeurs génotypiques dans la selection par index sur plusieurs caractères. *Biometrics* 25, 295-307. [245]

Rouvier, R. and Ricard, F. H. (1965). Etude de conformation du poulet. II. Recherche des composantes de la variabilité morphologique du poulet vivant. *Ann. zootech.* 14, 213-227. [142, 241, 244]

Rubin, J. (1967). Optimal classification into groups; an approach to solving the taxonomy problem. *J. theoret. Biol.* 15, 103-144. [295]

Russel, J. S., *see* Horton, I. F.

Sailer, S. B. and Flowers, J. M. (1969). Geographic variation in the American lobster. *Syst. Zool.* 18, 330-338. [270]

Sandberg, P., *see* Reyment, R. A.

Sanghvi, L. D., *see* Balakrishnan, V.

Sarkar, P. K., Bidwell, O. W. and Marcus, L. F. (1966). Selection of characteristics for numerical classification of soils. *Proc. Soil Sci. Am.* 30, 269-272. [267]

Saxena, S. K. (1969). Silicate solid solutions and geothermometry. 3. Distribution of Fe and Mg between coexisting Garnet and Biotite. *Contr. Mineral Petrol.* 22, 259-267. [162]

Schreider, E. (1960). "La Biométrie (Collection 'Que sais-je?')", 126 pp. Presses Universitaires, Paris. [243]

Scopoli, G. A. (1777). Introductio ad historiam naturalem, Praga. [7]

Scossiroli, R. E. (1959). Selezione artificiale per caratteri quantitativi e selezione naturale in *Drosophila melanogaster. Ric. Sci.* 29, 1-10. [243]

Scott, J. H. (1957). Muscle growth and function in relation to skeletal morphology. *Am. J. phys. Anthrop., n. ser.* 15, 277-234. [244]

Scott, W., *see* Moser, C. A.

Scott Blair, G. W. and Coppen, F. M. V. (1940). The subjective judgement of the elastic and plastic properties of soft bodies. *Brit. J. Psychol.* 31, 61-68. [340]

Seal, H. (1964). "Multivariate Statistical Analysis for Biologists", 207 pp. Methuen, London. [38, 43, 108, 140, 202, 339]

Selander, R. K., *see* Yang, S. H.

Shankarnarayan, D., *see* Sinha, R. N.

Sheals, J. G. (1964). The application of computer techniques to acarine taxonomy: a preliminary examination with species of the *Hypoaspis-Androlaelaps* complex (Acarina). *Proc. Linn. Soc. Lond.* 176, 11-21. [172, 264, 302]

Shepard, R. N. (1962). The analysis of proximities: multidimensional scaling with an unknown distance function. *Psychometrika* 27, 125-139, 219-246. [259, 303]

Sibson, R., *see* Jardine, C. J.

Siehl, A. (1970). Zur numerischen Klassifikation von Fusuliniden. Habilitation-sschrift. Univ. Bonn (unpublished). [306]

Simon, J. P. and Goodall, D. W. (1968). Relationship in annual species of *Medicago*. VI. Two dimensional chromatography of the phenolics and analysis of the results by probabilistic similarity methods. *Aust. J. Bot.* 16, 89-100. [261]

Sims, R. (1969). A numerical classification of megasceloid earthworms. *In* "The Soil Ecosystem" (J. G. Sheals, ed.), pp. 143-153. Syst. Assn. publn., London, No. 8. [342]

Sinha, R. N. and Shankarnarayan, D. (1955). Concepts of insect taxonomy in Ancient India. *Entom. News* 46, 243-247. [3]

Slater, P., *see* Rao, C. R.

Smirnov, E. S. (1963). The problem of taxonomic affinity in systematics. *J. gen. Biol.* 24, 172-181 (in Russian). [279]

Smirnov, E. S. (1968). On exact methods in systematics. *Syst. Zool.* 17, 1-13. [279]

Smirnov, E. S. (1969). "Taxonomic Analysis", 188 pp., University Publishing House, Moscow (in Russian). [279, 340]

Smith, H. F. (1936). A discriminant function for plant selection. *Ann. Eugen. Lond.* 7, 240-250. [245]

Smith, J. E. K. and Klem, L. (1961). Vowel recognition using a multiple discriminant function. *J. acoust. Soc. Am.* 33, 358 [54]

Sneath, P. H. A. (1966). A method for curve seeking from scattered points. *Computer J.* 8, 383-391. [93]

Sneath, P. H. A. (1967a). Trend-surface analysis of transformation grids. *J. Zool. Lond.* 151, 65-122. [325]

Sneath, P. H. A. (1967b). Some statistical problems in numerical taxonomy. *Statistician* 17, 1-12. [324]

Sneath, P. H. A., *see* Carmichael, J. W., Hodson, F. R., Jones, D., Lysenko, O., Sokal, R. R.

Sokal, R. R. (1961). Distance as a measure of taxonomic similarity. *Syst. Zool.* 10, 70-79. [38, 246]

Sokal, R. R. (1962). Variation and covariation of characters of alate *Pemphigus populi-transversus* in Eastern North America. *Evolution* 16, 227-245. [11, 243]

Sokal, R. R. (1963). The principles and practice of numerical taxonomy. *Taxon* 12, 190-199. [305]

Sokal, R. R. (1969). Animal taxonomy, theory and practice. *Q. Rev. Biol.* 44, 209-211. [272]

Sokal, R. R., *see* Camin, J. H. Gabriel, K. R., Michener, C. D., Rohlf, F. J.

Sokal, R. R. and Michener, C. D. (1958). A statistical method for evaluating systematic relationships. *Kans. Univ. Sci. Bull.* 38, 1409-1438. [282, 290]

Sokal, R. R. and Michener, C. D. (1967). The effects of different numerical techniques on the phenetic classification of bees of the *Hoplitis* complex (Megachilidae). *Proc. Linn. Soc. Lond.* 178, 59-74. [291]

Sokal, R. R. and Rinkel, R. C. (1963). Geographic variation of alate *Pemphigus populi-transversus* in Eastern North America. *Kans. Univ. Sci. Bull.* 44, 467-507. [214]

Sokal, R. R. and Sneath, P. H. A. (1963). "Principles of Numerical Taxonomy", 359 pp. W. H. Freeman & Co., San Francisco. [7, 34, 152, 164, 220, 242, 275, 278, 281, 283, 285, 286, 289, 311, 313, 317, 321, 343]

Sokal, R. R., Daly, H. V. and Rohlf, F. J. (1961). Factor analytical procedures in a biological model. *Kans. Univ. Sci. Bull.* 42, 1099-1121. [24, 202]

Solignac, M., *see* Bocquet, C.

Soria, J., *see* Heiser, C. B.

Spearman, C. (1904). General intelligence objectively determined and measured. *Am. J. Psychol.* 15, 201-293. [25, 202]

Spence, T. F., *see* Ashton, E. H.

Sporne, K. R. (1960). The correlation of biological characters. *Proc. Linn. Soc. Lond.* 171, 83-88. [35]

Sprent, P. (1968). Linear relationships in growth and size studies. *Biometrics* 24, 639-656. [28]

Spuhler, J. N., *see* Clark, P. J.

Spurnell, D. J. (1963). Some metallurgical applications of principal components. *Appl. Statist.* 12, 180-188. [160]

Stalker, H. D. and Carson, H. L. (1947). Morphological variation in natural populations of *Drosophila robusta* Sturtevant. *Evolution* 1, 237-248. [252]

Stapleton, H. E. (1956). The hand, with its five fingers, as the primitive basis of geometry, arithmetic and algebra. *Actes du 8-me Cong. int. Hist. Sci.* 1103 [1]

Stapleton, H. E. (1958). Ancient and modern aspects of Pythagoreanism. *Osiris* 30, 12-53. [1]

Storer, R. W. (1950). Geographic variation in the pigeon guillemots of North America. *Condor* 52, 28-31. [265]

Storms, L. H. (1958). Discrepancies between factorial and multivariate discrimination among groups as applied to personality theory. *J. ment. Sci.* 104, 713-721. [203]

Stower, W. J., Davies, D. E. and Jones, I. B. (1960). Morphometric studies of the Desert Locust *Schistocerca gregaria* (Forsk.). *J. anim. Ecol.* 29, 309-339. [56]

Strauch, F., *see* Jux, U.

Stroud, C. P. (1953). An application of factor analysis to the systematics of *Kalotermes*. *Syst. Zool.* 2, 76-92. [209]

Stuart, A. (1964). Review of Lawley and Maxwell's "Factor Analysis as a Statistical Method". *Biometrika* 51, 533. [142]

Stuart, A., *see* Kendall, M. G.

Sturrock, R. F. (1963). Observation on the use of *Trichostrongylus colubriformis* (Nematoda) infections of guinea pigs for laboratory experiments. *Parasitology* 53, 189-199. [258, 259]

Sylvester-Bradley, P. C. (1968). The science of diversity. *Syst. Zool.* 17, 176-181. [8]

Symmons, P. M. (1969). A morphometric measure of phase in the desert locust *Schistocerca gregaria* (Forsk.). *Bull. ent. Res.* 58, 803-809. [57]

Szczotka, F., *see* Perkal, J.

Talbot, P. A. and Mulhall, H. (1962). "The Physical Anthropology of Southern Nigeria", 127 pp. Cambridge University Press, Cambridge. [38, 109]

Teissier, G. (1938). Un essai d'analyse factorielle. Les variants sexuels de *Maia squinada. Biotypologie* 7, 73-96. [6]

Teissier, G. (1960). Relative growth. *In* "The Physiology of the Crustacea" (T. H. Waterman, ed.), Vol. 1, pp. 537-560. Academic Press, New York and London. [251]

Teissier, G., *see* Huxley, J. S.

Temple, J. T. (1968). The Lower Llandovery (Silurian) brachiopods from Keisley, Westmorland. *Palaeontol. Soc. Publ. No. 521*, 58 pp. [154, 155]

Thomas, B. A. M. (1961). Some industrial applications of multivariate analysis. *Appl. Statist.* 10, 1-8. [159]

Thomas, P. A. (1968). Geographic variation of the rabbit tick *Haemaphysalis leporispalustris* in North America. *Kans. Univ. Sci. Bull.* 47, 787-828. [266]

Thomson, D'A. W. (1942). "On Growth and Form", 1116 pp. 2nd Edn. Cambridge University Press, Cambridge. [6, 326]

Throckmorton, L. H. (1965). Similarity *versus* relationship in *Drosophila, Syst. Zool.* 14, 221-236. [307]

Thurstone, L. L. (1947). "Multiple Factor Analysis: a Development and Expansion of 'Factors of the Mind'," 535 pp. Chicago Univ. Press, Chicago, Ill. [201, 202, 327]

Tracey, J. G., *see* Webb, L. J.

Trevor, J. G., *see* Mukherjee, R.

Tryon, R. C. and Bailey, D. E. (1966). The BCTRY computer system of cluster and factor analysis. *Multivariate Behavioral Research* 1, 95-111. [322]

Turner, B., *see* Flake, R. H.

Tyler, S. A., *see* Laird, A. K.

Udvardy, M. D. F. (1969). "Dynamic Zoogeography", 445 pp. Van Nostrand Reinhold, New York. [232]

Van der Waerden, H. (1963). "Science Awakening", 306 pp. John Wiley & Sons, New York. [2]

Van Groenewoud, H. (1965). Ordination and classification of Swiss and Canadian coniferous forests by various biometric and other methods. *Ber. geobot. Inst. ETH, Stift. Rubel, Zurich* 36, 28-102. [225]

Valand, R. S., *see* Mantel, N.

Varadarajan, V. S., *see* Rao, C. R.

Vartio-Vaara, U., *see* Gyllenberg, H.

Veitch, L. G. (1963). Multiple linear regression techniques in the examination of the structure of layer lattice silicates. CSIRO, Melbourne, Divn. math. Statist. Tech. paper No. 14. [Preface]

Vencovsky, R., *see* Brieger, F. G.

Venekamp, J. T. N., *see* Ferrari, T. S.

Verdier, M., *see* Blackith, R. E.

Vinck, L. (1962). Le diagnostic formel. *Biométr.-Praxim* 3, 169-193 and 4, 3-22. [13, 296, 302]

Wahlstedt, W. J., *see* Reyment, R. A.

Walker, D., Milne, P., Guppy, J. and Williams, J. (1968). The computer-assisted storage and retrieval of pollen morphological data. *Pollen Spores* 10, 251-262. [293]

Waloff, N. (1966). Scotch broom (*Sarothamnus scoparius* (L.) Wimmer) and its insect fauna introduced into the Pacific north-west of America. *J. appl. Ecol.* 3, 293-311. [93, 94]

Warburton, F. W. (1962). The practical value of factor analysis in education. *Statistician* 12, 172-188. [202]

Warburton, F. W. (1964). A reply to Mr. Ehrenberg's criticism. *Statistician* 14, 49-51. [202]

Ward, L. K. (1968). The validity of the separation of *Thrips physapus*, L. and *T. hukkineni* Priesner (Thysanoptera, Thripidae). *Trans. R. ent. Soc. Lond.* 120, 395-416. [147]

Watson, L., Williams, W. T. and Lance, G. N. (1967). A mixed data numerical approach to angiosperm taxonomy; the classification of the Ericales. *Proc. Linn. Soc. Lond.* 178, 25-35. [263]

Webb, L. J., Tracey, J. G., Williams, W. T. and Lance, G. (1967). Studies in the numerical analysis of complex rain-forest communities. I. A comparison of methods applicable to site/species data. *J. Ecol.* 55, 171-191. [229]

Weber, E. (1959). The genetic analysis of characters with continuous variability on a Mendelian basis. I. Monohybrid segregation. *Genetics* 44, 1131-1139. [246]

Weber, E. (1960a). The genetical analysis of characters with continuous variability on a Mendelian basis. II. Monohybrid segregation and linkage analysis. *Genetics* 45, 459-466. [246]

Weber, E. (1960b). The genetical analysis of characters with continuous variability on a Mendelian basis. III. Dihybrid segregation. *Genetics* 45, 567-572. [246]

Webster, W. A., *see* Moss, W. W.

Weiner, J. M. and Dunn, O. J. (1966). Elimination of variates in linear discrimination problems. *Biometrics* 22, 268-275. [160]

Welch, P. D. and Wimpress, R. S. (1961). Two multivariate statistical computer programmes and their application to the vowel recognition problem. *J. acoust. Soc. Am.* 33, 426-434. [54]

Weresub, L. K., *see* Kendrick, W. B.

Westermann, G. E. C. (ed.) (1969). Sexual dimorphism in fossil Metazoa and taxonomic implications. International Union of Geological Sciences, ser. A(I). E. Schweizerbart'sche Verlagsbuchhandlung, 251 pp. [335]

White, F., *see* Prance, G. T.

White, J., *see* Moore, J. J.

White, M. J. D. and Andrew, L. E. (1959). Cytogenetics of the grasshopper *Moraba scurra*. V. Biometric effects of chromosomal inversions. *Evolution* 14, 284-292. [242]

Whitehead, F. H. (1955). Taxonomic studies in the genus *Cerastium:* I. *C. atrovirens, C. pumilum, C. semicandrum. Watsonia* 3, 213-227. [263]

Whitehead, F. H. (1956). Taxonomic studies in the genus *Cerastium:* II. *Cerastium subtetrandrum* (Lange) Murb. *Watsonia* 3, 324-326. [263]

Whitehead, F. H. (1959). Vegetational changes in response to alterations of surface roughness on M. Maiella, Italy. *J. Ecol.* 47, 603-606. [211]

Whitmore, T. E., *see* Greig-Smith, P.

Whittaker, H. M., *see* Delany, M. J.

Wicken, A. J., *see* Buck, S. F.

Wilkinson, C. (1967). A taxonomic revision of the genus *Teldenia* Moore (Lepidoptera; Drepanidae; Drepaninae). *Trans. R. ent. Soc. Lond.* 119, 303-362. [265, 309]

Wilkinson, C. (1970). Numerical taxonomic methods applied to some Indo-Australian Drepanidae: Lepidoptera. *J. nat. Hist.* 4, 269-288. [266, 342]

Wilkinson, E. M. (1951). Goethe's conception of form. *Proc. Br. Acad.* 37, 175-197. [5]

Williams, C. M., *see* Overall, J. E.

Williams, D. A., Beatty, R. A. and Burgoyne, P. S. (1970). Multivariate analysis in the genetics of spermatozoan dimensions in mice. *Proc. R. Soc. Lond.* B 175, 313-331. [247]

Williams, E. J. (1958). Prediction of clinical effects of Rh incompatibility. *Biometrics* 14, 143-144. [264]

Williams, J., *see* Walker, D.

Williams, W. T., *see* Ducker, S. C., Lambert, J. M., Watson, L., Webb, L. J.

Williams, W. T. and Lambert, J. M. (1959). Multivariate methods in plant ecology. I. Association analysis in plant communities. *J. Ecol.* 47, 83-101. [226, 290]

Williamson, M. H. (1961). A method for studying the relation of plankton variations to hydrography. *Bull. mar. Ecol.* 5, 224-229. [303]

Wilson, E. O. (1965). A consistency test for phylogenies based on contemporaneous species. *Syst. Zool.* 14, 214-220. [307]

Wimpress, R. S., *see* Welch, P. D.

Wirth, M., Estabrook, G. F. and Rogers, D. J. (1966). A graph theory model for systematic biology, with an example for the Oncodiinae (Orchidaceae). *Syst. Zool.* 15, 59-69. [296]

Wishart, D. (1969). Fortran II programs for 8 methods of cluster analysis (Clustan I). Computer contr. No. 38, 112 pp. State Geol. Survey, Lawrence, Kansas. [291]

Yang, S. H. and Selander, R. K. (1968). Hybridization in the grackle *Quiscalus quiscalus* in Louisiana. *Syst. Zool.* 17, 107-143. [245]

Yates, F. (1950). The place of statistics in the study of growth and form. *Proc. R. Soc. Lond.* B 137, 479-489. [257]

Yates, F. and Healy, M. J. R. (1964). How should we reform the teaching of statistics? *J. R. statist. Soc.* A 127, 199-210. [145]

Subject Index

Note: Persons listed in the author index are not included here; other persons mentioned are included in this index. An indication of the nature of each organism is added after the entry of a generic name, species are not included as such. Computer subroutines are not indexed.

A

Abakaliki people, anthropometry of, 109

Abstract morphometrics
analysis of literary styles as, 158
examples of, 150, 156, 225, 271

Abundance,
of Foraminifera, studied by canonical variates, 218
of liverworts, along first principal components, 229
replacing size vector, 150, 229
used in faunal comparisons, 234

Acacia (brigalow bush), canonical analysis of soil under, 267

Acarology, 266

Acceptance, as unproved taxonomic statement, 27

Acrididae, *see* Grasshoppers

Actinocamax (belemnite),
in coordinated multivariate study, 250
in time series, 59

Adansonian taxonomy,
equal weighting of characters in, 7
linked to Linnaean ideas by those of Goethe, 5

Adaptation,
and convergence, 316
skeletal, in bats, 103
to arboreal life, 102
to salinity, 212

Afrobolivina (Foraminifera), studied using principal components, 149

Agglomerative techniques,
as Achilles' heel of numerical taxonomy, 283
in classification, 283, 341

Agricultural field trials, 91

Allometry,
and correlated characters, 37
effect of, on morphometric ratios, 108
expressed by angles between vectors, 44
leading to matrix distortion, 251
multivariate equation of, 28
negative, 149
theories of, 28, 29
used in echinoid morphometrics, 256

Ammonites,
sexual dimorphism in, 335
studied using principal components, 148
studied using principal coordinates, 98

Anacridium (grasshopper), response to crowding of, 42

Analysis of variance,
manova, in canonical variate analysis, 112
related to canonical variate analysis, 88
to distinguish real from arbitrary clustering, 285

Anderson's hybrid index, 262

Angles between vectors,
as signposts to evolution, 254
representing geographical variation, 215
to study Collembola, 238
to study locusts, 41
to study nematodes, 258
use of, 16

Anopheles (mosquito), correlated characters in, 35

Ant-eaters studied using canonical variates, 106

Principal coordinates—*cont.*
  in paleoecology, 167
  in pedology, 268
  in petrology, 167
  in sedimentology, 167, 270
  residual variation in, 167
  situations appropriate for the use of, 141, 151
  to study ammonites, 98, 337
  to study evolution in heart-urchins, 253
  to study mites, 266
  to study ostracods, 166
  to study sponges, 270
Probabilistic approach,
  to identification, 280
  to similarity index, 261
  to weighting systems, 280
Promicroceras (ammonite) studied using principal coordinates, 98
Proper vector, 22
Propithecus (ape), shoulder structure of, 102
Prosimii, shoulder structure of, 100
Proximity analysis, in archaeology, 259
Psychological theory, in factor analysis, 25
Psychometry, 25, 99
Pythagoras, 2, 3, 4, 10

Q

Q-techniques,
  defined, 209
  information lost in, 210
  in marine ecology, 270
  in phytosociology, 226
  in principal component analysis, 209
  in principal coordinate analysis, 209
  interchangeability with R-techniques, 209
  to study collembola, 238
Quadratic discriminant functions,
  compared with linear, in theory, 31
  ease of computation of, 54
  in trend-surface analysis, 325
  in vowel recognition, 54
  use of rankings with, 304

Qualitative characters,
  generalized distances and, 172
  in higher taxonomy, 144
  in taxonomic practice, 274
  numbers needed of, 34
  special matrices for, 171
Quartimax procedure in factor analysis, 204

R

R-techniques,
  defined, 209
  information lost in, 210
  in marine ecology, 269
  in phytosociology, 226
  to study collembola, 238
Race identification, with discriminant functions, 58
Racial likeness, coefficient of, 38
Rana (frog), sexual dimorphism of, 335
Random variation,
  canalization of, 140
  importance of studying, 170,
  meanings of, 140
Rank correlation coefficients,
  in ecology, 235
  to study plankton, 303
Rankings,
  as replacements for measures, 105, 303
  in ecology, 236
  in multivariate analysis, 303
  in proximity analysis, 260
  in sociology, 156
  with quadratic discriminants, 304
Rashevsky, N., 13
Rate of evolution, measured by morphological change, 247
Ratios of characters,
  redundancy of, when added to character suite, 34
  related to discriminants, 30
  shortcomings of, in morphometrics, 27, 108, 338
Rats, allometric growth of, 28
Rearing conditions, influencing body form, 42, 43

410   SUBJECT INDEX